Sir Archibald Geikie

Class-Book of Geology

Sir Archibald Geikie

Class-Book of Geology

ISBN/EAN: 9783337059309

Printed in Europe, USA, Canada, Australia, Japan

Cover: Foto ©berggeist007 / pixelio.de

More available books at **www.hansebooks.com**

CLASS-BOOK OF GEOLOGY

OF

GEOLOGY

BY

Sir ARCHIBALD GEIKIE, F.R.S.

D.SC. CAMB., DUB. ; LL.D. EDIN., ST. AND.

CORRESPONDENT OF THE INSTITUTE OF FRANCE,
DIRECTOR-GENERAL OF THE GEOLOGICAL SURVEY OF THE UNITED KINGDOM, AND
DIRECTOR OF THE MUSEUM OF PRACTICAL GEOLOGY, LONDON ; FORMERLY
MURCHISON PROFESSOR OF GEOLOGY AND MINERALOGY IN THE
UNIVERSITY OF EDINBURGH.

ILLUSTRATED WITH WOODCUTS

London

MACMILLAN AND CO., Ltd.

NEW YORK: MACMILLAN & CO.

1896

First Edition 1886
Second Edition 1890
Reprinted 1891, 1892, 1893, 1894, 1896

PREFACE

THE present volume completes a series of educational works on Physical Geography and Geology, projected by me many years ago. In the *Primers*, published in 1873, the most elementary facts and principles were presented in such a way as I thought most likely to attract the learner, by stimulating at once his faculties of observation and reflection. The continued sale of large editions of these little books in this country and in America, and the translation of them into most European languages, leads me to believe that the practical methods of instruction adopted in them have been found useful. They were followed in 1877 by my *Class-Book of Physical Geography*, in which, upon as far as possible the same line of treatment, the subject was developed with greater breadth and fulness. This volume was meant to be immediately succeeded by a corresponding one on Geology, but pressure of other engagements has delayed till now the completion of this plan.

So many introductory works on Geology have been written that some apology or explanation seems required from an author who adds to their number. Experience of the practical work of teaching science long ago convinced me that what the young learner primarily needs is a class-book which will awaken his curiosity and interest. There should be enough of detail to enable him to understand how conclusions are arrived at. All through its chapters he should see how observation, generalisation, and induction go hand in hand in the progress of scientific research. But it should not be overloaded with technical details which, though of the highest importance, cannot be adequately understood until considerable advance has been made in the study. It ought to present a broad, luminous picture of each branch of the subject, necessarily, of course, incomplete, but perfectly correct and intelligible as far as it goes. This picture should be amplified in detail by a skilful teacher. It may, however, so arrest the attention of the learner himself as to lead him to seek, of his own accord, in larger treatises, fuller sources of information. To this ideal standard of a class-book I have striven in some measure to approach.

Originally, I purposed that this present volume should be uniform in size with the *Class-Book of Physical Geography*. But, as the illustrations were in progress, the advantage of adopting a larger page became evident, and with this greater scope and my own enthusiasm for the subject the book has gradually grown into what it now is. With few exceptions, the woodcuts have been drawn and engraved expressly for this volume. Mr. Sharman has kindly made for me most of the drawings of the fossils. The landscape sketches are chiefly from my own note-books. I have to thank Messrs. J. D. Cooper and M. Lacour for the skill with which they have given in wood-engraving the expression of the originals.

In preparing the Second Edition, I have thoroughly revised this Class-book, so as to keep it abreast of the onward progress of Geology. The sale of a large impression, and the numerous communications received from teachers and others, have led to the belief that the book might be made still more useful if printed in such a form as to admit of its being sold at a greatly reduced price. This change has now been effected; but the volume, though diminished in bulk, contains rather more matter than the first edition. Care has been taken to make the Index full and accurate.

30th June 1890.

CONTENTS

CHAPTER VI

CHAPTER VII

CHAPTER VIII

CHAPTER IX

PART II

ROCKS, AND HOW THEY TELL THE HISTORY OF THE EARTH

CHAPTER X

CHAPTER XI

PART III

THE STRUCTURE OF THE CRUST OF THE EARTH

CHAPTER XII

CHAPTER XIII

CHAPTER XIV

CHAPTER XV

PART IV

THE GEOLOGICAL RECORD OF THE HISTORY
OF THE EARTH

CHAPTER XVI

CHAPTER XXVI

•

CHAPTER XXVII

LIST OF ILLUSTRATIONS

CHAPTER I

THE main features of the dry land on which we live seem to remain unchanged from year to year. The valleys and plains familiar to our forefathers are still familiar to us, bearing the same meadows and woodlands, the same hamlets and villages, though generation after generation of men has meanwhile passed away. The hills and mountains now rise along the sky-line as they did long centuries ago, catching as of old the fresh rains of heaven and gathering them into the brooks and rivers which, through unknown ages, have never ceased to flow seawards. So steadfast do these features appear to stand, and so strong a contrast do they offer to the shortness and changeableness of human life, that they have become typical in our minds of all that is ancient and durable. We speak of the firm earth, of the everlasting hills, of the imperishable mountains, as if, where all else is fleeting and mutable, these forms at least remain unchanged.

And yet attentive observation of what takes place from day to day around us shows that the surface of a country is not now exactly as it used to be. We notice various changes of its topography going on now, which have doubtless been in progress for a long time, and the accumulated effect of which may ultimately transform altogether the character of a landscape. A strong gale, for instance, will level thousands of trees in its pathway, turning a tract of forest or woodland into a bare space, which may become a quaking morass, until perhaps changed into arable ground by the farmer. A flooded river will in a few hours cut away large slices from its banks, and spreading over fields and meadows, will bury many acres of fertile land under a covering of barren sand and shingle. A long-continued, heavy rain, by loosening masses

℃   B

of earth or rock on steep slopes, causes destructive landslips. A hard frost splinters the naked fronts of crags and cliffs, and breaks up bare soil. In short, every shower of rain and gust of wind, if we could only watch them narrowly enough, would be found to have done something towards modifying the surface of the land. Along the sea-margin, too, how ceaseless is the progress of change ! In most places, the waves are cutting away the land, sometimes even at so fast a rate as two or three feet in a year. Here and there, on the other hand, they cast sand and silt ashore so as to increase the breadth of the dry land.

These are ordinary everyday causes of alteration, and though singly insignificant enough, their united effect after long centuries cannot but be great. From time to time, however, other less frequent but more powerful influences come into play. In most large regions of the globe, the ground is often convulsed by earth-quakes, many of which leave permanent scars upon the surface of the land. Volcanoes, too, in many countries pour forth streams of molten rock and showers of dust and cinders that bury the surrounding districts and greatly alter their appearance.

Turning to the pages of human history, we find there the records of similar changes in bygone times. Lakes, on which our rude forefathers paddled their canoes and built their wattled island-dwellings, have wholly disappeared. Bogs, over whose treacherous surface these early hunters could not follow the chase of red deer or Irish elk, have become meadows and fields. Forests, where they hunted the wild boar, have been turned into grassy pastures. Cities have been entirely destroyed by earthquakes or have been entombed under the piles of ashes discharged from a burning mountain. So great have been the inroads of the sea that, in some instances, the sites of what a few hundred years ago were farms and hamlets, now lie under the sea half a mile or more from the modern shore. Elsewhere the land has gained upon the sea, and the harbours of an earlier time are now several miles distant from the coast-line.

But man has naturally kept note only of the more impressive changes ; in other words, of those which had most influence upon his own doings. We may be certain, however, that there have been innumerable minor alterations of the surface of the land within human history, of which no chronicler has made mention, either because they seemed too trivial, or because they took place so imperceptibly as never to be noticed. Fortunately, in many cases, these mutations of the land have written their own memorials,

which can be as satisfactorily interpreted as the ancient manu-
scripts from which our early national history is compiled.

In illustration of the character of these natural chronicles, let us
for a moment consider the subsoil beneath cities that have been
inhabited for many centuries.   In London, for example, when
excavations are made for drainage, building, or other purposes,
there are sometimes found, many feet below the level of the present
streets, mosaic pavements and foundations, together with earthen
vessels, bronze implements, ornaments, coins, and other relics of
Roman time.   Now, if we knew nothing, from actual authentic
history, of the existence of such a people as the Romans, or of
their former presence in England, these discoveries, deep beneath
the surface of modern London, would prove that long before the
present streets were built, the site of the city was occupied by a
civilised race which employed bronze and iron for the useful purposes of life, had a metal coinage, and showed not a little artistic
skill in its pottery, glass, and sculpture.   But down beneath the
rubbish wherein the Roman remains are embedded, lie gravels and
sands from which rudely-fashioned human implements of flint have
been obtained.   Whence we further learn that, before the civilised
metal-using people appeared, an earlier race had been there,
which employed weapons and instruments of roughly chipped
flint.

That this was the order of appearance of the successive peoples
that have inhabited the site of London is, of course, obvious.
But let us ask ourselves why it is obvious.   We observe that there
are, broadly speaking, three layers or deposits from which the
evidence is derived.   The upper layer is that which contains the
foundations and rubbish of modern London.   Next comes that
which encloses the relics of the Roman occupation.   At the
bottom lies the layer that preserves the scanty traces of the early
flint-folk.   The upper deposit is necessarily the newest, for it could
not be laid down until after the accumulation of those below it,
which must, of course, be progressively older, as they are traced
deeper from the surface.   By the mere fact that the layers lie one
above another, we are furnished with a simple clue which enables
us to determine their relative time of formation.   We may know
nothing whatever as to how old they are measured by years or
centuries.   But we can be absolutely certain of what is termed
their " order of superposition," or chronological sequence ; in other
words, we can be confident that the bottom layer came first and
the top layer last.

This kind of observation and reasoning will enable us to detect almost everywhere proofs that the surface of the land has not always been what it is to-day. In some districts, for example, when the dark layer of vegetable soil is turned up which supports the plants that keep the land so green, there may be found below it sand and gravel, full of smooth well-rounded stones. Such materials are to be seen in the course of formation where water keeps them moving to and fro, as on the beds of rivers, the margins of lakes, or the shores of the sea. Wherever smoothed rolled pebbles occur, they point to the influence of moving water ; so that we conclude, even though the site is now dry land, that the sand and gravel underneath it prove it to have been formerly under water. Again, below the soil in other regions, lie layers of oysters and other sea-shells. These remains, spread out like similar shells on the beach or bed of the sea at the present day, enable us to infer that where they lie the sea once rolled.

Pits, quarries, or other excavations that lay open still deeper layers of material, bring before us interesting and impressive testimony regarding the ancient mutations of the land. Suppose, by way of further illustration, that underneath a bed of sand full of oyster-shells, there lies a dark brown band of peat. This substance, composed of mosses and other water-loving plants, is formed in boggy places by the growth of marshy vegetation. Below the peat there might occur a layer of soft white marl full of lake-shells, such as may be observed on the bottoms of many lakes at the present time (compare Fig. 30). These three layers—oyster-bed, peat, and marl—would present a perfectly clear and intelligible record of a curious series of changes in the site of the locality. The bottom layer of white marl with its peculiar shells would show that at one time the place was occupied by a lake. The next layer of peat would indicate that, by the growth of marshy vegetation, the lake was gradually changed into a morass. The upper layer of oyster-shells would prove that the ground was then submerged beneath the sea. The present condition of the ground shows that subsequently the sea retired and the locality passed into dry land as it is to-day.

It is evident that by this method of examination information may be gathered regarding early conditions of the earth's surface, long before the authentic dates of human history. Such inquiries form the subject of Geology, which is the science that investigates the History of the Earth. The records in which this history is chronicled are the soils and rocks under our feet. It is the task of the geologist so to arrange and interpret these records as to show through what successive changes the globe has passed, and how the dry land has come to wear the aspect which it presents at the present time.

Just as the historian would be wholly unable to decipher the inscriptions of an ancient race of people unless he had first discovered a key to the language in which they are written, so the geologist would find himself baffled in his efforts to trace backward the history of the earth if he were not provided with a clue to the interpretation of the records in which that history is contained. Such a clue is furnished to him by a study of the operations of nature now in progress upon the earth's surface. Only in so far as he makes himself acquainted with these modern changes, can he hope to follow intelligently and successfully the story of earlier phases in the earth's progress. It will be seen that this truth has already been illustrated in the instances above given of the evidence that the surface of the land has not been always as it is now. The beds of sand and gravel, of oyster-shells, of peat and of marl, would have told us nothing as to ancient geography had we not been able to ascertain their origin and history by finding corresponding materials now in course of accumulation. To one ignorant of the peculiarities of fresh-water shells, the layer of marl would have conveyed no intelligible meaning. But knowing and recognising these peculiarities, we feel sure that the marl marks the site of a former lake. Thus the study of the Present supplies a key that unlocks the secrets of the Past.

In order, therefore, to trace back the history of the Earth, the geologist must begin by carefully watching the changes that now take place, and by observing how nature elaborates the materials that preserve more or less completely the record of these changes. In the following pages, I propose to follow this method of inquiry, and, as far as the subject will permit, to start with no assumptions which the learner cannot easily verify for himself. We shall begin with the familiar everyday operations of the air, rain, frost, and other natural agents. As these have been fully described in my *Class-Book of Physical Geography*, it will not be needful here to consider them again in detail. We shall rather pass on to inquire in what various ways they are engaged in contributing to the formation of new mineral accumulations, and in thereby providing fresh materials for the preservation of the facts on which geological history is founded. Having thus traced how new rocks are formed, we may then proceed to arrange the similar rocks of older time, marking what are the peculiarities of each and how they may best be classified.

If the labours of the geologist were concerned merely with the former mutations of the earth's surface,—how sea and land have changed places, how rivers have altered their courses, how lakes have been filled up, how valleys have been excavated, how mountains, peaks, and precipices have been carved, how plains have been spread out, and how the story of these revolutions has been written in enduring characters upon the very framework of the land,—he would feel the want of one of the great sources of interest in the study of the present face of nature. We naturally connect all modern changes of the earth's surface with the life of the plants and animals that flourish there, and more especially with their influence on the progress of Man himself. If there were no similar connection of the ancient changes with once living things—if the history of the earth were merely one of dead inert matter—it would lose much of its interest for us. But happily that history includes the records of successive generations of plants and animals which, from early times, have peopled land and sea. The remains of these organisms have been preserved in the deposits of different ages, and can be compared and contrasted with those of the modern world.

To realise how such preservation has been possible, and how far the forms so retained afford an adequate picture of the life of the time to which they belonged, we must turn once more to watch how nature deals with this matter at the present time. Of the millions of flowers, shrubs, and trees which year after year clothe the land with beauty, how many relics are preserved? Where are the successive generations of insect, bird, and beast which have appeared in this country since man first set foot upon its soil? They have utterly vanished. If all their living descendants could suddenly be swept away, how could we tell that such plants and animals ever lived at all? It must be confessed that the vast majority of them leave no trace behind. Nevertheless we should be able to recover relics of some of them by searching in the comparatively few places where, at the present day, dead plants and animals are entombed and preserved. From the alluvial terraces of rivers, from the silt of lake-bottoms, from the depths of peat-mosses, from the floors of subterranean caverns, from the incrustations left by springs, we might recover traces of some at least of the living things that people the land. And from these fragmentary and incomplete records we might conjecture what

may have been the general character of the life of the time.   By
searching the similar records of earlier ages the geologist has
brought to light many profoundly interesting vestiges of vegetation
and of animal life belonging to types that have long since passed
away.

It must be evident, however, that were we to confine our inquiries
merely to its surface, we should necessarily gain a most imperfect
view of the general history of the Earth.   Beneath that surface,
as volcanoes show, there lies a hot interior, which must have profoundly influenced the changes of the outer parts or crust of the
planet.   The study of volcanoes enables us to penetrate, as it
were, a little way into that interior, and to understand some of
the processes in progress there.   But our knowledge of the inside
of the Earth can obviously be based only to a very limited extent
on direct observation, for man cannot penetrate far below the
surface.   The deepest mines do not go deep enough to reach
materials differing in any essential respect from those visible above
ground.   Nevertheless, by inference from such observations as
can be made, and by repeated and varied experiments in labora-
tories, imitating as closely as can be devised what may be supposed to be the conditions that exist deep within the globe, some
probable conclusions can be drawn even as to the changes that
take place in those deeper recesses that lie for ever concealed
from our eyes.   These conclusions will be stated in later chapters
of this book, and the rocks will be described, on the origin of
which they appear to throw light.

I have compared the soils and rocks with which geology deals
to the records out of which the historian writes the chronicles of
a nation.   We might vary the simile by likening them to the
materials employed in the construction of a great building.   It
is of course interesting enough to know what kinds of marble,
granite, mortar, wood, brass, or iron, have been chosen by an
architect.   But much more important is it to inquire how these
various substances have been grouped together so as to form such
a building.   In like manner, besides the nature and mode of
origin of the various rocks of which the visible and accessible
part of the earth consists, we ought to know how these varied
substances have been arranged so as to build up what we can see
of the outer part or crust of our globe.   In short, we should try
to trace what may be called the architecture of the planet, noting
how each variety of rock occupies its own characteristic place, and
how they are all grouped and braced together in the solid framework

of the land.   This then will be the next subject for consideration in this volume.

But in a great historical edifice, like one of the Gothic minsters of Europe, for example, there are often several different styles. A student of architecture can detect these distinctions, and by their means can show that a cathedral has not been completed in one age ; that it may even have been partially destroyed and re-built during successive centuries, only finally taking its present form after many political vicissitudes and many changes of architectural taste.   Each edifice has thus a separate history, which is recorded by the way the materials have been shaped and put together in the various parts of the masonry.   So it is with the architecture of the Earth.   We have evidence of many demolitions and rebuild-ings, and the story of their general progress can still be deciphered among the rocks.   It is the business of Geology to trace out that story, to put all the scattered materials together, and to make known by what a long succession of changes the Earth has reached its present state.   An outline of what science has accom-plished in this task will form the last and concluding part of this book.

In the following chapters I wish two principles to be kept steadily in view.   In the first place, looking upon Geology as the study of the Earth's history, we need not at first concern ourselves with any details, save those that may be needed to enable us clearly to understand what the general character and progress of this history have been.   In a science which embraces so vast a range as Geology, the multiplicity of facts to be examined and remembered may seem at first to be almost overwhelming.   But a selection of the essential facts is sufficient to give the learner a clear view of the general principles and conclusions of the science, and to enable him to enter with intelligence and interest into more detailed treatises.   In the second place, Geology is essentially a science of observation.   The facts with which it deals should, as far as possible, be verified by our own personal examination.   We should lose no opportunity of seeing with our own eyes the actual progress of the changes which it investigates, and the proofs which it adduces of similar changes in the far past.   To do this will lead us into the fields and hills, to the banks of rivers and lakes, and to the shores of the sea.   We can hardly take any country walk, indeed, in which with duly observant eye we may not detect either some geological operation in actual progress, or the evidence of one which was completed long ago.   Having learnt what to look for and how to interpret it when seen, we are as it were gifted with a new sense. Every landscape comes to possess a fresh interest and charm, for we carry about with us everywhere an added power of enjoyment, whether the scenery has long been familiar or presents itself for the first time. I would therefore seek at the outset to impress upon those who propose to read the following pages, that one of the main objects with which this book is written is to foster a habit of observation, and to serve as a guide to what they are themselves to look for, rather than merely to relate what has been seen and determined by others. If they will so learn these lessons, I feel sure that they will never regret the time and labour they may spend over the task.

PART I

THE MATERIALS FOR THE HISTORY OF THE EARTH

CHAPTER II

THE INFLUENCE OF THE ATMOSPHERE IN THE CHANGES OF THE EARTH'S SURFACE

IN the history of mankind no sharp line can be drawn between the events that are happening now or have happened within the last few generations, and those that took place long ago, and which are sometimes, though inaccurately, spoken of as historical. Every people is enacting its history to-day just as fully as it did many centuries ago. The historian recognises this continuity in human progress. He knows that the feelings and aspirations which guided mankind in old times were essentially the same influences that impel them now, and therefore that the wider his knowledge of his fellowmen of the present day, the broader will be his grasp in dealing with the transactions of former generations. So too is it with the history of the Earth. That history is in progress now as really as it has ever been, and its events are being recorded in the same way and by the same agents as in the far past. Its continuity has never been broken. Obviously, therefore, if we would explore its records "in the dark backward and abysm of time," we should first make ourselves familiar with the manner in which these records are being written from day to day before our eyes.

In this first Part, attention will accordingly be given to the changes in progress upon the Earth at the present time, and to the various ways in which the passing of these changes is chronicled in natural records. We shall watch the actual transaction of geological history, and mark in what way its incidents inscribe themselves on the page of the earth's surface.[1] Every day and hour witness the enacting of some geological event, trifling and transient or stupendous and durable. Sometimes the event leaves behind it only an imperceptible trace of its passage, at other times it graves itself almost imperishably in the annals of the globe. In tracing the origin and development of these geological annals of the present time, we shall best qualify ourselves for deciphering the records of the early revolutions of the planet. We are thereby led to study the various chronicles compiled respectively by the air, rain, rivers, springs, glaciers, the sea, plants and animals, volcanoes and earthquakes—in other words, all the deposits left by the operations of these agents, the scars or other features made by them upon the earth's surface, and all other memorials of geological change. Having learnt how modern deposits are produced, and how they preserve the story of their origin, we shall then be able to group with them the corresponding deposits of earlier times, and to embrace all the geological records, ancient as well as modern, in one general scheme of classification. Such a scheme will enable us to see the continuity of the materials of geological history, and will fix definitely for us the character and relative position of all the chief rocks out of which the visible part of the globe is composed.

Weathering.—The gradual change that overtakes everything on the face of the earth is expressed in all languages by familiar phrases which imply that the mere passing of time is the cause of the change. As Sir Thomas Browne quaintly said more than two hundred years ago, "time antiquates antiquities, and hath an art to make dust of all things." We speak of the dust of antiquity and the gnawing tooth of time. We say that things are time-eaten, worn with age, crumbling under a weight of years. Nothing suggests such epithets so strikingly as an old building. We know that the masonry at first was smooth and fresh ; but now we describe it as weather-beaten, decayed, corroded. So distinctive is this appearance that it is always looked for in an ancient piece of stone-work ; and if not seen, its absence at once suggests a doubt whether the masonry can really be old. No matter of what

---

[1] For descriptions of the ordinary operations of geological agents the reader is referred to my *Class-Book of Physical Geography.* My object now is to direct attention to what is most enduring in these operations, and in what various ways they form permanent geological records.

varieties of stone the edifice may have been built, a few generations may be enough to give them this look of venerable antiquity.   The surface that was left smoothly polished by the builders grows rough and uneven, with scars and holes eaten into it.   Portions of the original polish that may here and there have escaped, serve as a measure of how much has actually been removed from the rest of the surface.

Now, if in the lapse of time, stone which has been artificially dressed is wasted away, we may be quite certain that the same stone in its natural position on the slope of a hill or valley, or by the edge of a river or of the sea, must decay in a similar way. Indeed, an examination of any crumbling building will show that, in proportion as the chiselled surface disappears, the stone puts on the ordinary look which it wears where it has never been cut by man, and where only the finger of time has touched it. Could we remove some of the decayed stones from the building and insert them into a natural crag or cliff of the same kind of stone, their peculiar time-worn aspect would be found to be so exactly that of the rest of the cliff that probably no one would ever suspect that a mason's tools had once been upon them.

FIG. 1.—Weathering of rock, as shown by old masonry. (The "false-bedding" and other original structures of the stone are revealed by weathering.)

From this identity of surface between the time-worn stones of an old building and the stone of a cliff we may confidently infer that the decay so characteristic of ancient masonry is as marked upon natural faces of rock.   The gradual disappearance of the artificial smoothness given by the mason, and its replacement by the ordinary natural rough surface of the stone, shows that this natural surface must also be the result of decay.   And as the peculiar crumbling character is universal, we may be sure that the decay with which it is connected must be general over the globe.

But the mere passing of time obviously cannot change anything, and to say that it does is only a convenient figure of speech. It is not time, but the natural processes which require time for

their work, that produce the widespread decay over the surface of
the earth.   Of these natural processes, there are four that specially
deserve consideration—changes of temperature, saturation and
desiccation, frost, and rain.

(1) Changes of Temperature.—In countries where the
days are excessively hot, with nights correspondingly cool, the
surfaces of rocks heated sometimes, as in parts of Africa, up to
more than 130° Fahr. by a tropical sun, undergo considerable
expansion in consequence of this increase of temperature.   At
night, on the other hand, the rapid radiation quickly chills the
stone and causes it to contract.   Hence the superficial parts, being
in a perpetual state of strain, gradually crack up or peel off.   The
face of a cliff is thus worn slowly backward, and the prostrate
blocks that fall from it are reduced to smaller fragments and
finally to dust.   Where, as in Europe and the settled parts of
North America, the contrasts of temperature are not so marked,
the same kind of waste takes place in a less striking manner.

(2) Saturation and Desiccation.—Another cause of the
decay of the exposed surfaces of rocks is to be sought in the alternate soaking of them with rain and drying of them in sunshine,
whereby the component particles of the stone are loosened and
fall to powder.   Some kinds of stone freshly quarried and left to
this kind of action are rapidly disintegrated.   The rock called
shale (see p. 153) is peculiarly liable to decay from this cause.
The cliffs into which it sometimes rises show at their base long
trails of rubbish entirely derived from its waste.

(3) Frost.—A third and familiar source of decay in stone
exposed to the atmosphere is to be found in the action of Frost
The water that falls from the air upon the surface of the land
soaks into the soil and into · the pores of rocks.   When the
temperature of the air falls below the freezing point, the imprisoned
moisture expands as it passes into ice, and in expanding pushes
aside the particles between which it is entangled.   Where this
takes place in soil, the pebbles and the grains of sand and earth
are separated from each other by the ice that shoots between them.
They are all frozen into a solid mass that rings like stone under
our feet ; but, as soon as a thaw sets in, the ice that formed the
binding cement passes into water which converts the soil into soft
earth or mud.   This process, repeated winter after winter, breaks
up the materials of the soil, and enables them to be more easily
made use of by plants and more readily blown away by wind or
washed off by rain.   Where the action of frost affects the surface

of a rock, the particles separated from each other are eventually
blown or washed away, or the rock peels off in thin crusts or
breaks up into angular pieces, which are gradually disintegrated
and removed.

(4) Rain.— One further cause of decay may be sought in the remarkable power possessed by Rain of chemically corroding stones.
In falling through the atmosphere, rain absorbs the gases of the air,
and with their aid attacks surfaces of rock.   With the oxygen thus
acquired, it *oxidises* those substances which can still take more of
this gas, causing them to *rust* (pp. 117, 123).   As a consequence
of this alteration, the cohesion of the particles is usually weakened,
and the stone crumbles down.   With the carbon-dioxide, or carbonic acid, it dissolves and removes some of the more soluble
ingredients in the form of carbonates, thereby also usually loosening the component particles of the stone.   In general, the influence
of rain is to cause the exposed parts of rocks to rot from the surface inward.   Where the ground is protected with vegetation, the
decay is no doubt retarded ; but in the absence of vegetation, the
outer crust of the decayed layer is apt to be washed off by rain, or
when dried to powder may be blown away and scattered by wind.
As fast as it is removed from the surface, however, it is renewed
underneath by the continued soaking of rain into the stone.

Effects of Weathering.—Hence one of the first lessons to
be learnt—when from the common evidence around us we seek to
know what has been the history of the ground on which we live—
is one of ceaseless decay.   All over the land, in all kinds of
climates, and from various causes, bare surfaces of soil and rock
yield to the influences of the atmosphere or weather.   The decay
thus set in motion is commonly called "weathering."   That it
may often be comparatively rapid is familiarly and instructively
shown in buildings or open-air monuments of which the dates are
precisely known.   Marble tombstones in the graveyards of large
towns, for example, hardly keep their inscriptions legible for even
so long as a century.   Before that time, the surface of the stone
has crumbled away into a kind of sand.   Everywhere the weather-
eaten surfaces, the crumbling crust of decayed stone, and the
scattered blocks and trains of rubbish, tell their tale of universal
waste.

It is well to take numerous opportunities of observing the process of this decay in different situations and on various kinds of
materials.   We can thus best realise the important part which
weathering must play in the changes of the earth's surface, and

we prepare ourselves for the consideration of the next question that arises, What becomes of all the rotted material ?—a question to answer which leads us into the very foundations of geological history.

Openings from the soil down into the rock underneath often afford instructive lessons regarding the decay of the surface of the land. Fig. 2, for instance, is a drawing of one of these sections, in which a gradual passage may be traced from solid sandstone (*a*) underneath up into broken-up sandstone (*b*), and thence into the earthy layer (*c*) that supports the vegetation of the surface. Traced from below upwards, the rock is found to become more and more broken and crumbling, with an increasing number of rootlets that strike freely through it in all directions, until it passes insensibly into the uppermost dark layer of vegetable soil or humus. This dark layer owes its characteristic brown or black colour to the decaying remains of vegetation diffused through it.

FIG. 2.—Passage of sandstone upwards into soil.

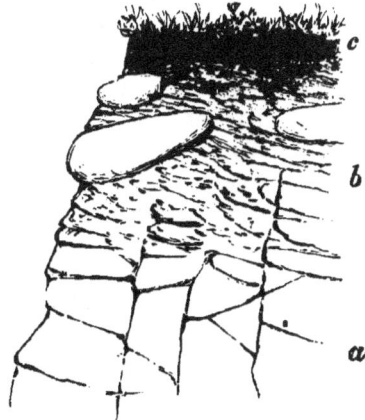

Again, granite in its unweathered state is a hard, compact, crystalline rock that may be quarried out in large solid blocks (*a* in Fig. 3), yet when traced upward to within a few feet from the surface it may be seen to have been split by innumerable rents into fragments which are nevertheless still lying in their original position. As these fragments are attacked by percolating moisture, their surfaces decay, leaving the still unweathered parts as rounded blocks (*b*), which might at first be mistaken for transported boulders. They are, however, parts of the rock broken up in place, and not fragments that have been carried from a distance. The little quartz veins that traverse the solid granite can be recognised running through the

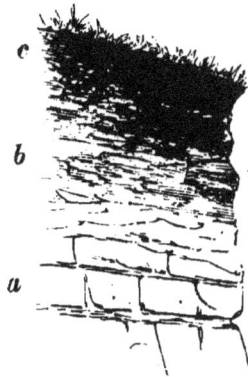

FIG. 3.—Passage of granite upwards into soil.

decayed and fresh parts alike. But, besides being broken into pieces, the granite rots away and loses its cohesion. Some of the smaller pieces can be crumbled down between the fingers, and this decay increases upwards, until the rock becomes a mere sand or sandy clay in which a few harder kernels are still left. Into this soft layer roots may descend from the surface, and, like the sandstone, the granite merges above into the overlying soil (*c*).

Soil and Subsoil.—In such sections as the foregoing, three distinct layers can be recognised which pass into each other. At the bottom lies the *rock*, either undecayed or at least still fresh enough to show its true nature. Next comes the broken-up crumbling layer through which stray roots descend, and which is known as the *subsoil*. At the top lies the dark band, crowded with rootlets and forming the true *soil*. These three layers obviously represent successive stages in the decay of the surface of the land. The soil is the layer of most complete decay. The subsoil is an intermediate band where the progress of decomposition has not advanced so far, while the shattered rock underneath shows the earlier stages of disintegration. Vegetation sends its roots and rootlets through the rotted rock. As the plants die, they are succeeded by others, and the rotted remains of their successive generations gradually darken the uppermost decomposed layer. Worms, insects, and larger animals that may die on the surface, likewise add their mouldering remains to this uppermost deposit. And thus from animals and plants there is furnished to the soil that *organic matter* on which its fertility so much depends. The very decay of the vegetation helps to promote that of the underlying rock, for it supplies various organic acids ready to be absorbed by percolating rain-water, the power of which to decompose rocks is thereby increased (p. 24).

It is obvious, then, that in answer to the question, What becomes of the rotted material produced by weathering? we may confidently assert that, over surfaces of land protected by a cover of vegetation, this material in large measure accumulates where it is formed. Such accumulation will naturally take place chiefly on flat or gently inclined ground. Where the slope is steep, the decomposed layer will tend to travel down-hill by mere gravitation, and to be further impelled downward by descending rain-water.

If there is so intimate a connection between the soil at the surface and the rock underneath, we can readily understand that soils should vary from one district to another, according to the nature of the underlying rocks. Clays will produce clayey soil, sandstones, sandy soil, or, where these two kinds of rock occur together, they may give rise to sandy clay or loam. Hence, knowing what the underlying rock is, we may usually infer what must be the character of the overlying soil, or, from the nature of the soil, we may form an opinion respecting the quality of the rock that lies below.

But it will probably occur to the thoughtful observer that when once a covering of soil and subsoil has been formed over a level piece of ground, especially where there is also an overlying carpet of verdure, the process of decay should cease—the very layer of rotted material coming eventually to protect the rock from further disintegration. Undoubtedly, under these circumstances, weathering is reduced to its feeblest condition. But that it still continues will be evident from some considerations, the force of which will be better understood a few pages further on. If the process were wholly arrested, then in course of time plants growing on the surface would extract from the soil all the nutriment they could get out of it, and with the increasing impoverishment of the soil, they would dwindle away and finally die out, until perhaps only the simpler forms of vegetation would grow on the site. Something of this kind not improbably takes place where forests decay and are replaced by scrub and grass. But the long-continued vigorous growth of the same kind of plants upon a tract of land doubtless indicates that in some way the process of weathering is not entirely arrested, but that, as generation succeeds generation, the plants are still able to draw nutriment from fresh portions of decomposed rock. A cutting made through the soil and subsoil shows that roots force their way downward into the rock, which splits up and allows percolating water to soak downwards through it. The subsoil thus gradually eats its way into the solid rock below. Influences are at work also, whereby there is an imperceptible removal of material from the surface of the soil. Notable among these influences are Rain, Wind, and Earthworms.

Wherever soil is bare of vegetation it is directly exposed to removal by **Rain**. Ground is seldom so flat that rain may not flow a little way along the surface before sinking underneath. In its flow, it carries off the finer particles of the soil. These may travel each time only a short way, but as the operation is repeated, they are in the course of years gradually moved down to lower ground or to some runnel or brook that sweeps them away seaward. Both on gentle and on steep slopes, this transporting power of rain is continually removing the upper layer of bared soil.

C

Where soil is exposed to the sun, it is liable to be dried into mere dust, which is borne off by **Wind.** How readily this may happen is often strikingly seen after dry weather in spring-time. The earth of ploughed fields becomes loose and powdery, and clouds of its finer particles are carried up into the air and transported to other farms, as gusts of wind sweep across. "March dust," which is a proverbial expression, may be remembered as an illustration of one way in which the upper parts of the soil are removed.

Even where a grassy turf protects the general surface, bare places may always be found whence this covering has been removed. Rabbits, moles, and other animals throw out soil from their burrows. Mice sometimes lay it bare by eating the pasture down to the roots. The common **Earthworms** bring up to daylight in the course of a year an almost incredible quantity of it in their castings. Mr. Darwin estimated that this quantity is in some places not less than 10 tons per annum over an acre of ground. Only the finest particles of mould are swallowed by worms and conveyed by them to the surface, and it is precisely these which are most apt to be washed off by rain or to be dried and blown away as dust by the wind. Where it remains on the ground, the soil brought up by worms covers over stones and other objects lying there, which consequently seem to sink into the earth. The operation of these animals causes the materials of the soil to be thoroughly mixed. In tropical countries, the termite or "white ant" conveys a prodigious amount of fine earth up into the open air. With this material it builds hills sometimes 60 feet high and visible for a distance of several miles ; likewise tunnels and chambers, which it plasters all over the stems and branches of trees, often so continuously that hardly any bark can be seen. The fine soil thus exposed is liable to be blown away by the wind or washed off by the fierce tropical rains.

Although, therefore, the layer of vegetable soil which covers the land appears to be a permanent protection, it does not really prevent a large amount of material from being removed even from grassy ground. It forms the record of the slow and almost imperceptible geological changes that affect the regions where it accumulates,—the quiet fall of rain, the gradual rotting away of the upper part of the underlying rock, the growth and decay of a long succession of generations of plants, the ceaseless labours of the earthworm, the scarcely appreciable removal of material from the surface by the action of rain and wind, and the equally insensible descent of the crumbling subsoil farther and farther into the solid stone below.   Having learnt how all this is told by the soil beneath our feet, we should be ready to recognise in the soil of former ages a similar chronicle of quiet atmospheric disintegration.

Talus.—Besides soil and subsoil, there are other forms in which decomposed rock accumulates on the surface of the land. Where a large mass of bare rock rises up as a steep bank or cliff, it is liable to constant degradation, and the materials detached from its surface accumulate down the slopes, forming what is known as a *Talus* (Fig. 4).   In mountainous or hilly regions,

Fig. 4.—Talus-slopes at the foot of a line of cliffs.

where rocky precipices rise high into the air, there gather at their feet and down their clefts long trails or *screes* of loose blocks that have been split off from them by the weather.   Such slopes, especially where they are not too steep, and where the rubbish that forms them is not too coarse, may be more or less covered with vegetation, which in some measure arrests the descent of the debris.   But from time to time, during heavy rains, deep gullies are torn out of them by rapidly formed torrents, which sweep down their materials to lower levels (Fig. 10).   The sections laid bare in these gullies show that the rubbish is arranged in more or less distinct layers which lie generally parallel with the surface of the slope ; in other words, it is rudely stratified, and its layers or strata are inclined at the angle of the declivity which seldom exceeds 35°.

Rain-wash, Brick-earth. — On more gentle slopes, even where no bare rock projects into the air, the fall of rain gradually washes down the upper parts of the soil to lower levels.   Hence arise thick accumulations of what is known as *rain-wash*—soil mixed often with angular fragments of still undecomposed rock, and not infrequently forming a kind of brick-earth (Fig. 5). Deposits of this nature are still gathering now, though their lower portions may be of great antiquity. In the south-east of England, for instance, the brick-earths contain the bones of animals that have long since passed away.

Dust.—By the action of wind, above referred to, a vast amount of fine dust and sand is carried up into the air and strewn far and wide over the land. In dry countries, such as large tracts of Central Asia, the air is often thick with a fine yellow dust which may entirely obscure the sun at mid-day, and which settles over everything. After many centuries, a deposit, which may be hundreds of feet deep, is thus accumulated on the surface of the land. Some of the ancient cities of the Old World, Nineveh and Babylon for example, after being long abandoned by man, have gradually been buried under the fine soil drifted over them by the wind and intercepted and protected by the weeds that grew up over the ruins. Even in regions where, as in Britain, there is a large annual rainfall, seasons of drought occur, during which there may be a considerable drifting of the finer particles of soil by the wind. We probably hardly realise how much the soil may be removed here and heightened there from this cause.

FIG. 5.—Section of rain-wash or brick-earth. 7. Vegetable soil. 6 Brick-earth. 5.White sand. 4. Brick-earth. 3. White sand. 2. Brick-earth. 1. Gravel with seams of sand.

Sand-dunes.—Some of the most striking and familiar examples of the accumulation of loose deposits by the wind are those to which the name of Dunes is given. On sandy shores, exposed to winds that blow landwards, the sand is dried and then carried away from the beach, gathering into long mounds or ridges which run parallel to the coast-line. These ridges are often 50 or 60 feet, sometimes even more than 250 feet high, with deep troughs and irregular circular hollows between them, and they occasionally form a strip several miles broad, bordering the sea. The particles of sand are driven inland by the wind, and the dunes gradually bury fields, roads, and villages, unless their progress is arrested by the growth of vegetation over their shifting surfaces. On many parts of the west coast of Europe, the dunes are marching into the interior at the rate of 20 feet in a year. Hence large tracts of land have within historic times been entirely lost under them.   In the north of Scotland, for example, an ancient and extensive barony, so noted for its fertility that it was called " the granary of Moray," was devastated about the middle of the seventeenth century by the moving sands, which now rise in barren ridges more than 100 feet above the site of the buried land.   In the interior of continents also, where with great dryness of climate there is a continual disintegration of the surface of rocks, wide wastes of sand accumulate, as in the deserts of Libya, Arabia,

FIG. 6.—Sand-dunes.

and Gobi, in the heart of Australia, and in many of the western parts of the United States.

There can be no doubt, however, that though the layer of vegetable soil, the heaps of rubbish that gather on slopes and at the base of rocky banks and precipices, and the widespread drifting of dust and sand over the land, afford evidence that much of the material arising from the general decay of the surface of the land accumulates under various forms upon that surface, nevertheless its stay there is not permanent.   Wind and rain are continually removing it, sometimes in vast quantities, into the sea. Every brook, made muddy by heavy rain, is an example of this transport, for the mud that discolours the water is simply the finer material of the soil washed off by rain.   When we reflect upon the multitude of streams, large and small, in all parts of the globe, and consider that they are all busy carrying their freights of mud to the sea, we can in some measure appreciate how great must be the total annual amount of material so removed. What becomes of this material will form the subject of succeeding chapters.

Summary.—The first lesson to be learnt from an examination of the surface of the land is, that everywhere decay is in progress upon it. Wherever the solid rock rises into the air, it breaks up and crumbles away under the various influences combined in the process of Weathering. The wasted materials caused by this universal disintegration partly accumulate where they are formed, and make soil. But in large measure, also, they are blown away by wind and washed off by rain. Even where they appear to be securely protected by a covering of vegetation, the common earth-worm brings the finer parts of them up to the surface, where they come within reach of rain and wind, so that on tracts permanently grassed over, there may be a continuous and not inconsiderable removal of fine soil from the surface. In proportion as the upper layers of soil are removed, roots and percolating water are enabled to reach down farther into the solid rock which is broken up into subsoil, and thus the general surface of the land is insensibly lowered.

Besides accumulating *in situ* as subsoil and soil, the debris of decomposed rock forms talus-slopes and screes at the foot of crags, and a layer of rain-wash or brick-earth over gentler slopes. Where the action of wind comes markedly into play, tracts of sand-dunes may be piled up along the borders of the sea and of lakes, or in the arid interior of continents ; and wide regions have been in course of time buried under the fine dust which is some-times so thick in the air as to obscure the noonday sun. But in none of these forms can the accumulation of decomposed material be regarded as permanent. So long as it is exposed to the influences of the atmosphere, this material is still liable to be swept away from the surface of the land and borne outwards into the sea.

CHAPTER III

It appears, then, that from various causes all over the globe, there is a continual decay of the surface of the land ; that the decomposed material partly accumulates as soil, subsoil, and sheets or heaps of loose earth or sand, but that much of it is washed off the land by rain or blown into the rivers or into the sea by wind. We have now to consider the part taken by Running Water in this transport. From the single rain-drop up to the mighty river, every portion of the water that flows over the land is busy with its own share of the work. When we reflect on the amount of rain that falls annually over the land, and on the number of streams, large and small, that are ceaselessly at work, we realise how difficult it must be to form any fit notion of the entire amount of change which, even in a single year, these agents work upon the surface of the earth.

The influence of rain in the decay of the surface of the land was briefly alluded to in the last chapter. As soon as a drop of rain reaches the ground, it begins its appointed geological task, dissolving what it can carry off in solution, and pushing forward and downward whatever it has power to move. As the rain-drops gather into runnels, the same duty, but on a larger scale, is performed by them ; and as the runnels unite into large streams, and these into yet mightier rivers, the operations, though becoming colossal in magnitude, remain essentially the same in kind. In the operations of the nearest brook, we see before us in miniature a sample of the changes produced by the thousands of rivers which, in all quarters of the globe, are flowing from the mountains to the sea. Watching these operations from day to day, we discover that they may all be classed under two heads. In the first place, the brook hollows out the channel in which it flows and thus aids in the general waste of the surface of the land ; and in the second place, it carries away fine silt and other material resulting from that waste, and either deposits it again on the land or carries it out to sea.   Rivers are thus at once agents that themselves directly degrade the land, and that sweep the loosened detritus towards the ocean.   An acquaintance with each of these kinds of work is needful to enable us to understand the nature of the records which river-action leaves behind it.

i. Erosive and transporting Power of Running Water.

Chemical Action.—We have seen that rain in its descent from the clouds absorbs air, and that with the oxygen and carbonic acid which it thus obtains it proceeds to corrode the surfaces of rock on which it falls.   When it reaches the ground and absorbs the acids termed " humous," which are supplied by the decomposing vegetation of the soil, it acquires increased power of eating into the stones over which it flows.   When it rolls along as a runnel, brook, or river, it no doubt still attacks the rocks of its channel, though its action in this respect is not so easily detected.   In some circumstances, however, the solvent influence of river-water upon solid rocks is strikingly displayed.   Where the water contains a large proportion of the acids of the soil, and flows over a kind of rock specially liable to be eaten away by these acids, the most favourable conditions are presented for observing the change. Thus, a stream which issues from a peat-bog is usually dark brown in colour, from the vegetable solutions which it extracts from the moss.   Among these solutions are some of the organic acids referred to, ready to eat into the surface of the rocks or loose stones which the stream may encounter in its descent.   No kind of rock is more liable than limestone to corrosion under such circumstances.   Peaty water flowing over it eats it away with comparative rapidity, while those portions of the rock that rise above the stream escape solution, except in so far as they are attacked by rain.   Hence arise some curious features in the scenery of limestone districts.   The walls of limestone above the water, being attacked only by the atmosphere, are not eaten away so fast as their base, over which the stream flows.   They are consequently undermined, and are sometimes cut into dark tunnels and passages (Fig. 7).   Even where the solvent action of the water of rivers is otherwise inappreciable, it can be detected by

means of chemical analysis. Thus rivers, partly by the action of
their water upon the loose stones and solid rocks of their channels,
and partly by the contributions they receive from Springs (which
will be afterwards described), convey a vast amount of dissolved
material into the sea. The mineral substance thus invisibly
transported consists of various salts. One of the most abundant
of these—carbonate of lime—is the substance that forms limestone, and furnishes the mineral matter required for the hard parts
of a large proportion of the lower animals. It is a matter of some
interest to know that this substance, so indispensable for the

Fig. 7.—Erosion of limestone by the solvent action of a peaty stream,
Durness, Sutherlandshire.

formation of the shells of so large a number of sea-creatures,
is constantly supplied to the sea by the streams that flow into it.[1]
The rivers of Western Europe, for instance, have been ascertained
to convey about 1 part of dissolved mineral matter in every 5000
parts of water, and of this mineral matter about a half consists
of carbonate of lime. It has been estimated that the Rhine bears
enough carbonate of lime into the sea every year to make three
hundred and thirty-two thousand millions of oysters of the usual
size. Another abundant ingredient of river-water is gypsum or

[1] There is now reason, however, to suspect that the carbonate of lime in
marine organisms is not derived so much from the comparatively minute
proportion of that substance present in solution in sea-water, as from the
much more abundant sulphate of lime which undergoes apparently a process
of chemical transformation into carbonate within the living animals.

sulphate of lime, of which the Thames is computed to carry annually past London not less than 180,000 tons. The total quantity of carbonate of lime, removed from the limestones of its basin by this river in a year, amounts, on an average, to 140 tons from every square mile, which is estimated to be equal to the lowering of the general surface to the extent of $\frac{9}{100}$ of an inch from each square mile in a century, or one foot in 13,200 years.

Mechanical Action—(1) Transport.—The dissolved material forms but a small proportion of the total amount of mineral substances conveyed by rivers from land to sea. A single shower of rain washes off fine dust and soil from the surface of the ground into the nearest brook which then rolls along with a discoloured current. An increase in the volume of the water enables a stream to sweep along sand, gravel, and blocks of stone lying in its channel, and to keep these materials moving until, as the declivity lessens and the rain ceases, the current becomes too feeble to do more than lazily carry onward the fine silt that discolours it. Every stream, large or small, is ceaselessly busy transporting mud, sand, or gravel. And as the ultimate destination of all this sediment is the bottom of the sea, it is evident that if there be no compensating influences at work to repair the constant loss, the land must in the end be all worn away.

Some of the most instructive lessons regarding the work of running water on land are afforded by the beds of mountain-torrents. Huge blocks, detached from the crags and cliffs on either side, may there be seen cumbering the pathway of the water, which seems quite powerless to move such masses and can only sweep round them or find a passage beneath them. But visit such a torrent when it is swollen with heavy rains or rapidly melted snow, and you will hear the stones knocking against each other or on the rocky bottom, as they are driven downwards by the flood. Or when the stream is at its lowest, in dry summer weather, follow its course a little way down hill, and you will see that by degrees the blocks, losing their sharp edges, have become rounded boulders, and that these are gradually replaced by coarse shingle and then by finer gravel. In the quieter reaches of the water, sheets of sand begin to make their appearance, and at last when the stream reaches the plains, no sediment of coarser grain than mere silt may be seen in its channel. It is thus obvious that in the constant transport maintained by watercourses, the carried materials, by being rolled along rocky channels and continually ground against each other, diminish in size as they descend. A river flowing from a range of mountains to the distant ocean may be likened to a mill, into which large angular masses of rock are cast at the upper end, and out of which only fine sand and silt are discharged at the lower.

Partly, therefore, owing to the fine dust and soil swept into them by wind and rain from the slowly decomposing surface of the land, and partly to the friction of the detritus which they sweep along their channels, rivers always contain more or less mineral matter suspended in their water or travelling with the current on the bottom. The amount of material thus transported varies greatly in different rivers, and at successive seasons even in the same river. In some cases, the rainfall is spread so equably through the year that the rivers flow onward with a quiet monotony, never rising much above nor sinking much below their average level. In such circumstances, the amount of sediment they carry downward is proportionately small. On the other hand, where either from heavy periodical rains or from rapid melting of snow, rivers are liable to floods, they acquire an enormously increased power of transport, and their burden of sediment is proportionately augmented. In a few days or weeks of high water, they may convey to the sea a hundredfold the amount of mineral matter which they could carry in a whole year of their quieter mood.

Measurements have been made of the proportions of sediment in the waters of different rivers at various seasons of the year. The results, as might be expected, show great variations. Thus the Garonne, rising among the higher peaks of the Pyrenees, drains a large area of the south of France, and is subject to floods by which an enormous quantity of sediment is swept down from the mountains to the plains. Its proportion of mud has been estimated to be as much as 1 part in 100 parts of water. The Durance, which takes its source high on the western flank of the Cottian Alps, is one of the rapidest and muddiest rivers in Europe. Its angle of slope varies from 1 in 208 to 1 in 467, the average declivity of the great rivers of the globe being probably not more than 1 in 2600, while that of a navigable stream ought not to exceed 10 inches per mile or 1 in 6336. The Durance is, therefore, rather a torrent than a river. With this rapidity of descent is conjoined an excessive capacity for transporting sediment. In floods of exceptional severity, the proportion of mud in the stream has been estimated at one-tenth by weight of the water, while the average proportion for nine years from 1867 to 1875 was about $\frac{1}{850}$. Probably the best general average is to be obtained from a river which drains a wide region exhibiting considerable diversities of climate, topography, rocks, and soils.   The Mississippi presents a good illustration of these diversities, and has accordingly been taken as a kind of typical river, furnishing, so to speak, a standard by which the operations of other rivers may be compared, and which may perhaps be assumed as a fair average for all the rivers of the globe.   Numerous measurements have been made of the proportion of sediment carried into the Gulf of Mexico by this vast river, with the result of showing that the average amount of sediment is by weight 1 part in every 1500 parts of water, or little more than one-third of the proportion in the water of the Durance.

If now we assume that, all over the world, the general average proportion of sediment floating in the water of rivers is 1 part in every 1500 of water, we can readily understand how seriously in the course of time must the land be lowered by the constant removal of so much decomposed rock from its surface.   Knowing the area of the basin drained by a river, and also the proportion of sediment in its water, we can easily calculate the general loss from the surface of the basin.   The ratio of the weight or "specific gravity" of the silt to that of solid rock may be taken to be as 19 is to 25.   Accordingly the Mississippi conveys annually from its drainage basin an amount of sediment equivalent to the removal of $\frac{1}{6000}$ part of a foot of rock from the general surface of the basin.   At this rate, one foot of rock will be worn away every 6000 years.   If we take the general height of the land of the whole globe to be 2120 feet, and suppose it to be continuously wasted at the same rate at which the Mississippi basin is now suffering, then the whole dry land would be carried into the sea in 12,720,000 years.   Or if we assume the mean height of Europe to be 973 feet and that this continent is degraded at the Mississippi rate of waste until the last vestige of it disappears, the process of destruction would be completed in rather less than 6,000,000 years.   Such estimates are not intended to be close approximations to the truth.   As the land is lowered, the rate of decay will gradually diminish, so that the later stages of decay will be enormously protracted.   But by taking the rate of operation now ascertained to be in progress in such a river basin as the Mississippi, we obtain a valuable standard of comparison, and learn that the degradation of the land is much greater and more rapid than might have been supposed.

(2) Erosion.—But rivers are not merely carriers of the mud, sand, and gravel swept into their channels by other agencies.   By

keeping these materials in motion, the currents reduce them in size, and at the same time employ them to hollow out the channels wherein they move. The mutual friction that grinds down large blocks of rock into sand and mud, tells also upon the rocky beds along which the material is driven. The most solid rocks are worn down; deep long gorges are dug out, and the water-courses, when they have once chosen their sites, remain on them and sink gradually deeper and deeper beneath the general level of the country. The surfaces of stone exposed to this attrition assume the familiar smoothed and rounded appearance which is known as *water-worn*. The loose stones lying in the channel of a stream, and the solid rocks as high up as floods can scour them, present this characteristic aspect. Here and there, where a few stones have been caught in an eddy of the current, and are kept in constant gyration, they reduce each other in dimensions, and at the same time grind out a hollow in the underlying rock. The sand and mud produced by the friction are swept off ·by the current, and the stones when sufficiently reduced in size are also carried away. But their places are eventually taken by other blocks brought down by floods, so that the supply of grinding material is kept up until the original hollow is enlarged into a wide deep caldron, at the bottom of which the stones can only be stirred by the heaviest floods. Cavities of this kind, known as *pot-holes*, are of frequent occurrence in rocky watercourses as well as on rocky shores, in short, wherever eddies of water can keep shingle rotating upon solid rock. As they often coalesce by the wearing away of the intervening wall of rock, they greatly aid in the deepening of a watercourse. In most rocky gorges, a succession of old pot-holes may be traced far above the present level of the stream (Fig. 8).

That it is by means of the gravel and other detritus pushed along the bottom by the current, rather than by the mere friction of the water on its bed, that a river excavates its channel, is most strikingly shown immediately below a lake. In traversing a lake, the tributary streams deposit their sediment on its bottom, because the still water checks their current and, by depriving the water of its more rapid movement, compels it to drop its burden of gravel, sand, and silt (see p. 42). Filtered in this way, the various streams united in the lake escape at its lower end as a clear trans-parent river. The Rhone, for instance, flows into the Lake of Geneva as a turbid stream; it issues from that great reservoir at Geneva as a rushing current of the bluest, most translucent water which, though it sweeps over ledges of rock, has not yet been able to grind them down into a deep gorge. The Niagara, also, filtered by Lake Erie, has not acquired sediment enough to enable it to cut deeply into the rocks over which it foams in its rapids before throwing itself over the great Falls.

One of the most characteristic features of streams is the singularly sinuous courses they follow. As a rule, too, the flatter

FIG. 8.—Pot-holes worn out by the gyration of stones in the bed of a stream.

the ground over which they flow, the more do they wind. Not uncommonly they form loops, the nearest bends of which in the end unite, and as the current passes along the now straightened channel, the old one is left to become by degrees a lake or pond of stagnant water, then a marsh, and lastly, dry ground. We might suppose that in flowing off the land, water would take the shortest and most direct road to the sea. But this is far from being the case. The slightest inequalities of level have originally determined sinuosities of the channels, while trifling differences in the hardness of the banks, in the accumulation of sediment, and in the direction of the currents and eddies, have been enough to turn a stream now to one side now to another, until it has assumed its present meandering course. How easily this may be done can be instructively observed on a roadway or other bare surface of ground. When quite dry and smooth, hardly any depressions in which water would flow might be detected on such a surface. But after a heavy shower of rain, runnels of muddy water will be seen coursing down the slope in serpentine channels that at once recall the winding rivers of a great drainage-system. The slightest differences of level have been enough to turn the water from side to side. A mere pebble or projecting heap of earth or tuft of grass has sufficed to cause a bend. The water, though always descending, has only been able to reach the bottom by keeping the lowest levels, and turning from right to left as these guided it.

When a river has once taken its course and has begun to excavate its channel, only some great disturbance, such as an earthquake or volcanic eruption, can turn it out of that course. If its original pathway has been a winding one, it goes on digging out its bed which, with all its bends, gradually sinks below the level of the surrounding country. The deep and picturesque gorge in which the Moselle winds from Trèves to Coblenz has in this way been slowly eroded out of the undulating tableland across which the river originally flowed.

In another and most characteristic way, the shape of the ground and the nature and arrangement of the rocks over which they flow, materially influence rivers in the forms into which they carve their channels. The Rhone and the Niagara, for instance, though filtered by the lakes through which they flow, do not run far before plunging into deep ravines. Obviously such ravines cannot have been dug out by the same process of mechanical attrition whereby river-channels in general are eroded. Yet the frequency of gorges in river scenery shows that they cannot be due to any exceptional operation. They may generally be accounted for by some arrangement of rocks wherein a bed of harder material is underlain by one more easily removable. Where a stream, after flowing over the upper bed, encounters the decomposable bed below, it eats away the latter more rapidly. The overlying hard rock is thus undermined, and, as its support is destroyed, slice after slice is cut away from it. The waterfall which this kind of structure produces continues to eat its way backward or up the course of the stream, so long as the necessary conditions are maintained of hard rocks lying upon soft. Any change of structure which would bring the

hard rocks down to the bed of the channel, and remove the sc
rocks from the action of the current and the dash of the spra
would gradually destroy the waterfall.   It is obvious that, by cu
ting its way backward, a waterfall excavates a ravine.

The renowned Falls of Niagara supply a striking illustration
the process now described.   The vast body of water which issu
from Lake Erie, after flowing through a level country for a fe
miles, rushes down its rapids and then plunges over a precipi
of solid limestone.   Beneath this hard rock lies a band of cor
paratively easily eroded shale.   As the water loosens and remov
the lower rock, large portions of the face of the precipice behir
the Falls are from time to time precipitated into the boiling flo
below.   The waterfall is thus slowly prolonging the ravine belc
the Falls.   The magnificent gorge in which the Niagara, after i
tumultuous descent, flows sullenly to Lake Ontario is not less th:
7 miles long, from 200 to 400 yards wide, and from 200 to 3
feet deep.   There is no reason to doubt that this chasm has be
entirely dug out by the gradual recession of the Falls from t
cliffs at Queenstown, over which the river at first poured.   V
may form some conception of the amount of rock thus remov
from the estimate that it would make a rampart about 12 feet hi
and 6 feet thick extending right round the whole globe at t
equator.   Still more gigantic are the gorges or cañons of t
Colorado and its tributaries in Western America.   The Grai
Cañon of the Colorado is 300 miles long, and in some places mc
than 6000 feet deep (Fig. 9).   The country traversed by it is
network of profound ravines, at the bottom of which the streai
flow that have eroded them out of the table-land.

ii. DEPOSITION OF MATERIALS BY RUNNING WATER.

Permanent Records of River-Action.—If, then, all t
streams on the surface of the globe are engaged in the double ta
of digging out their channels and carrying away the loose materi
that arise from the decomposition of the surface of the land, l
us ask ourselves what memorials of these operations they lea
behind them.   In what form do the running waters of the la
inscribe their annals in geological history ?   If these waters cou
suddenly be dried up all over the earth, how could we tell wh
changes they had once worked upon the surface of the land ?   C
we detect the traces of ancient rivers where there are no rivers no
From what has been said in this lesson it will be evident th

Fig. 9.—Grand Cañon of the Colorado. (Holmes.)

in answer to such questions as these, we may affirm that one unmistakable evidence of the former presence of rivers is to be found
in the channels which they have eroded. The gorges, rocky
defiles, pot-holes, and water-worn rocks which mark the pathway
of a stream would long remain as striking memorials of the work
of running water. In districts, now dry and barren, such as large
regions in the Levant, there are abundant channels (*wadies*) now
seldom or never occupied by a stream, but which were evidently
at one time the beds of active torrents.

Alluvium.—But more universal testimony to the work of
running water is to be found in the deposits which it has accumulated. To these deposits the general name of *alluvium* has been
given. Spreading out on either side, sometimes far beyond the
limits of the ordinary or modern channels, these deposits, even
when worn into fragmentary patches, retain their clear record of
the operations of the river. Let us in imagination follow the course
of a river from the mountains to the sea, and mark as we go the
circumstances under which the accumulation of sediment takes place.

The power possessed by running water to carry forward sediment depends mainly upon the velocity of the current. The more
rapidly a stream flows, the more sediment can it transport, and
the larger are the blocks which it can move. The velocity is
regulated chiefly by the angle of slope ; the greater the declivity,
the higher the velocity and the larger the capacity of the stream
to carry down debris. Any cause, therefore, which lessens the
velocity of a current diminishes its carrying power. If water,
bearing along gravel, sand, or mud, is checked in its flow, some
of these materials will drop and remain at rest on the bottom.
In the course of every stream, various conditions arise whereby
the velocity of the current is reduced. One of the most obvious
of these is a diminution in the slope of the channel. Another is
the union of a rapid tributary with a more gently flowing stream.
A third is the junction of a stream with the still waters of a lake
(see p. 42) or with the sea. In these circumstances, the flow of
the water being checked, the sediment at once begins to fall to
the bottom.

Tracing now the progress of a river, for illustrations of this law
of deposition, we find that among the mountains where the river
takes its rise, the torrents that rush down the declivities have torn
out of them such vast quantities of soil and rock as to seam them
with deep clefts and gullies. Where each of these rapid streamlets reaches the valley below, its rapidity of motion is at once

lessened, and with this slackening of speed and consequent loss of
carrying power, there is an accompanying deposit of detritus.
Blocks of rock, angular rubbish, rounded shingle, sand, and earth
are thrown down in the form of a cone of which the apex starts
from the bottom of the gully and the base spreads out over the
plain (Fig. 10).   Such cones vary in dimensions according to the

FIG. 10.—Gullies torn out of the side of a mountain by descending torrents, with
cones of detritus at their base.

size of the torrent and the comparative ease with which the rocks
of the mountain-side can be loosened and removed.   Some of
them, thrown down by the transient runnels of the last sudden
rain-storm, may not be more than a few cubic yards in bulk.   But
on the skirts of mountainous regions they may grow into masses
hundreds of feet thick and many miles in diameter.   The valleys
in a range of mountains afford many striking examples of these
alluvial cones or *fans*, as they are called.   Where the tributary

torrents are numerous, a succession of such cones or fans, nearly
or quite touching each other, spreads over the floor of a valley.
From this cause, so large an amount of detritus has within historic
times been swept down into some of the valleys of the Tyrol
that churches and other buildings are now half-buried in the
accumulation.   •

Looking more closely at the materials brought down by the
torrents, we find them arranged in rude irregular layers, sloping
downwards into the plain, the coarsest and most angular detritus
lying nearest to the mountains, while more rounded and water-worn
shingle or sand extends to the outer margin of the cone.   This
grouping of irregular layers of angular and half-rounded detritus
is characteristic of the action of torrents.   Hence, where it occurs,
even though no water may run there at the present day, it may
be regarded as indicating that at some former time a torrent swept
down detritus over that site.

Quitting the more abrupt declivities, and augmented by
numerous tributaries from either side, the stream whose course we
are tracing loses the character of a torrent and assumes that of a
river.   It still flows with velocity enough to carry along not only
mud and sand, but even somewhat coarse gravel.   The large
angular blocks of the torrential part of its course, however, are no
longer to be seen, and all the detritus becomes more and more
rounded and smoothed as we follow it towards the plains.   At
many places, deposits of gravel or sand take place, more especially at the inner side of the curves which the stream makes as it
winds down the valley.   Sweeping with a more rapid flow round
the outer side of each curve, the current lingers in eddies on the
inner side and drops there a quantity of sediment.   When the
water is low, strips of bare sand and shingle on the concave side
of each bend of the stream form a distinctive feature in river
scenery.   It is interesting to walk along one of these strips and
to mark how the current has left its record there.   The stones
are well smoothed and rounded, showing that they have been
rolled far enough along the bottom of the channel to lose
their original sharp edges, and to pass from the state of
rough angular detritus into that of thoroughly water-worn gravel.
Further, they will be found not to lie entirely at random, as might
at first sight be imagined.   A little examination will show that,
where the stones are oblong, they are generally placed with their
longer axis pointing across the stream.   This would naturally be
the position which they would assume where the current kept

rolling them forward along the channel.   Those which are flat in shape will be observed usually to slope up stream.   That the sloping face must look in the direction from which the current moves will be evident from Fig. 11, where a current, moving in the direction of the arrow and gradually diminishing in force, would no longer be able to overturn the stones which it had so

FIG. 11.—Flat stones in a bank of river-shingle, showing the direction of the current (indicated by the arrow) that transported and left them.

placed as to offer the least obstacle to its passage.   Had the current flowed from the opposite quarter, it would have found the upturned edges of the stones exposed to it, and would have readily overturned them until they found a position in which they again presented least resistance to the water.   In a section of gravel, it is thus often quite possible to tell from what quarter the current flowed that deposited the pebbles.

Yet another feature in the arrangement of the materials is well seen where a digging has been made in one of the alluvial banks, but better still in a section of one of the terraces to be immediately referred to.   The layers of gravel or sand in some bands may be observed to be inclined at a steeper angle than in others, as shown in the accompanying figure (Fig. 12).   In such cases, it will be noticed that the slope of the more inclined layers is down the stream, and hence that their direction gives a clue to that of the current which arranged them.   We may watch similar layers in the act of deposition among shallow pools into which currents

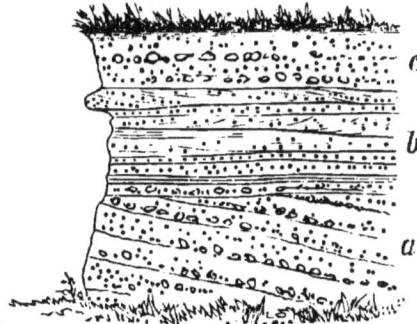

FIG. 12.—Section of alluvium showing direction of currents.

are discharging sediment.   The gravel or sand may be observed moving along the bottom, and then falling over the edge of a bank into the bottom of the pool.   As the sediment advances by successive additions to its steep slope in front, it gradually fills the pool up.   Its progress may be compared to that of a railway embankment formed by the discharge of waggon-loads of rubbish down its end.   A section through such an embankment would reveal a series of bands of variously coloured materials inclined steeply towards the direction in which the waggon-loads were thrown down. Yet the top of the embankment may be kept quite level for the permanent way. The nearly level bands (*b*, *c*) in Fig. 12 represent the general bottom on which the sediment accumulated, while the steeper lines in the lower gravel (*a*) point to the existence and direction of the currents by which sediment was pushed forward along that bottom. (Compare pp. 174, 175.)

As the river flows onward through a gradually expanding valley, another characteristic feature becomes prominent. Flanking each side of the flat land through which the stream pursues

FIG. 13.—River-terraces.

its winding course, there runs a steep slope or bank a few feet or yards in height, terminating above in a second or higher plain, which again may be bordered with another similar bank, above which there may lie a third plain. These slopes and plains form a group of terraces, rising step by step above and away from the river, sometimes to a height of several hundred feet, and occasionally to the number of 6 or 8 or even more (Fig. 13). Here and there, by the narrowing of the intervening strip of plain, two terraces merge into one, and at some places the river in winding down the valley has cut away great slices from the terraces, perhaps even entirely removing them and eating back into the rock out of which the valley has been excavated. Sections are thus exposed showing a succession of gravels, sands, and loams like those of the present river. From the line of the uppermost terrace down to the spits of shingle now forming in the channel, we have evidently a chronologically arranged series of river-deposits, the oldest being at the top and the youngest at the bottom. But how could the river have flowed at the level of these high gravels, so far above its present limits? An examination of the behaviour of the stream during floods will help towards an answer to this question.

When from heavy rains or melted snows the river overflows its banks, it spreads out over the level ground on either side. The tract liable to be thus submerged during inundations is called the *flood-plain*. As the river rises in flood, it becomes more and more turbid from the quantity of mud and silt poured into it by its tributaries on either side. Its increase in volume likewise augments its velocity, and consequently its power of scouring its bed and of transporting the coarser detritus resting there. Large quantities of shingle may thus be swept out of the ordinary channel and be strewn across the nearer parts of the flood-plain. As the current spreads over this plain, its velocity and transporting capacity diminish, and consequently sediment begins to be thrown down. Grass, bushes, and trees, standing on the flood-plain, filter some of the sediment out of the water. Fine mud and sand, for instance, adhere to the leaves and stems, whence they are eventually washed off by rain into the soil underneath. In this way, the flood-plain is gradually heightened by the river itself. At the same time, the bed of the river is deepened by the scour of the current, until, in the end, even the highest floods are no longer able to inundate the flood-plain. The differ- ence of level between that plain and the surface of the river gradually increases; by degrees the river begins to cut away the edges of the terrace which it cannot now overflow, and to form a new flood-plain at a lower level. In this manner, it slowly lowers its bed, and leaves on either side a set of alluvial terraces to mark successive stages in the process of excavation. If during this process the level of the land should be raised, the slope of the rivers, and consequently their scour, would be augmented, and they would thereby acquire greater capacity for the formation of terraces. There is reason to believe that this has taken place both in Europe and North America.

While it is obvious that the highest terraces must be the oldest, and that the series is progressively younger down to the terrace that is being formed at the present time, nevertheless, in the materials comprising any one terrace, those lying at the top must be the youngest. This apparent contradiction arises from the double action of the river in eroding its bed and depositing its sediment. If there were no lowering of the channel, then the deposits would follow the usual order of sequence, the oldest being below and the youngest above. This order is maintained in the constituents of each single terrace, for the lowermost layers of gravel must evidently have been accumulated before the deposit of those that overlie them. But when the level of the water is lowered, the next set of deposits must, though younger, lie at a lower level than those that preceded them. In no case, however, will the older beds, though higher in position, be found really to overlie the younger. They have been formed at different levels.

The gravel, sand, and loam laid down by a river are marked, as we have seen, by an arrangement in layers, beds, or strata lying one upon another. This *stratified* disposition indeed is characteristic of all sedimentary accumulations, and is best developed where currents have been most active in transporting and assorting the materials (p. 172). It is the feature that first catches the eye in any river-bank, where a section of the older deposits or "alluvium" is exposed. Beds of coarser and finer detritus alternate with each other, but the coarsest are generally to be observed below and the finest above. The "deltas" accumulated by rivers in lakes and in the sea will be noticed in Chapters IV and VII.

But besides the inorganic detritus carried forward by a river, we have also to consider the fate of the remains of plants and the carcases of animals that are swept down, especially during floods. Swollen by sudden and heavy rains, a river will rise above its ordinary level and uproot trees and shrubs. On such occasions, too, moles and rabbits are drowned and buried in their burrows on the alluvial flood-plain. Birds, insects, and even some of the larger mammals are from time to time drowned, swept away by floods and buried in the sediment, and their remains, where of a durable kind or where sufficiently covered over, may be preserved for an indefinite period. The shells and fishes living in the river itself may also be killed during the flood, and may be entombed with the other organisms in the sediment.

Summary. —The material produced by the universal decay of the surface of the land is washed off by rain and swept seawards by brooks and rivers. The rate at which the general level of the land is being lowered by the operation of running water may be approximately ascertained by measuring or estimating the amount of mineral matter carried seaward every year from a definite region,

such as a river-basin. Taking merely the matter in mechanical
suspension, and assuming that the proportion of it transported
annually in the water of the Mississippi may be regarded as an
average proportion for the rivers of Europe, we find that this
continent, at the Mississippi rate of degradation, might be reduced
to the sea-level in rather less than 6,000,000 years.

In pursuing their course over the land, running waters gradually deepen and widen the channels in which they flow, partly
by chemically dissolving the rocks and partly by rubbing them
down by the friction of the transported sand, gravel, and stones.
When they have once chosen their channels, they usually keep to
them, and the sinuous windings, at first determined by trifling
inequalities on the surface of country across which the streams
began to flow, are gradually deepened into picturesque gorges.
In the excavation of such ravines, waterfalls play an important
part by gradually receding up stream.   River-channels, especially
if cut deeply into the solid rock, remain as enduring monuments
of the work of running water.   ʼ

But still more important as geological records, because more
frequent and covering a larger area, are the deposits which rivers
leave as their memorials.   Whatever checks the velocity of a
current weakens its transporting power, and causes it to drop
some of its sediment to the bottom.   Accordingly, accumulations
of sediment occur at the foot of torrent slopes, along the lower
and more level ground, especially on the inner or concave side of
the loops, over the flood-plains, and finally in the deltas formed
where rivers enter lakes or the sea.   In these various situations,
thick stratified beds of silt, sand, and gravel may be formed,
enclosing the remains of the plants and animals living on the land
at the time.   As a river deepens its channel, it leaves on either
side alluvial terraces that mark successive flood-plains over which
it has flowed.

CHAPTER IV

Fresh-water Lakes.—According to the law stated in last chapter, that when water is checked in its flow, it must drop some of its sediment, lakes are pre-eminently places for the deposition and accumulation of mineral matter. In their quiet depths, the debris worn away from the surface of the land is filtered out of the water and allowed to gather undisturbed upon the bottom. The tributary streams may enter a large lake swollen and muddy, but the escaping river is transparent. It is evident, therefore, that lakes must be continually silting up, and that when this process is complete, the site of a lake will be occupied by a series of deposits comprising a record of how the water was made to disappear.

To those who know the aspect of lakes only in fine weather, they may seem places where geological operations are at their very minimum of activity. The placid surface of the water ripples upon beaches of gravel or spits of sand ; reeds and marshy plants grow out into the shallows ; the few streamlets that tumble down from the surrounding hills furnish perhaps the only sounds that break the stillness, but their music and motion are at once hushed when they lose themselves in the lake. The scene might serve as the very emblem of perfectly undisturbed conditions of repose. But come back to this same scene during an autumn storm, when the mists have gathered all round the hills, and the rain, after pouring down for hours, has turned every gully into the track of a roaring torrent. Each tributary brook, hardly visible perhaps in drought, now rushes foaming and muddy from its dell and sweeps out into the lake. The large streams bear along on their swift brown currents trunks of trees, leaves, twigs, with now and then the carcase of some animal that has been drowned by the rising

flood. Hour after hour, from every side, these innumerable
swollen waters bear their freights of gravel, sand, and mud into
the lake. Hundreds or thousands of tons of sediment must thus be
swept down during a single storm. When we multiply this result
by the number of storms in a year and by the number of years in
an ordinary human life, we need not be surprised to be told that
even within the memory of the present generation, and still more
within historic times, conspicuous changes have taken place in
many lakes.

Filling up of Lakes.—In the Lake of Lucerne, for example,
the River Reuss, which bears down the drainage of the huge
mountains round the St. Gothard, deposits about 7,000,000 cubic
feet of sediment every year. Since the year 1714 the Kander,

FIG. 14.—Alluvial terraces on the side of an emptied reservoir.

which drains the northern flanks of the centre of the Bernese
Oberland, is said to have thrown into the lower end of the Lake
of Thun such an amount of sediment as to form an area of 230
acres, now partly woodland, partly meadow and marsh. Since
the time of the Romans, the Rhone has filled up the upper end
of the Lake of Geneva to such an extent that a Roman harbour,
still called Port Valais, is now nearly two miles from the edge of
the lake, the intervening ground having been converted first into
marshes and then into meadows and farms.

It is at the mouths of streams pouring into a lake that the
process of filling up is most rapid and striking. But it may be
detected at many other places round the margin. Instructive
lessons on this subject may be learned at a reservoir formed by
damming back the waters of a steep-sided valley, and liable to
be sometimes nearly dry (Fig. 14). In such a situation, when
the water is low, it may be noticed that a series of parallel lines
runs all round the sides of the reservoir, and that these lines

consist of gravel, sand, or earth. Each of them marks a former level of the water, and they show that the reservoir was not drained off at once but intermittently, each pause in the diminution of level being marked by a line of sediment. It is easy to watch how these lines are formed along the present margin of the water. The loose debris from the bare slope above, partly by its own gravitation, partly by the wash of rain, slides down into the water. But as soon as it gets there, its further downward movement is

Fig. 15.—Parallel roads of Glen Roy.

arrested. By the ripple of the water it is gently moved up and down, but keeps on the whole just below the line to which the water reaches. So long as it is concealed under the water, its position and extent can hardly be realised. But as soon as the level of the reservoir sinks, the sediment is left as a marked shelf or terrace. In natural lakes, the same process is going on, though hardly recognisable, because concealed under the water. But if by any means a lake could be rapidly emptied, its former level would be marked by a shelf or alluvial terrace. In some cases, the barrier of a lake has been removed, and the sinking of the water has revealed the terrace. The famous "parallel roads" of Glen Roy, in the west of Scotland, are notable examples (Fig. 15).

The valleys in that region were anciently dammed up by large glaciers. The drainage accumulated behind the ice, filled up the valleys and converted them into a series of lakes or fresh-water "fjords." The former levels of these sheets of water and the successive stages of their diminution and disappearance are shown by the series of alluvial shelves known as "parallel roads." The highest of these is 1140 feet, the middle 1059 feet, and the lowest 847 feet above the level of the sea.

Thus, partly by the washing of detritus down from the adjoining slopes by rain, partly by the sediment carried into them by streams, and partly by the growth of marshy vegetation along their margins, lakes are visibly diminishing in size. In mountainous countries, every stage of this appearance may be observed (Fig. 16). Where the lakes are deep, the tongues of sediment or

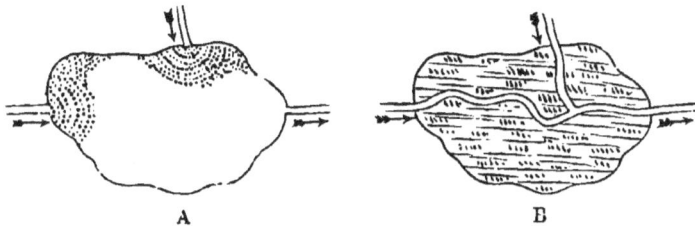

A   B

FIG. 16.—Stages in the filling up of a lake. In A two streamlets are represented as pouring their "deltas" into a lake. In B they have filled the lake up, converting it into a meadow across which they wind on their way down the valley.

"deltas" which the streams push in front of them have not yet been able to advance far from the shore. In other cases, every tributary has built up an alluvial plain which grows outwards and along the coast, until it unites with those of its neighbours to form a nearly continuous belt of flat meadow and marsh round the lake. By degrees, as this belt increases in width, the lake narrows, until the whole tract is finally converted into an alluvial plain, through which the river and its tributaries wind on their way to lower levels. The successive flat meadow-like expansions of valleys among hills and mountains were probably in most cases originally lakes which have in this manner been gradually filled up.

Lake Deposits.—The bottoms of lakes must evidently contain many interesting relics. . Dispersed through the shingle, sand, and mud that gather there, are the remains of plants and animals that lived on the surrounding land. Leaves, fruits, twigs, branches, and trunks embedded in the silt may preserve for an indefinite period their record of the vegetation of the time. The wings or wing-cases of insects, the shells of land-snails, the bones of birds and mammals, carried down into the depths of a lake and entombed in the silt there, will remain as a chronicle of the kind of animals that haunted the surrounding hills and valleys.

The layers of gravel, sand, and silt laid down on the floor of a lake differ in some respects from those deposited in the terraces of a river, being generally finer in grain, and including a larger proportion of silt, mud, or clay among them, especially away from the margin of the lake. They are, no doubt, further distinguished by the greater abundance of the remains of plants and animals preserved in them.

But lakes likewise serve as receptacles for a series of deposits which are peculiar to them, and which consequently have much interest and importance as they furnish a ready means of detecting the sites of lakes that have long disappeared. The molluscs that live in lacustrine waters are distinct from the snails of the adjoining shores. Their dead shells gather on the bottoms of some lakes in such numbers as to form there a deposit of the white crumbling marl, already referred to on p. 4. In course of time this deposit may grow to be many feet or yards in thickness. The shells in the upper parts may be quite fresh, some of the animals having only recently died ; but they become more and more decayed below until, towards the bottom of the deposit, the marl passes into a more compact chalk-like substance in which few or no shells may be recognisable (Fig. 17). On the sites of lakes that have been

FIG. 17.—Piece of shell-marl containing shells of *Limnæa peregra.*

naturally filled up or artificially drained, such marl has been extensively dug as a manure for land. Besides the shells from the decay of which it is chiefly formed, it sometimes yields the bones of deer, oxen, and other animals, whose carcases must originally have sunk to the bottom of the lake and been there gradually covered up in the growing mass of marl. Many examples of these marl-deposits are to be found among the drained lakes of Scotland and Ireland.

Yet another peculiar accumulation is met with on the bottom of some lakes, particularly in Sweden. In the neighbourhood of banks of reeds and on the sloping shallows of the larger lakes, a deposit of hydrated peroxide of iron takes place, in the form of concretions varying in size from small grains like gunpowder up to cakes measuring six inches across. The iron is no doubt dissolved out of the rocks of the neighbourhood by water containing organic acids or carbonic acid. In this condition, it is liable to be oxidised on exposure. As after oxidation it can no longer be retained in solution, it is precipitated to the bottom where it collects in grains which by successive additions to their surface become pellets, balls, or cakes. Possibly some of the microscopic plants (diatoms) which abound on the bottoms of the lakes may facilitate the accumulation of the iron by abstracting this substance from the water and depositing it inside their siliceous coverings. Beds of concretionary brown ironstone are formed in Sweden from 10 to 200 yards long, 5 to 15 yards broad, and from 8 to 30 inches thick. During winter when the lakes are frozen over, the iron is raked up from the bottom through holes made for the purpose in the ice, and is largely used for the manufacture of iron in the Swedish furnaces. When the iron has been removed, it begins to form again, and instances are known where, after the supply had been completely exhausted, beds several inches in thickness were formed again in twenty-six years.

Salt-Lakes.—The salt-lakes of desert regions present a wholly peculiar series of deposits. These sheets of water have no outlet; yet there is reason to believe that most of them were at first fresh, and discharged their outflow like ordinary lakes. Owing to geological changes of level and of climate, they have long ceased to overflow. The water that runs into them, instead of escaping by a river, is evaporated back into the air. But the various mineral salts carried by it in solution from rocks and soils are not evaporated also. They remain behind in the lakes, which are consequently becoming gradually salter. Among the salts thus introduced, common salt (sodium-chloride) and gypsum (calcium-sulphate) are two of the most important. These substances, as the water evaporates in the shallows, bays, and pools, are precipitated to the bottom where they form solid layers of salt and gypsum. The latter substance begins to be thrown down when 37 per cent of the water containing it has been evaporated. The sodium-chloride does not appear until 93 per cent of the water has disappeared. In the order of deposit, therefore, gypsum comes before the salt (see p. 136). Some bitter lakes contain sodium-carbonate, in others magnesium-chloride is abundant. The Dead Sea, the Great Salt Lake of Utah, and many other salt lakes and inland seas furnish interesting evidence of the way in which they have gradually changed. In the upper terraces of the Great Salt Lake, 1000 feet or more above the present level of the water, fresh-water shells occur, showing that the basin was at first fresh. The valley-bottoms around saline lakes are now crusted with gypsum, salt, or other efflorescence, and their waters are almost wholly devoid of life. Such conditions as these help us to understand how great deposits of gypsum and rock-salt were formed in England, Germany, and many other regions where the climate would not now permit of any such condensation of the water (Chapter XXII).

Summary.—The records inscribed by lakes in geological history consist of layers of various kinds of sediment. These deposits may form mere shelves or terraces along the margin of the water which, if drained off, will leave them as evidence of its former levels. By the long-continued operations of rain, brooks, and rivers, continually bringing down sediment, lakes are gradually filled up with alluvium, and finally become flat meadow-land with tributary streams winding through it. The deposits that thus replace the lacustrine water consist mainly of sand or gravel near shore, while finer silt occupies the site of the deeper water. They may also include beds of marl formed of fresh-water shells, and sheets of brown iron ore. Throughout them all, remains of the plants and animals of the surrounding land are likely to be entombed and preserved.

Salt lakes leave, as their enduring memorial, beds of rock-salt and gypsum, sometimes carbonate of soda and other salts. Many of them were at first fresh, as is shown by the presence of ordinary fresh-water shells in their upper terraces. But by change of climate and long-continued excess of evaporation over precipitation, the water has gradually become more and more saline, and has sometimes disappeared altogether, leaving behind it deposits of common salt, gypsum, and other chemical precipitates.

CHAPTER V

THE changes made by running water upon the land are not confined to that portion of the rainfall which courses along the surface.
Even when it sinks underground and seems to have passed out of
the general circulation, the subterranean moisture does not remain
inactive. After travelling for a longer or shorter distance through
the pores of rocks, or along their joints and other divisional planes,
it finds its way once more to daylight and reappears in *Springs*.[1]
In this underground journey, it corrodes rocks, somewhat in the
same way as rain attacks those that are exposed to the outer air,
and it works some curious changes upon the face of the land.
Subterranean water thus leaves distinct and characteristic memorials
as its contribution to geological history.

There are two aspects in which the work of underground water
may be considered here. In the first place, portions of the substance of subterranean rocks are removed by the percolating water
and in large measure carried up above ground ; in the second
place, some of these materials are laid down again in a new form
and take a conspicuous place among the geological monuments of
their time.

In the removal of mineral substance, water percolating through
rocks acts in two distinct ways, mechanical and chemical, each of
which shows itself in its own peculiar effects upon the surface.

(1) **Mechanical Action.** — While slowly filtering through
porous materials, water tends to remove loose particles and thus
to lessen the support of overlying rocks. But even where there
is no transport, the water itself, by saturating a porous layer that
rests upon a more or less impervious one, loosens the cohesion of
that porous layer. The overlying mass of rock is thus made to

[1] *Physical Geography Class-Book*, p. 222.

E

rest upon a watery and weakened platform, and if from its position it should have a tendency to gravitate in any given direction, it may at last yield to this tendency and slide downwards. Along the sides of sea-cliffs, on the precipitous slopes of valleys or river-gorges, or on the declivities of hills and mountains, the conditions are often extremely favourable for the descent of large masses of rock from higher to lower levels.

Remarkable illustrations of such *Landslips*, as they are called, have been observed along the south coast of England, where certain porous sandy rocks underlying a thick sheet of chalk rest upon more or less impervious clays, which, by arresting the water in its descent, throw it out along the base of the slopes. After much wet weather, the upper surface of these clays becomes, as it were, lubricated by the accumulation of water, and large slices of the overlying rocks, having their support thereby weakened, break off from the solid cliffs behind and slide down towards the sea. The most memorable example occurred at Christmas time, in the year 1839, on the coast of Devonshire not far from Axmouth. At that locality, the chalk-downs end off in a line of broken cliff some 500 feet above the sea. From the edge of the downs flanked by this cliff a tract about 800 yards long, containing not less than 30 acres of arable land, sank down with all its fields, hedgerows, and path-ways. This sunken mass, where it broke away from the upland, left behind it a new cliff, showing along the crest the truncated ends of the fields, of which the continuation was to be found in a chasm more than 200 feet deep. While the ground sank into this defile and was tilted steeply towards the base of the cliff, it was torn up by a long rent running on the whole in the line of the cliff, and by many parallel and transverse fissures. Half a century has passed away since this landslip occurred. The cliff remains much as it was at first, and the sunken fields with their bits of hedgerow still slope steeply down to the bottom of the declivity (Fig. 18). But the lapse of time has allowed the influence of the atmosphere to come into play. The outstanding dislocated frag-ments with their vertical walls and flat tops, showing segments of fields, have been gradually worn into tower-like masses with sloping declivities of debris. The long parallel rent has been widened by rain into a defile with shelving sides. Everywhere the rawness of the original fissures has been softened by the rich tapestry of verdure which the genial climate of that southern coast fosters in every sheltered nook. But the scars have not been healed, and they will no doubt remain still visible for many a year to come.

FIG. 18.—View of Axmouth landslip (as it appeared in April 1885).

Along the south coast of England, many landslips, of which there is no historical record, have produced some of the most picturesque scenery of that region. Masses that have slipped away from the main cliff have so grouped themselves down the slopes that hillocks and hollows succeed each other in endless confusion, as in the well-known Undercliff of the Isle of Wight. Some of the tumbled rocks are still fresh enough to show that they have fallen at no very remote period, or even that the slipping still continues ; others, again, have yielded so much to the weather that their date doubtless goes far back into the past, and some of them are crowned with what are now venerable ruins.

The most stupendous landslips on record have occurred in mountainous countries. Upwards of 150 destructive examples have been chronicled in Switzerland. Of these, one of the most memorable was that of the Rossberg, a mountain lying behind the Rigi, and composed of thick masses of hard red sandstone and conglomerate so arranged as to slope down into the valley of Goldau. The summer of the year 1806 having been particularly wet, so large an amount of water had collected in the more porous layers of rock as to weaken the support of the overlying mass ; consequently a large part of the side of the mountain suddenly gave way and rushed down into the valley, burying under the debris about a square German mile of fertile land, four villages containing 330 cottages and outhouses, and 457 inhabitants. To this day, huge angular blocks of sandstone lying on the farther side of the valley bear witness to the destruction caused by this land-slip, and the scar on the mountain-slope whence the fallen masses descended is still fresh.

(2) **Chemical Action**—(*a*) Solution.—But it is by its chemical action on the rocks through which it flows that subterranean water removes by far the largest amount of mineral matter, and produces the greatest geological change. Even pure water will dissolve a minute quantity of the substance of many rocks. But rain is far from being chemically pure water. In previous chapters it has been described as taking oxygen and carbonic acid out of the air in its descent, and abstracting organic acids and carbonic acid from the soil through which it sinks. By help of these ingredients, it is enabled to attack even the most durable rocks, and to carry some of their dissolved substance up to the surface of the ground.

One of the substances most readily attacked and removed even by pure water is the mineral known as carbonate of lime. Among other impurities, natural waters generally contain carbonic acid, which may be derived from the air or from the soil ; occasionally from some deeper subterranean source. The presence of this acid gives the water greatly increased solvent power, enabling it readily to attack carbonate of lime, whether in the form of limestone, or diffused through rocks composed mainly of other substances. Even lime, which is not in the form of carbonate, but is united with silica in various crystalline minerals (silicates, p. 130), may by this means be decomposed and combined with carbonic acid. It is then removed in solution as carbonate. So long as the water retains enough of free carbonic acid, it can keep the carbonate of lime in solution and carry it onward.

Limestone is a rock almost entirely composed of carbonate of lime. It occurs in most parts of the world, covering sometimes tracts of hundreds or thousands of square miles, and often rising into groups of hills, or even into ranges of mountains (see pp. 154, 158). The abundance of this rock affords ample opportunity for the display of the solvent action of subterranean water. Trickling down the vertical joints and along the planes between the limestone beds, the water dissolves and removes the stone, until in the course of centuries these passages are gradually enlarged into clefts, tunnels, and caverns. The ground becomes honeycombed with openings into dark subterranean chambers, and running streams fall into these openings and continue their course underground.

Every country which possesses large limestone tracts furnishes examples of the way in which such labyrinthine tunnels and systems of caverns are excavated. In England, for example, the Peak Cavern of Derbyshire is believed to be 2300 feet long, and in some places 120 feet high. On a much more magnificent scale are the caverns of Adelsberg near Trieste, which have been explored to a distance of between four and five miles, but are probably still more extensive. The river Poik has broken into one part of the labyrinth of chambers, through which it rushes before emerging again to the light. Narrow tunnels expand into spacious halls, beyond which egress is again afforded by low passages into other lofty recesses. The most stupendous chamber measures 669 feet in length, 630 feet in breadth, and 111 feet in height. From the roofs hang pendent white *stalactites* (p. 55), which, uniting with the floor, form pillars of endless varieties of form and size. Still more gigantic is the system of subterranean passages in the Mammoth Cave of Kentucky, the accessible parts of which are believed to have a combined length of about 150 miles. The largest cavern in this vast labyrinth has an area of two acres, and is covered by a vault 125 feet high.

Of the mineral matter dissolved by permeating water out of the rocks underground, by far the larger part is discharged by springs into rivers, and ultimately finds its way to the sea. The total amount of material thus supplied to the sea every year must be enormous. Much of it, indeed, is abstracted from ocean-water by the numerous tribes of marine plants and animals. In particular, the lime, silica, and organic matter are readily seized upon to build up the framework and furnish the food of these creatures. But probably more mineral matter is supplied in solution than is re-

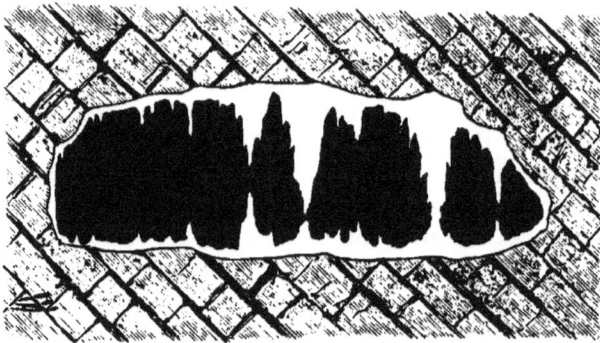

FIG. 19.—Section of cavern with stalactites and stalagmite.

quired by the organisms of the sea, in which case the water of the sea must be gradually growing heavier and salter.

(*b*) Deposition.—But it is the smaller proportion of the material not conveyed into the sea that specially demands attention. Every spring, even the purest and most transparent, contains mineral solutions in sufficient quantity to be detected by chemical analysis. Hence all plants and animals that drink the water of springs and rivers necessarily imbibe these solutions which, indeed, supply some of the mineral salts whereof the harder parts both of plants and animals are constructed. Many springs, however, contain so large a proportion of mineral matter, that when they reach the surface and begin to evaporate, they drop their solutions as a precipitate, which settles down upon the bottom or on objects within reach of the water. After years of undisturbed continuance, extensive sheets of mineral material may in this manner be accumulated, which remain as enduring monuments of the work of underground water, even long after the springs that formed them have ceased to flow.

Calcareous Springs.—Among the accumulations of this nature by far the most frequent and important are those formed by what are called Calcareous Springs. In regions abounding in lime-stone or rocks containing much carbonate of lime, the subterranean waters which, as we have seen, gradually erode such vast systems of tunnels, clefts, and caverns, carry away the dissolved rock, and retain it in solution only so long as they can keep their carbonic acid. As soon as they begin to evaporate and to lose some of this acid, they lose also the power of retaining so much carbonate of lime in solution. This substance is accordingly dropped as a fine white precipitate, which gathers on the surfaces over which the water trickles or flows.

The most familiar example of this process is to be seen under the arches of bridges and vaults. Long pendent white stalks or *stalactites* hang from between the joints of the masonry, while wavy ribs of the same substance run down the piers or walls, and even collect upon the ground (*stalagmite*). A few years may suffice to drape an archway with a kind of fringe of these pencil-like icicles of stone. Per-colating from above through the joints between the stones of the masonry, the rain-water, armed with its minute proportion of carbonic acid at once attacks the lime of the mortar and forms carbonate of lime, which is carried

FIG. 20.—Section show-ing successive layers of growth in a stalactite.

downward in solution. Arriving at the surface of the arch, the water gathers into a drop, which remains hanging there for a brief interval before it falls to the ground. That interval suffices to allow some of the carbonic acid to escape, and some of the water to evaporate. Consequently, round the outer rim of the drop a slight precipitation of white chalky carbonate of lime takes place. This circular pellicle, after the drop falls, is increased by a similar deposit from the next drop, and thus drop by drop the original rim or ring is gradually lengthened into a tube which may eventually be filled up inside, and may be thickened irregularly outside by the trickle of calcareous water (Fig. 20). But the deposition on the roof does not exhaust the stock of dissolved carbonate. When the drops reach the ground the same process of evaporation and precipitation continues. Little mounds of the same white substance are built up on the floor, and, if the place remain undisturbed, may grow until they unite with the stalactites from the roof, forming white pillars that reach from floor to ceiling (Fig. 19, and p. 154).

It is in limestone caverns that stalactitic growth is seen on the most colossal scale. These quiet recesses having remained undisturbed for many ages, the process of solution and precipitation has advanced without interruption until, in many cases, vast caverns have been transformed into grottoes of the most marvellous beauty. White glistening fringes and curtains of crystalline carbonate of lime, or *spar*, as it is popularly called, hang in endless variety and beauty of form from the roof. Pillars of every dimension, from slender wands up to thick-ribbed columns like those of a cathedral, connect the roof and the pavement. The walls, projecting in massive buttresses and retiring into alcoves, are everywhere festooned with a grotesque drapery of stone. The floor is crowded with mounds and bosses of strangely imitative forms which recall some of the oddest shapes above ground. Wandering through such a scene, the visitor somehow feels himself to be in another world, where much of the architecture and ornament belongs to styles utterly unlike those which can be seen anywhere else.

The material composing stalactite and stalagmite is at first, as already stated, a fine white chalky pulp-like substance which dries into a white powder. But as the deposition continues, the older layers, being impregnated with calcareous water, receive a precipitation of carbonate of lime between their minute pores and crevices, and assume a crystalline structure. Solidifying and hardening by degrees, they end by becoming a compact crystalline stone (spar) which rings under the hammer.

The numerous caverns of limestone districts have offered ready shelter to various kinds of wild animals and to man himself. Some of them (*Bone-Caves*) have been hyæna-dens, and from under their hard floor of stalagmite the bones of hyænas and of the creatures they fed upon are disinterred in abundance. Rude human implements have likewise been obtained from the same deposits, showing that man was contemporary with animals which have long been extinct. The solvent action of underground water has thus been of the utmost service in geological history, first, in forming caverns that could be used as retreats, and then in providing a hard incrustation which should effectually seal up and preserve the relics of the denizens left upon the cavern-floors.

Calcareous springs, issuing from limestone or other rock abounding in lime, deposit carbonate of lime as a white precipitate. So large is the proportion of mineral contained by some waters that thick and extensive accumulations of it have been formed. The substance thus deposited is known by the name of *Calcareous Tufa, Calc-sinter,* or *Travertine.* It varies in texture, some kinds being loose and crumbling, others hard and crystalline. In many places it is composed of thin layers or laminæ, of which sixty may be counted in the thickness of an inch, but bound together into a solid stone. These laminæ mark the successive layers of deposit. They are formed parallel to the surface over which the water flows or trickles, hence they may be observed not only on the flat bottoms of the pools, but irregularly enveloping the walls of the channel as far up as the dash of water or spray can reach. Rounded bosses may thus be formed above the level of the stream, and the recesses may be hung with stalactites.

The calcareous springs of Northern and Central Italy have long been noted for the large amount of their dissolved lime, the rapidity with which it is deposited, and the extensive masses in which it has accumulated. Thus at San Filippo in Tuscany, it is deposited in places at the rate of one foot in four months, and it has been piled up to a depth of at least 250 feet, forming a hill a mile and a quarter long, and a third of a mile broad. So compact are many of the Italian travertines that they have from time immemorial been extensively used as a building stone, which can be dressed and is remarkably durable. Many of the finest buildings of ancient and modern Rome have been constructed of travertine.

A familiar feature of many calcareous springs deserves notice here. The precipitation of calc-sinter is not always due merely to evaporation. In many cases, where the proportion of carbonate of lime in solution is so small that under ordinary circumstances no precipitation of it would take place, large masses of it have been deposited in a peculiar fibrous form. On examination, this precipitation will be found to be caused by the action of plants, particularly bog-mosses which, decomposing the carbonic acid in the water, cause the lime-carbonate to be deposited along their stems and leaflets. The plants are thus incrusted with sinter which, preserving their forms, looks as if it were composed of heaps of moss turned into stone. Hence the name of *petrifying* springs often given to waters where this process is to be seen. There is, however, no true petrifaction or conversion of the actual substance of the plants into stone. The fibres are merely incrusted with travertine, inside of which they eventually die and decay. But as the plants continue to grow outward, they increase the sinter by fresh layers, while the inner and dead parts of the mass are filled up and solidified by the deposit of the precipitate within their cavities.

A growing accumulation of travertine presents a special

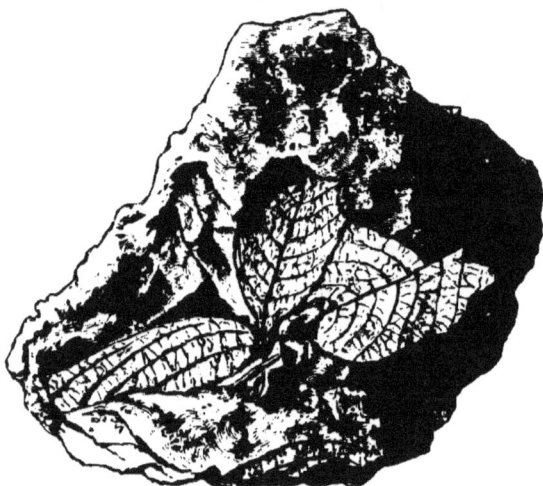

FIG. 21.—Travertine with impressions of leaves.

interest to the geologist from the fact that it offers exceptional facilities for the preservation of remains of the plants and animals of the neighbourhood. Leaves from the surrounding trees and shrubs are blown into pools or fall upon moist surfaces where the precipitation of lime is actively going on (Fig. 21). Dead insects, snail-shells, birds, small mammals, and other denizens of the district may fall or be carried into similar positions. These remains may be rapidly enclosed within the stony substance before they have time to decay, and even if they should afterwards moulder into dust, the sinter enclosing them retains the mould of their forms, and thus preserves for an indefinite period the record of their former existence.

Chalybeate Springs.—A second but less abundant deposit from springs is found in regions where the rocks below ground contain decomposing sulphide of iron (p. 137). Water percolating through such rocks and oxidising the sulphur of that mineral, forms sulphate of iron (ferrous sulphate) which it removes in solution. The presence of any notable quantity of this sulphate is at once revealed by the marked inky taste of the water and by the yellowish-brown precipitate on the sides and bottom of the channel. Such water is termed *Chalybeate.* When it mixes with other water containing dissolved carbonates (which are so generally present in running water), the sulphate is decomposed, the sulphuric acid passing over to the lime or alkali of the carbonate, while the iron takes up oxygen and falls to the bottom as a yellowish-brown precipitate (limonite, p. 129). This interchange of combinations, with the consequent precipitation of iron-oxide, may continue for a considerable distance from the outflow of the chalybeate water. Nearest the source the deposit of hydrated ferric oxide or ochre is thickest. It encloses leaves, stems, and other organic remains, and preserves moulds or casts of their forms. It also cements the loose sand and shingle of a river-bottom into solid rock.

Siliceous Springs.—One other deposit from spring-water may be enumerated here. In volcanic regions, hot springs (geysers) rise to the surface which, besides other mineral ingredients, contain a considerable proportion of *silica* (p. 117). This substance is deposited as *Siliceous Sinter* round the vents whence the water is discharged, where it forms a white stone rising into mounds and terraces with fringes and bunches of coral-like growth. Where many springs have risen in the same district, their respective sheets of sinter may unite, and thus extensive tracts are buried under the deposit. In Iceland, for example, one of the sheets is said to be two leagues long, a quarter of a league wide, and a hundred feet thick. In the Yellowstone Park of North America, many valleys are floored over with heaps of sinter, and in New Zealand other extensive accumulations of the same material are to be found. It is obvious that, like travertine, siliceous sinter may readily entomb and preserve a record of the plants and animals that lived at the time of its deposition.

Summary.—The underground circulation of water produces changes that leave durable records in geological history. These changes are of two kinds. (1) Landslips are caused, by which the forms of cliffs, hills, and mountains are permanently altered.

Vast labyrinths of subterranean tunnels, galleries, and caverns are dissolved out of calcareous rocks, and openings are made from these passages up to the surface whereby rivers are engulfed. Many of the caves thus hollowed out have served as dens of wild beasts and dwelling-places for man, and the relics of these inhabitants have been preserved under the stalagmite of the floors. (2) An enormous quantity of mineral matter is brought up to the surface by springs. Most of the solutions are conveyed ultimately to the sea where they partly supply the substances required by the teeming population of marine plants and animals. But, under favourable circumstances, considerable deposits of mineral matter are made by springs, more especially in the form of travertine, siliceous sinter, and ochre. In these deposits the remains of terrestrial vegetation, also of insects, birds, mammals, and other animals, are not infrequently preserved, and remain as permanent memorials of the life of the time when they flourished.

CHAPTER VI

ICE in various ways alters the surface of the land. By disintegrating and eroding even the most durable rocks, and by removing loose materials and piling them up elsewhere, it greatly modifies the details of a landscape. As it assumes various forms, so it accomplishes its work with considerable diversity. The action of frost upon soil and bare surfaces of rock has already (p. 13) been described. We have now to consider the action of frozen rivers and lakes, snow and glaciers, which have each their own characteristic style of operation, and leave behind them their distinctive contribution to the geological history of the earth.

Frozen Rivers and Lakes.—In countries with a severe winter climate, the rivers and lakes are frozen over, and the cake of ice that covers them may be more than two feet thick. When this cake is broken up in early summer, large masses of it are driven ashore, tearing up the littoral boulders, gravel, sand, or mud, and pushing them to a height of many feet above the ordinary level of the water. When the ice melts, huge heaps of detritus are found to have been piled up by it, which remain as enduring monuments of its power. Not only so, but large fragments of the ice that has been formed along shore and has enclosed blocks of stone, gravel, and sand, are driven away and may travel many miles before they melt and drop their freight of stones. On the St. Lawrence and on the coast of Labrador, there is a constant transportation of boulders by this means. Further, besides freezing over the surface, the water not infrequently forms a loose spongy kind of ice on the bottom (*Anchor-ice*, *Ground-ice*) which encloses stones and gravel, and carries them up to the surface where it joins the cake of ice there. This bottom-ice is formed abundantly on some parts of the Canadian rivers. Swept down by the current, it accumulates

against the bars or banks, or is pushed over the upper ice, and
from time to time gathers into temporary barriers, the bursting of
which may cause destructive floods.   In the river St. Lawrence,
banks and islets have been to a large extent worn down by the
grating of successive ice-rafts upon them.

Snow.—On level or gently inclined ground, whence snow disappears merely by melting or evaporation, it exercises, while it
remains, a protective influence upon the soil and vegetation,
shielding them from the action of frost.   On slopes of sufficient declivity, however, the sheet of snow acquires a tendency to
descend by gravitation, as we may often see on house-roofs in winter.
In many cases, it creeps or slides down the side of a hill or valley,
and in so doing pushes forward bare soil, loose stones, or other
objects lying on the surface.   By this means, the debris of
weathered rock in exposed situations is gradually thrust down-hill
and the rock is bared for further disintegration.   But where the
declivities are steep enough to allow the snow to break off in
large sheets and to rush rapidly down, the most striking changes
are observable.   Such descending masses are known as *Avalanches.*   Varying from 10 to 50 feet or more in thickness and
several hundred yards broad and long, they sweep down the
mountain sides with terrific force, carrying away trees, soil, houses,
and even large blocks of rock.   The winter of 1884-85 was
especially remarkable for the number of avalanches in the valleys
of the Alps, and for the enormous loss of life and property which
they caused   In such mountain ground, not only are declivities
bared of their trees, soil, and boulders, but huge mounds of debris
are piled up in the valleys below.   Frequently, also, such a
quantity of snow, ice, and rubbish is thrown across the course of
a stream as to dam back the water, which accumulates until it
overflows or sweeps away the barrier.   In another but indirect
way, snow may powerfully affect the surface of a district where,
by rapid melting, it so swells the rivers as to give rise to destructive floods.

While, therefore, the influence of snow is on the whole to
protect the surface of the land, it shows itself in mountainous
regions singularly destructive, and leaves as chief memorials of
this destructiveness the mounds and rough heaps of earth and
stones that mark where the down-rushing avalanches have come
to rest.

Glaciers and Ice-Sheets leave their record in characters so
distinct that they cannot usually be confounded with those of any

other kind of geological agent.   The changes which they produce
on the surface of the land may be divided into two parts : (1) the
transport of materials from the high grounds to lower levels, and
(2) the erosion of their beds.

(1) Transport.—As a glacier descends its valley, it receives
upon its surface the earth, sand, mud, gravel, boulders, and blocks
of rock that roll or are washed down from the slopes on either
side. Most of this rubbish accumulates on the edges of the
glacier, where it is slowly borne to lower levels as the ice creeps
downwards. But some of it falls into the crevasses or rents by

Fig. 22.—Glacier with medial and lateral moraines.

which the ice is split, and may either be imprisoned within the
glacier, or may reach the rocky floor over which the ice is sliding.
The rubbish borne onward upon the surface of the glacier is known
as *moraine-stuff*.   The mounds of it running along each side of
the glacier form *lateral moraines*, those on the right-hand side as
we look down the length of the valley being the right lateral
moraine, those on the other side the left lateral moraine.   Where
two glaciers unite, the left lateral moraine of the one joins the
right lateral moraine of the other, forming what is called a *medial
moraine* that runs down the middle of the united glacier.   Where
a glacier has many tributaries bearing much moraine-stuff, its

surface may be like a bare plain covered with earth and stones, so that, except where a yawning crevasse reveals the clear blue gleam of the ice below, nothing but earth and stones meets the eye. When the glacier melts, the detritus is thrown in heaps upon the valley, forming there the *terminal moraine*.

Glaciers, like rivers, are subject to variations of level. Even from year to year they slowly sink below their previous limit or rise above it. The glacier of La Brenva, for example, on the Italian side of Mont Blanc, subsided no less than 300 feet in the first half of the present century. One notable consequence of such diminution is that the blocks of rock lying on the edges of a glacier are stranded on the side of the valley, as the ice shrinks

Fig. 23.—Perched blocks scattered over ice-worn surface of rock.

away from them. Such *Perched Blocks* or *Erratics* (Fig. 23), as they are called, afford an excellent means of noting how much higher and longer a glacier has once been than it is now. Their great size (some of them are as large as good-sized cottages) and their peculiar positions make it quite certain that they could not have been transported by any current of water. They are often poised on the tops of crags, on the very edges of precipices, or on steep slopes where they could never have been left by any flood, even had the flood been capable of moving them. The agent that deposited them in such positions must have been one that acted very quietly and slowly, letting the blocks gently sink into the sites they now occupy. The only agent known to us that could have done this is glacier-ice. We can actually see similar blocks on the glaciers now, and others which have only recently been stranded on the side of a valley from which the ice has sunk. In the Swiss valleys, the scattered ice-borne boulders may be observed by hundreds far above the existing level of the glaciers and many miles beyond where these now end. If the origin of the dispersed erratics is self-evident in a valley where a glacier is still busy transporting them, those that occur in valleys which are now destitute of glaciers can offer no difficulty ; they become, indeed, striking monuments that glaciers once existed there.

Scattered erratic blocks offer much interesting evidence of the movements of the ice by which they were transported. In a glacier-valley, the blocks that fall upon the ice remain on the side from which they have descended. Hence, if there is any notable difference between the rocks of the two sides, this difference will be recognisable in the composition of the moraines, and will remain distinct even to the end of the glacier. If, therefore, in a district from which the glaciers have disappeared, we can trace up the scattered blocks to their sources among the mountains, we thereby obtain evidence of the actual track followed by the vanished glaciers. The limits to which these blocks are traceable do not, of course, absolutely fix the limits of the ice that transported them. They prove, however, that the ice extended at least as far as they occur, but it may obviously have risen higher and advanced farther than the space within which the blocks are now confined. In Europe, some striking examples occur of the use of this kind of evidence. Thus the peculiar blocks of the Valais can be traced all the way to the site of the modern city of Lyons. There can therefore be no doubt that the glacier of the Rhone once extended over all that intervening country and reached at least as far as Lyons,—a distance. of not less than 170 miles from where it now ends. Again, from the occurrence of blocks of some of the char- acteristic rocks of Southern Scandinavia, in Northern Germany, Belgium, and the east of England, we learn that a great sheet of ice once filled up the bed of the Baltic and the North Sea, carrying with it immense numbers of northern erratics. In Britain, where there are now neither glaciers nor snow-fields, the abundant dis- persion of boulders from the chief tracts of high ground shows that this country was once in large part buried under ice, like modern Greenland. The evidence for these statements will be more fully given in a later part of this volume (Chapter XXVII).

Besides the moraine-stuff carried along on the surface, loose detritus and blocks of rock are pushed onwards under the ice.

F

When a glacier retires, this earthy and stony debris, where not swept away by the escaping river, is left on the floor of the valley. One remarkable feature of the stones in it is that a large proportion of them are smoothed, polished, and covered with fine scratches or ruts, such as would be made by hard sharp-pointed fragments of stone or grains of sand. These markings run for the most part along the length of each oblong stone, but not infrequently cross each other, and sometimes an older may be noticed partially effaced by a newer set. This peculiar striation is a most characteristic mark of the action of glaciers. The stones under the ice are fixed in the line of least resistance—that

FIG. 24.—Stone smoothed and striated by glacier-ice.

is, end on. In this position, under the weight of hundreds of feet of ice, they are pressed upon the floor over which the glacier is travelling. Every sharp edge of stone or grain of sand, pressed along the surface of a block, or over which the block itself is slowly drawn, engraves a fine scratch or a deeper rut. As the block moves onward, it is more and more scratched, losing its corners and edges, and becoming smaller and smoother till, if it travel far enough, it may be entirely ground into sand or mud (Fig. 24).

(2) Erosion.—The same process of erosion is carried on upon the solid rocks over which the ice moves. These are smoothed, striated, and polished by the friction of the grains of sand, pebbles, and blocks of stone crushed against them by the slowly creeping mass of ice. Every boss of rock that looks toward the quarter from which the overlying ice is moving is ground away, while those that face to the opposite side are more or less

sharp and unworn.   The striation is especially noteworthy.   From
the fine scratches, such as are made by grains of sand, up to deep
ruts like those of cart-wheels in unmended roadways, or to still
wider and deeper hollows, all the friction-markings run in a
general uniform direction, which is that of the motion of the
glacier.   Such striated surfaces could only be produced by some
agent with rigidity enough to hold the sand-grains and stones in
position, and press them steadily onward upon the rocks.   A

25.—Ice-striation on the floor and side of a valley.

river polishes the rocks of its channel by driving shingle and sand
across them ; but the currents are perpetually tossing these
materials now to one side, now to another, so that smoothed and
polished surfaces are produced, but with nothing at all resembling
striation.   A glacier, however, by keeping its grinding materials
fixed in the bottom of the ice, engraves its characteristic parallel
striæ and groovings, as it slowly creeps down the valley.   All the
surfaces of rock within reach of the ice are smoothed, polished,
and striated.   Such surfaces present the most unmistakable
evidence of glacier-action, for they can be produced by no other
known natural agency.   Hence, where they occur in glacier

valleys, far above and beyond the present limits of the ice, they prove how greatly the ice has sunk.   In regions also where there are now no glaciers, these rock-markings remain as almost imperishable witnesses that glaciers once existed.   By means of their evidence, for example, we can trace the march of great ice-sheets which once enveloped the whole of Scandinavia and lay deep upon nearly the whole of Britain.

The river that escapes from the end of a glacier is always milky or muddy.   The fine sand and mud that discolour the water are not supplied by the thawing of the clear ice, nor by the sparkling brooks that gush out of the mountain-slopes, nor by the melting of the snows among the peaks that rise on either side. This material can only come from the rocky floor of the glacier itself.   It is the fine sediment ground away from the rocks and loose stones by their mutual friction under the pressure of the overlying ice.   It serves thus as a kind of index or measure of the amount of material worn off the rocky bed by the grinding action of the glacier.   We can readily see that as this erosion and transport are continually in progress, the amount of material removed in the course of time must be very great.   It has been estimated, for example, that the Justedal glacier in Norway removes annually from its bed 2,427,000 cubic feet of sediment. At this rate the amount removed in a century would be enough to fill up a valley or ravine 10 miles long, 100 feet broad, and 40 feet deep.

In arctic and antarctic latitudes, where the land is buried under a vast ice-sheet, which is continually creeping seaward and break-ing off into huge masses that float away as icebergs, there must be a constant erosion of the terrestrial surface.   Were the ice to retire from these regions, the ground would be found to wear what is called a *glaciated* surface ; that is to say, all the bare rocks would present a characteristic ice-worn aspect, rising into smooth rounded bosses like dolphins' backs (*roches moutonnées*), and sinking into hollows that would become lake-basins.   Every-where these bare rocks would show the striæ and groovings graven upon them by the ice, radiating generally from the central high grounds, and thus indicating the direction of flow of the main streams of the ice-sheet.   Piles of earth, ice-polished stones, and blocks of rock would be found strewn over the country, especially in the valleys and over the plains.   These materials would still further illustrate the movements of the ice, for they would be found to be singularly local in character, each district having supplied its own contribution of detritus. Thus from a region of red sandstone, the rubbish would be red and sandy; from one of black slate, it would be black and clayey (see Chapter XXVI).

Summary.—In this chapter we have seen that Ice in various ways affects the surface of the land and leaves its mark there. Frost, as already explained in Chapter II, pulverises soil, disintegrates exposed surfaces of stone, and splits open bare rocks along their lines of natural joint. On rivers and lakes, the disrupted ice wears down banks and pushes up mounds of sand, gravel, and boulders along the shores. Snow lying on the surface of the land protects that surface from the action of frost and air. In the condition of avalanches, snow causes large quantities of earth, soil, and blocks of rock to be removed from the mountain-slopes and piled up on the valleys. In the form of glaciers, ice transports the debris of the mountains to lower levels, bearing along and sometimes stranding masses of rock as large as cottages, which no other known natural agent could transport. Moving down a valley, a glacier wears away the rocks, giving them a peculiar smoothed and striated surface which is thoroughly characteristic. By this grinding action, it erodes its bed and produces a large amount of fine sediment, which is carried away by the river that escapes at the end of the ice-stream. Land-ice thus leaves thoroughly distinctive and enduring memorials of its presence in polished and grooved rocks, in masses of earth, clay, or gravel, with striated stones, and in the dispersal of erratic blocks from principal masses of high ground. These memorials may remain for ages after the ice itself has vanished. By their evidence we know that the present glaciers of the Alps are only a shrunk remnant of the great ice-fields which once covered that region; that the Scandinavian glaciers swept across what is now the bed of the North Sea as far as the mouth of the Thames; and that Scotland, Ireland, Wales, and the greater part of England were buried under great sheets of ice which crept downwards into the North Sea on the one side, and into the Atlantic on the other (Chapter XXVII).

CHAPTER VII

WE have now to inquire how the work of the Sea is registered in geological history. This work is broadly of two kinds. In the first place, the sea is engaged in wearing away the edges of the land, and in the second place, being the great receptacle into which all the materials, worn away from the land, are transported, it arranges these materials over its floor, ready to be raised again into land at some future time.

i. **Demolition of the Land.**—In its work of destruction along the coasts of the land, the sea acts to some extent (though we do not yet know how far) by chemically dissolving the rocks and sediments which it covers. Cast-iron bars, for example, have been found to be so corroded by sea-water as to lose nearly half their strength in fifty years. Doubtless many minerals and rocks are liable to similar attacks.

But it is by its mechanical effects that the sea accomplishes most of its erosion. The mere weight with which ocean-waves fall upon exposed coasts breaks off fragments of rock from cliffs. Masses, 13 tons in weight, have been known to be quarried out of the solid rock by the force of the breakers in Shetland, at a height of 70 feet above sea-level. As a wave may fall with a blow equal to a pressure of 3 tons on the square foot, it compresses the air in every cleft and cranny of a cliff, and when it drops it allows the air instantly to expand again. By this alternate compression and expansion, portions of the cliff are loosened and removed. Where there is any weaker part in the rock, a long tunnel may be excavated, which may even be drilled through to the daylight above, forming an opening at some distance inland from the edge of the cliff. During storms, the breakers rush through such a tunnel, and spout forth from the opening (or *blow-hole*) in clouds of spray (Fig. 26).

Probably the most effective part of the destructive action of the sea is to be found in the battery of gravel, shingle, and loose blocks of stone which the waves discharge against cliffs exposed to their fury. These loose materials, caught up by the advancing breakers and thrown with great force upon the rocks of a coast-line, are dragged back in the recoil of the water, but only to be again lifted and swung forward. In this loud turmoil, the loose stones are reduced in size and are ground smooth by friction against each

FIG. 26.—Buller of Buchan—a caldron-shaped cavity or blow-hole worn out of granite by the sea on the coast of Aberdeenshire.

other and upon the solid cliff. The well-rounded and polished aspect of the gravel on such storm-beaten shores is an eloquent testimony to the work of the waves. But still more striking, because more measurable, is the proof that the very cliffs themselves cannot resist the blows dealt upon them by the wave-borne stones. Above the ordinary limit reached by the tides, the rocks rise with a rough ragged face, bearing the scars inflicted on it by the ceaseless attacks of the air, rain, frost, and the other agencies that waste the surface of the land. But all along the base of the cliff, within reach of the waves, the rocks have been smoothed and polished by the ceaseless grinding of the shingle upon them, while arches, tunnels, solitary pillars, half-tide skerries, creeks, and caves attest the steady advance of the sea and the gradual demolition of the shore.

Every rocky coast-line exposed to a tempestuous sea affords illustrations of these features of the work of waves. Even where the rocks are of the most durable kind, they cannot resist the ceaseless artillery of the ocean. They are slowly battered down, and every stage in their demolition may be witnessed, from the sunken reef, which at some distance from the shore marks where the coast-line once ran, up to the tunnelled cliff from which a huge mass was detached during the storms of last winter. But where the materials composing the cliffs are more easily removed, the progress of the waves may be comparatively rapid. Thus on the east coast of Yorkshire between Spurn Point and Flamborough Head, the cliffs consist of boulder-clay, and vary up to more than 100 feet in height. At high water, the tide rises against the base of these cliffs, and easily scours away the loose debris which would otherwise gather there and protect them. Hence, within historic times, a large tract of land, with its parishes, farms, villages, and sea-ports, has been washed away, the rate of loss being estimated at not less than $2\frac{1}{4}$ yards in a year. Since the Roman occupation a strip of land between 2 and 3 miles broad is believed to have disappeared.

It is evident that to carry on effectively this mechanical erosion, the sea-water must be in rapid motion. But in the deeper recesses of the ocean, where there is probably no appreciable movement of the water, there can hardly be any sensible erosion. In truth, it is only in the upper parts of the sea, which are liable to be affected by wind, that the conditions for marine erosion can be said to exist. The space within which these conditions are to be looked for is that comprised between the lowest depth to which the influence of waves and marine currents extends, and the greatest height to which breakers are thrown upon the land. These limits, no doubt, vary considerably in different regions. In some parts of the open sea, as off the coast of Florida, the disturbing action of the waves has been supposed to reach to a depth of 600 feet, though the average limit is probably greatly less. On exposed promontories in stormy seas, such as those of the north of Scotland, breakers have been known to hurl up stones to a height of 300 feet above sea-level. But probably the zone, within which the erosive work of the sea is mainly carried on, does not as a rule exceed 300 feet in vertical range.

FIG. 27.—The Stacks of Duncansby, Caithness, a wave-beaten coast-line.

Within some such limits as these, the sea is engaged in gnawing away the edges of the land. A little reflection will show us
that, if no counteracting operation should come into play, the prolonged erosive action of the waves would reduce the land below
the sea-level. If we suppose the average rate of demolition to be
10 feet in a century, then it would take not less than 52,800 years
to cut away a strip one mile broad from the edge of the land. But
while the sea is slowly eating away the coast-line, the whole surface
of the land is at the same time crumbling down, and the wasted
materials are being carried away by rivers into the sea at such a
rate that, long before the sea could pare away more than a mere
narrow selvage, the whole land might be worn down to the sea-
level by air, rain, and rivers (p. 28).

But there are counteracting influences in nature that would
probably prevent the complete demolition of the land. What
these influences are will be more fully considered in a later chapter.
In the meantime, it will be enough to bear in mind that while the
land is constantly worn down by the forces that are acting upon
its surface, it is liable from time to time to be uplifted by other
forces acting from below. And the existing relation between the
amount and height of land, and the extent of sea, on the face of
the globe, must be looked upon as the balance between the working of both these antagonistic classes of agencies.

FIG. 28.—Section of submarine plain. *l*, Land cut into caves, tunnels, sea-stacks, reefs,
and skerries by the waves, and reduced to a platform below the level of the sea (*s s*) on
which the gravel, sand, and mud (*d*) produced by the waste of the coast may accumulate.

But without considering for the present whether the results of
the erosion performed by the sea will be interrupted or arrested,
we can readily perceive that their tendency is toward the reduc-
tion of the level of the land to a submarine plain (Fig. 28). As
the waves cut away slice after slice from a coast-line, the portion
of land which they thus overflow, and over which they drive the
shingle to and fro, is worn down until it comes below the lower
limit of breaker-action, where it may be covered up with sand or

mud. When the abraded land has been reduced to this level, it reaches a limit where erosion ceases, and where the sea, no longer able to wear it down further, protects it from injury by other agents of demolition. This lower limit of destruction on the surface of the earth has been termed "the base-level of erosion."

We see, then, that the goal toward which all the wear and tear of a coast-line tends, is the formation of a more or less level platform cut out of the land. Yet an attentive study of the process will convince us that in the production of such a platform the sea has really had less to do than the atmospheric agents of destruction. An ordinary sea-cliff is not a vertical wall. In the great majority of cases it slopes seaward at a steep angle ; but if it had been formed, and were now being cut away, mainly by the sea, it ought obviously to have receded fastest where the waves attack it—that is, at its base. In other words, if sea-cliffs retired chiefly because they are demolished by the sea, they ought to be most eroded at the bottom, and should therefore be usually overhanging precipices. That this is not the case shows that some other agency is concerned which causes the higher parts of a cliff to recede faster than those below. This agency can be no other than that of the atmospheric forces—air, frost, rain, and springs. These cause the face of the cliff to crumble down, detaching mass after mass, which, piled up below, serve as a breakwater, and must be broken up and removed by the waves before the solid cliff behind them can be attacked.

ii. **Accumulations formed by the Sea.**—It is not its erosive action that constitutes the most important claim of the sea to the careful study of the geologist. After all, the mere marginal belt or fringe within which this action is confined forms such a small fraction of the whole terrestrial area of the globe, that its importance dwindles down when we compare it with the enormously vaster surface over which the operations of the air, rain, rivers, springs, and glaciers are displayed. But when we regard the sea as the receptacle into which all the materials worn off the land ultimately find their way, we see what a large part it must play in geological history.

During the last fifteen years great additions have been made to our knowledge of the sea-bottom all over the world. Portions of the deposits accumulating there have been dredged up even from the deepest abysses, so that it is now possible to construct charts, showing the general distribution of materials over the floor of the ocean.

Beginning at the shore, let us trace the various types of marine deposits outward to the floors of the great ocean-abysses. In many places, the sea is more or less barred back by the accumulation of sediment worn away from the land. In estuaries, for example, there is often such an amount of mud in the water that the bottom on either side is gradually raised above the level of tide-mark, and forms eventually a series of meadows which the sea can no longer overflow. At the mouths of rivers with a considerable current, a check is given to the flow of the water when it reaches the sea, and there is a consequent arrest of its detritus. Hence, a bar is formed across the outflow of a river, which during floods is swept seawards, and during on-shore gales is driven again inland. Even where there is no large river, the smaller streams flowing off the surface of a country may carry down sediment enough to be arrested by the sea, and to be thrown up as a long bank or bar running parallel with the coast. Behind this bar, the drainage of the interior accumulates in long lagoons, which find an outflow through some breach in the bar, or by soaking through the porous materials of the bar itself. A large part of the eastern coast of the United States is fringed with such bars and lagoons. A space several hundred miles long on the east coast of India is similarly bordered.

But the most remarkable kind of accumulation of terrestrial detritus in the sea is undoubtedly that of *river-deltas*. Where the tidal scour is not too great, the sediment brought down by a large river into a marine bay or gulf gradually sinks to the bottom as the fresh spreads over and mingles with the salt water. During floods, coarse sediment is swept along, while during low states of the river nothing but fine mud may be transported. Alternating sheets of different kinds of sediment are thus laid down one upon another on the sea-floor, until by degrees they reach the surface, and thus gradually increase the breadth of the land. Some deltas are of enormous size and depth. That of the Ganges and Brahmaputra covers an area of between 50,000 and 60,000 square miles—that is, about as large as England and Wales. It has been bored through to a depth of 481 feet, and has been found to consist of numerous alternations of fine clays, marls, and sands or sandstones, with occasional layers of gravel. In all this great thickness of sediment, no trace of marine organisms was found, but land-plants and bones of terrestrial and fluviatile animals occurred. Lower Egypt has been formed by the growth of the delta of the Nile, whereby a wide tract of alluvial land has not only filled up the bottom of the valley, but has advanced into the Mediterranean.

Turning now to the deposits that are more distinctively those of the sea itself, we find that ridges of coarse shingle, gravel, and sand are piled up along the extreme upper limit reached by the waves. The coarsest materials are for the most part thrown highest, especially in bays and narrow creeks where the breakers are confined within converging shores. In such situations, during heavy gales, *storm-beaches* of coarse rounded shingle are formed sometimes several yards above ordinary high-tide mark (Fig. 29).

Fig. 29.—Storm-beach ponding back a stream and forming a lake ; west coast of Sutherlandshire.

Where a barrier of this kind is thrown across the mouth of a brook, the fresh water may be ponded back to form a small lake, of which the outflow usually escapes by percolation through the shingle. In sheltered bays, behind headlands, or on parts of a coast-line where tidal currents meet, detritus may accumulate in spits or bars. Islands have in this way been gradually united to each other or to the mainland, while the mainland itself has gained considerably in breadth. At Romney Marsh, on the south-east coast of England, for instance, a tract of more than 80 square miles, which in Roman times was in great part covered by the sea at high water, is now dry land, having been gained partly by the natural increase of shingle thrown up by the waves and partly by the barriers artificially erected to exclude the sea.

While the coarsest shingle usually accumulates towards the upper part of the beach, the materials generally arrange themselves according to size and weight, becoming on the whole finer as they are traced towards low-water mark. But patches of coarse gravel may be noticed on any part of a beach, and large boulders may be seen even below the limits of the lowest tides. As a rule, the deposits formed along a beach, and in the sea immediately beyond, include the coarsest kinds of marine sediment. They are also marked by frequent alternations of coarse and fine detritus, these rapid interchanges pointing to the varying action of the waves and strong shore-currents. Towards the lower limit of breaker-action, fine gravel and sand are allowed to settle down, and beyond these, in quiet depths where the bottom is not disturbed, fine sand and mud washed away from the land slowly accumulate.

The distance to which the finer detritus of the land is carried by ocean-currents before it finds its way to the bottom, varies up to about 200 miles or more. Within this belt of sea, the land-derived materials are distributed over the ocean-floor. Coarse and fine gravel and sand are the most common materials in the areas nearest the land. Beyond these lie tracts of fine sand and silt with occasional patches of gravel. Still farther from the land, at depths of 600 feet and upwards, fine blue and green muds are found, composed of minute particles of such minerals as form the ordinary rocks of the land. But traced out into the open ocean, these various deposits of recognisable terrestrial origin give place to thoroughly oceanic accumulations, especially to widespread sheets of exceedingly fine red and brown clay. This clay, the most generally diffused deposit of the deeper or abysmal parts of the sea, appears to be derived from the decomposition of volcanic fragments either washed away from volcanic islands or supplied by submarine eruptions. That it is accumulated with extreme slowness is shown by two curious and interesting kinds of evidence. Where it occurs farthest removed from land, great numbers of sharks' teeth, with ear-bones and other bones of whales, have been dredged up from it, some of these relics being quite fresh, others partially coated with a crust of brown peroxide of manganese, some wholly and thickly enveloped in this substance. The same haul of the dredge has brought up bones in all these conditions, so that they must be lying side by side on the red clay floor of the ocean abysses. The deposition of manganese is no doubt an exceedingly slow process, but it is evidently faster than the deposition of the red clay. The bones dredged up probably represent a long succession of generations of animals. Yet so tardily does the red clay gather over them, that the older ones are not yet covered up by it, though they have had time to be deeply encased in oxide of manganese. The second kind of evidence of the extreme slowness of deposit in the ocean abysses is supplied by minute spherules of metallic iron, which occurring in numbers dispersed through the red clay, have been identified as portions of meteorites or falling stars. These particles no doubt fall all over the ocean, but it is only where the rate of deposition of sediment is exceedingly slow that they may be expected to be detected.

Besides the sediments now enumerated, the bottom of the sea receives abundant accumulations of the remains of shells, corals, foraminifera and other marine creatures; but these will be described in the next chapter, where an account is given of the various ways in which plants and animals, both upon the land and in the sea, inscribe their records in geological history. It must also be borne in mind that throughout all the sediments of the sea-floor, from the upper part of the beach down to the bottom of the deepest and remotest abyss, the remains of the plants, sponges, corals, shells, fishes and other organisms of the ocean may be entombed and preserved. It will suffice here to remember that various depths and regions of the sea have their own characteristic forms of life, the remains of which are preserved in the sediments accumulating there, and that although gravel, sand, and mud laid down beneath the sea may not differ in any recognisable detail from similar materials deposited in a lake or river, yet the presence of marine organisms in them would be enough to prove that they had been formed in the sea. It is evident, also, that if the sea-floor over a wide area were raised into land, the extent of the deposits would show that they could not have been accumulated in any mere river or lake, but must bear witness to the former presence of the sea itself.

Summary.—The sea records its work upon the surface of the earth in a twofold way. In the first place, in co-operation with the atmospheric agents of disintegration, it eats away the margin of the land and planes it down. The final result of this process if uninterrupted would be to reduce the level of the land to that of a submarine platform, the position of the surface of which would be determined by the lower limit of effective breaker-action. In the second place, the sea gathers over its floor all the detritus worn by every agency from the surface of the land. This material

is not distributed at random ; it is assorted and arranged by the
waves and currents, the coarsest portions being laid down nearest
the land, and the finest in stiller and deeper water.   The belt of
sea-floor within which this deposition takes place probably does
not much exceed a breadth of 200 miles.   Beyond that belt, the
bottom of the ocean is covered to a large extent with deposits of
red clay derived from the decomposition of volcanic material and
laid down with extreme slowness.   These and the widespread
deposits of dead sea-organisms (to be described in next chapter)
are truly oceanic accumulations, recognisably distinct from those
derived from terrestrial sources within the narrow zone of deposition near the land.

CHAPTER VIII

HOW PLANTS AND ANIMALS INSCRIBE THEIR RECORDS IN GEOLOGICAL HISTORY

BROADLY considered, there are two distinct ways in which Plants and Animals leave their mark upon the surface of the earth. In the first place, they act directly by promoting or arresting the decay of the land, and by forming out of their own remains deposits which are sometimes thick and extensive. In the second place, their remains are transported and entombed in sedimentary accumulations of many different kinds, and furnish important evidence as to the conditions under which these accumulations were formed. Each of these two forms of memorial deserves our careful attention, for, taken together, they comprise the most generally interesting departments of geology, and those in which the history of the earth is principally discussed.[1]

i. **Direct action of living things upon the surface of the globe.**—This action is often of a destructive kind, both plants and animals taking their part in promoting the general disintegration of rocks and soils. Thus, by their decay they furnish to the soil those organic acids referred to on pp. 16, 24, as so important in increasing the solvent power of water, and thereby promoting the waste of rocks. By thrusting their roots into crevices of cliffs, plants loosen and gradually wedge off pieces of rock, and by sending their roots and rootlets through the soil, they open up the subsoil to be attacked by air and descending moisture (p. 16). The action of the common earthworm in bringing up fine soil to be exposed to the influences of wind and rain was referred to at p. 18. Many burrowing animals also, such as the mole and

---

[1] In the Appendix a Table of the Vegetable and Animal Kingdoms is given, from which the organic grade of the plants and animals referred to in this and subsequent chapters may be understood.

rabbit, throw up large quantities of soil and subsoil which are liable to be blown or washed away.

On the other hand, the action may be conservative, as, for instance, where, by forming a covering of turf, vegetation protects the soil underneath from being rapidly removed, or where sand-loving plants bind together the surface of dunes, and thereby arrest the progress of the sand, or where forests shield a mountain-side from the effects of heavy rains and descending avalanches.

(1) *Deposits formed of the remains of Plants.*—But it is chiefly by the aggregation of their own remains into more or less extensive deposits that plants and animals leave their most prominent and enduring memorials. As examples of the way in which this is done by plants, reference may be made to peat-bogs, mangrove-swamps, infusorial earth, and calcareous sea-weeds.

Peat-bogs.—In temperate and arctic countries, marshy vegetation accumulates in peat-bogs over areas from an acre or two to many square miles, and to a depth of sometimes 50 feet. These deposits are largely due to the growth of bog-mosses and other aquatic plants which, dying in their lower parts, continue to grow upward on the same spot. On flat or gently-inclined moors, in hollows between hills, on valley-bottoms, and in shallow lakes, this marshy vegetation accumulates as a wet spongy fibrous mass, the lower portions of which by degrees become a more or less compact dark brown or black pulpy substance, wherein the fibrous texture, so well seen in the upper or younger parts, in large measure disappears. In a thick bed of peat, it is not infrequently possible to detect a succession of plant remains, showing that one kind of vegetation has given place to another during the accumulation of the mass. In Europe, as already mentioned on p. 4, peat-bogs often rest directly upon fresh-water marl containing remains of lacustrine shells (1 in Fig. 30). In every such case, it is evident that the peat has accumulated on the site of a shallow lake which has been filled up, and converted into a morass by the growth of marsh-plants along its edges and over its floor. The lowest parts of the peat may contain remains of the reeds, sedges, and other aquatic plants which choked up

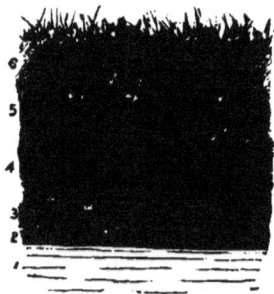

FIG. 30.—Section of a peat-bog.

the lake (2, 3). Higher up, the peat consists almost entirely of the matted fibres of different mosses, especially of the kind known as Bog-moss or *Sphagnum* (4). The uppermost layers (5, 6) may be full of roots of different heaths which spread over the surface of the bog.

The rate of growth of peat has been observed in different situations in Central Europe to vary from less than a foot to about two feet in ten years ; but in more northern latitudes the growth is probably slower. Many thousand square miles of Europe and North America are covered with peat-bogs, those of Ireland being computed to occupy a seventh part of the surface of the island, or upwards of 4000 square miles.

As the aquatic plants grow from the sides toward the centre of a shallow lake, they gradually cover over the surface of the water with a spongy layer of matted vegetation. Animals, and man himself, venturing on this treacherous surface sink through it, and may be drowned in the black peaty mire underneath. Long afterwards, when the morass has become firm ground, and openings are made in it for digging out the peat to be used as fuel, their bodies may be found in an excellent state of preservation. The peaty water so protects them from decay that the very skin and hair sometimes remain. In Ireland, numerous skeletons of the great Irish elk have been obtained from the bogs, though the animal itself has been extinct since before the beginning of the authentic history of the country.

Mangrove-swamps.—Along the flat shores of tropical lands, the mangrove trees grow out into the salt water, forming a belt of jungle which runs up or completely fills the creeks and bays. So dense is the vegetation that the sand and mud, washed into the sea from the land, are arrested among the roots and radicles of the trees, and thus the sea is gradually replaced by firm ground. The coast of Florida is fringed with such mangrove-swamps for a breadth of from 5 to 20 miles. In such regions, not only does the growth of these swamps add to the breadth of the land, but the sea is barred back, and prevented from attacking the newly-formed ground inside.

Infusorial earth.—A third kind of vegetable deposit to be referred to here is that known by the names of infusorial earth, diatom-earth, and tripoli-powder. It consists almost entirely of the minute frustules of microscopic plants called diatoms, which are found abundantly in lakes and likewise in some regions of the ocean (Fig. 31). These lowly organisms are remarkable for

secreting silica in their structure. As they die, their singularly
durable siliceous remains fall like a fine dust on the bottom of the
water, and accumulate there as a pale gray or straw-coloured
deposit, which, when dry, is like flour, and in its pure varieties is
made almost entirely of silica (90 to 97 per cent). Underneath
the peat-bogs of Britain a layer of this material is sometimes met
with. One of the most famous examples is that of Richmond,
Virginia, where a bed of it occurs 30 feet thick. At Bilin in
Bohemia also an important bed has long been known. The
bottom of some parts of the Southern Ocean is covered with a
diatom-ooze made up mainly of siliceous diatoms, but containing

FIG. 31.—Diatom-earth from floor of Antarctic Ocean, magnified 300 diameters
(*Challenger* Expedition).

also other siliceous organisms (radiolarians) and calcareous foraminifera (Fig. 31).

Accumulations of sea-weeds.—Yet one further illustration
of plant-action in the building up of solid rock may be given. As
a rule the plants of the sea form no permanent accumulations,
though here and there under favourable conditions, such as in
bays and estuaries, they may be thrown up and buried under
sand so as eventually to be compressed into a kind of peat. Some
sea-weeds, however, abstract from sea-water carbonate of lime,
which they secrete to such an extent as to form a hard stony
structure, as in the case of the common nullipore. When the
plants die, their remains are thrown ashore and pounded up
by the waves, and being durable they form a white calcareous
sand. By the action of the wind, this sand is blown inland and
may accumulate into dunes. But unlike ordinary sand, it is liable
to be slightly dissolved by rain-water, and as the portion so dis-

solved is soon redeposited by the evaporation of the moisture, the little sand-grains are cemented together, and a hard crust is formed which protects the sand underneath from being blown away. Meanwhile rain-water percolating through the mounds gradually solidifies them by cementing the particles of sand to each other, and thick masses of solid white stone are thus produced. Changes of this kind have taken place on a great scale at Bermuda, where all the dry land consists of limestone formed of compacted calcareous sand, mainly the detritus·of sea-weeds.

(2) *Deposits formed of the remains of Animals.*—Animals are, on the whole, far more successful than plants in leaving enduring

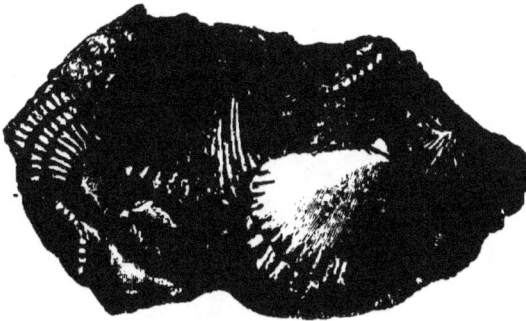

FIG. 32.—Recent limestone (Common Cockle, etc., cemented in a matrix of broken shells).

memorials of their life and work. They secrete hard outer shells and internal skeletons endowed with great durability, and capable of being piled up into thick and extensive deposits which may be solidified into compact and enduring stone. On land, we have an example of this kind of accumulation in the lacustrine marl already (pp. 4, 46) described as formed of the congregated remains of various shells. But it is in the sea that animals, secreting carbonate of lime, build up thick masses of rock, such as shell-banks, ooze, and coral reefs (see Chapter XI, p. 158).

Shell-banks.—Some molluscs, such as the oyster, live in populous communities upon submarine banks. In the course of generations, thick accumulations of their shells are formed on these banks. By the action of currents also large quantities of broken shells are drifted to various parts of the sea-bottom not far from land. Such deposits of shells, *in situ* or transported, may be more or less mixed with or buried under sand and silt, according as the currents vary in direction and force. On the other hand, they may be gradually cemented into a solid calcareous mass, as has been observed off the coast of Florida, where they form on the sea-bottom a sheet of limestone, made up of their remains.

Ooze.—From observations made during the great expedition of the *Challenger*, it has been estimated that in a square mile of the tropical ocean down to a depth of 100 fathoms there are more than 16 tons of carbonate of lime in the form of living animals. A continual rain of dead calcareous organisms is falling to the bottom, where their remains accumulate as a soft chalky *ooze.*

FIG. 33.—Globigerina ooze dredged up by *Challenger* Expedition from a depth of 1900 fathoms in the North Atlantic (²⁄₁).

Wide tracts of the ocean-floor are covered with a pale grey ooze of this nature, composed mainly of the remains of the shells of the foraminifer *Globigerina* (Fig. 33). In the north Atlantic this deposit probably extends not less than 1300 miles from east to west, and several hundred miles from north to south.

Here and there, especially among volcanic islands, portions of the sea-bed have been raised up into land, and masses of modern limestone have thereby been exposed to view. Though they are full of the same kind of shells as are still living in the neighbouring sea, they have been cemented into compact and even somewhat crystalline rock, which has been eaten into caverns by percolating water, like limestones of much older date. This cementation, as above remarked, is due to water permeating the stone, dissolving from its outer parts the calcareous matter of shells, corallines, and other organic remains, and redepositing it again lower down, so as to cement the organic detritus into a compact stone.

Coral-reefs offer an impressive example of how extensive masses of solid rock may be built up entirely of the aggregated remains of animals. In some of the warmer seas of the globe, and notably in the track of the great ocean-currents, where marine life is so abundant, various kinds of coral take root upon the edges and summits of submerged ridges and peaks, as well as on the shelving sea-bottom facing continents or encircling islands (1 in Fig. 34). These creatures do not appear to flourish at a greater depth than 15 or 20 fathoms, and they are killed by exposure to sun and air. The vertical space within which they live may therefore be stated broadly as about 100 feet. They grow in colonies, each composed of many individuals, but all united into one mass, which at first may be merely a little solitary clump on the sea-floor, but which, as it grows, joins other similar clumps to form what is

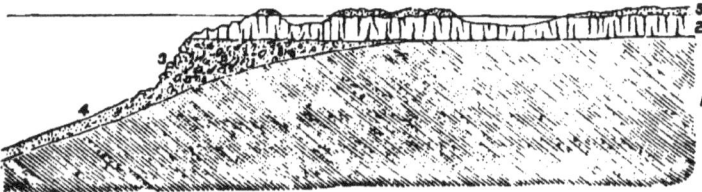

FIG. 34.—Section of a coral-reef. 1. Top of the submarine ridge or bank on which the corals begin to build. 2. Coral-reef. 3. Talus of large blocks of coral-rock on which the reef is built outward. 4. Fine coral sand and mud produced by the grinding action of the breakers on the edge of the reef. 5. Coral sand thrown up by the waves and gradually accumulating above their reach to form dry ground.

known as a reef. Each individual secretes from the sea-water a hard calcareous skeleton inside its transparent jelly-like body, and when it dies, this skeleton forms part of the platform upon which the next generation starts. Thus the reef is gradually built upward as a mass of calcareous rock (2), though only its upper surface is covered with living corals. These creatures continue to work upward until they reach low-water mark, and then their further upward progress is checked. But they are still able to grow outward. On the outer edges of the reef they flourish most vigorously, for there, amid the play of the breakers, they find the food that is brought to them by the ocean-currents. From time to time fragments are torn off by breakers from the reef and roll down its steep front (3). There, partly by the chemical action of the sea-water, and partly by the fine calcareous mud and sand (4), produced by the grinding action of the waves and washed into their crevices, these loose blocks are cemented into a firm, steep slope, on the top of which the reef continues to grow outwards. Blocks of coral and quantities of coral-sand are also thrown up on the surface of the reef, where by degrees they form a belt of low land above the reach of the waves (5).   On the inside of the reef, where the corals cannot find the abundant food-supply afforded by the open water outside, they dwindle and die.   Thus the tendency of all reefs must be to grow seawards, and to increase in breadth. Perhaps their breadth may afford some indication of their relative age.

Where a reef has started on a shelving sea-bottom near the coast of a continent, or round a volcanic island, the space of water inside is termed the *Lagoon Channel*.   Where the reef has been built up on some submarine ridge or peak, and there is consequently no land inside, the enclosed space of water is called a *Lagoon,* and the circular reef of coral is known as an *Atoll*.   If no subsidence of the sea-bottom takes place, the maximum thickness of a reef must be limited by the space within which the corals can thrive—that is, a vertical depth of about 100 feet from the surface of the sea.   But the effect of the destruction of the ocean-front of the reef, and the piling up of a slope of its fragments on the sea-bottom outside, will be to furnish a platform of the same materials on which the reef itself may grow outward, so that the united mass of calcareous rock may attain a very much greater thickness than 100 feet.   On the other hand, if the sea-bottom were to sink at so slow a rate that the reef-building corals could keep pace with the subsidence, a mass of calcareous rock many thousand feet thick might obviously be formed by them.   It is a disputed question in which of these two ways atolls have been formed.

It is remarkable how rapidly and completely the structure of the coral-skeleton is effaced from the coral-rock, and a more or less crystalline and compact texture is put in its place.   The change is brought about partly by the action of both sea-water and rain-water in dissolving and redepositing carbonate of lime among the minute interstices of the rock, and partly also by the abundant mud and sand produced by the pounding action of the breakers on the reef, and washed into the crevices.   On the portion of a reef laid dry at low water, the coral-rock looks in many places as solid and old as some of the ancient white limestones and marbles of the land.   There, in pools where a current or ripple of water keeps the grains of coral-sand in motion, each grain may be seen to have taken a spherical form unlike that of the ordinary irregularly rounded or angular particles.   This arises because carbonate of lime in solution in the water is deposited round each grain as it moves along. A mass of such grains aggregated together is called *oolite*, from its resemblance to fish-roe. In many limestones, now forming wide tracts of richly culti-vated country, this oolitic structure is strikingly exhibited. There can be no doubt that in these cases it was produced in a similar way to that now in progress on coral-reefs (see pp. 141, 155).

In the coral tracts of the Pacific Ocean there are nearly 300 coral islands, besides extensive reefs round volcanic islands. Others occur in the Indian Ocean. Coral-reefs abound in the West Indian Seas, where, on many of the islands, they have been upraised into dry land, in Cuba to a height of 1100 feet above sea-level. The Great Barrier Reef that fronts the north-eastern coast of Australia is 1250 miles long, and from 10 to 90 miles broad.

There are other ways in which the aggregation of animal remains forms more or less extensive and durable rocks. To some of these reference will be made in later chapters. Enough has been said here to show that by the accumulation of their hard parts animals leave permanent records of their presence both on land and in the sea.

ii. **Preservation of remains of Plants and Animals in sedimentary deposits.**—But it is not only in rocks formed out of their remains that living things leave their enduring records. These remains may be preserved in almost every kind of deposit, under the most wonderful variety of conditions. And as it is in large measure from their occurrence in such deposits that the geologist derives the evidence that successive tribes of plants and animals have peopled the globe, and that the climate and geo-graphy of the earth have greatly varied at different periods, we shall find it useful to observe the different ways in which the remains both of plants and animals are at this moment being entombed and preserved upon the land and in the sea. With the knowledge thus gained, it will be easier to understand the lessons taught by the organic remains that lie among the various solid rocks around us.

It is evident that in the vast majority of cases, the plants and animals of the land leave no perceptible trace of their presence. Of the forests that once covered so much of Central and Northern Europe, which is now bare ground, most have disappeared, and unless authentic history told that they had once flourished, we should never have known anything about them. There were also herds of wild oxen, bears, wolves, and other denizens contemporaneous with the vanished forests.   But they too have passed away, and we might ransack the soil in vain for any trace of them.

If the remains of terrestrial vegetation and animals are anywhere preserved it must obviously be only locally, but the favourable circumstances for their preservation, although not everywhere to be found, do present themselves in many places if we seek for them.   The fundamental condition is that the relics should, as soon as possible after death, be so covered up as to be protected from the air and from too rapid decomposition.   Where this condition is fulfilled, the more durable of them may be preserved for an indefinite series of ages.

(*a*) On the Land there are various places where the remains both of plants and animals are buried and shielded from decay. To some of these reference has already been made.   Thus amid the fine silt, mud, and marl gathering on the floors of lakes, leaves, fruits, and branches, or tree-trunks, washed from the neighbouring shores, may be imbedded, together with insects, birds, fishes, lizards, frogs, field-mice, rabbits, and other inhabitants. These remains may of course often decay on the lake-bottom, but where they sink into or are quickly covered up by the sediment, they may be effectually preserved from obliteration.   They undergo a change, indeed, being gradually turned into stone, as will be described in Chapter XV.   But this conversion may be effected so gently as to retain the finest microscopic textures of the original organisms.

In peat-bogs also, as already stated (p. 83), wild animals are often engulfed, and their soft parts are occasionally preserved as well as their skeletons.   The deltas of river-mouths must receive abundantly the remains of animals swept off by floods.   As the carcases float seawards, they begin to fall to pieces and the separate bones sink to the bottom, where they are soon buried in the silt.   Among the first bones to separate from the rest of the skeleton are the lower jaws (pp. 308, 311).   We should therefore expect that in excavations made in a delta these bones would occur most frequently.   The rest of the skeleton is apt to be carried farther out to sea before it can find its way to the bottom. The stalagmite floor of caverns has already been referred to (p. 56) as an admirable material for enclosing and preserving organic remains.   The animals that fell into these recesses, or used them as dens in which they lived or into which they dragged their prey, have left their bones on the floors, where, encased in or covered by solid stalagmite, these relics have remained for ages. Most of our knowledge of the animals which inhabited Europe at the time when man appeared, is derived from the materials disinterred from these Bone-caves. Allusion has also been made to the travertine formed by mineral-springs and to the facility with which leaves, shells, insects, and small birds, reptiles, or mammals may be enclosed and preserved in it (p. 58). Thus, while the plants and animals of the land for the most part die and decay into mere mould, there are here and there localities where their remains are covered up from decay and preserved as memorials of the life of the time.

(*b*) On the bottom of the Sea the conditions for the preservation of organic remains are more general and favourable than on land. Among the sands and gravels of the shore, some of the stronger shells that live in the shallower waters near land may be covered up and preserved, though often only in rolled fragments. It is below tide-mark, however, and more especially beneath the limit to which the disturbing action of breakers descends, that the remains of the denizens of the sea are most likely to be buried in sediment and to be preserved there as memorials of the life of the sea. It is evident that hard and therefore durable relics have the best chance of escaping destruction. Shells, corals, corallines, spicules of sponges, teeth, vertebræ, and ear-bones of fishes may be securely entombed in successive layers of silt or mud. But the vast crowds of marine creatures that have no hard parts must almost always perish without leaving any trace whatever of their existence. And even in the case of those which possess hard shells or skeletons, it will be easily understood that the great majority of them must be decomposed upon the sea-bottom, their component elements passing back again into the sea-water from which they were originally derived. It is only where sediment is deposited fast enough to cover them up and protect them before they have time to decay, that they may be expected to be preserved.

In the most favourable circumstances, therefore, only a very small proportion of the creatures living in the sea at any time leave a tangible record of their presence in the deposits of the sea-bottom. It is in the upper waters of the ocean, and especially in the neighbourhood of land, that life is most abundant. The same region also is that in which the sediment derived from the waste of the land is chiefly distributed. Hence it is in these marginal parts of the ocean that the conditions for preserving memorials of the animals that inhabit the sea are best developed.

As we recede from the land, the rate of deposit of sediment on the sea-floor gradually diminishes, until in the central abysses it reaches that feeble stage so strikingly brought before us by the evidence of the manganese nodules (p. 78). The larger and thinner calcareous organisms are attacked by the sea-water and dissolved, apparently before they can sink to the bottom ; at least their remains are comparatively rarely found there. It is such indestructible objects as sharks' teeth and vertebræ and ear-bones of whales that form the most conspicuous organic relics in these abysmal deposits.

Summary.—Plants and animals leave their records in geological history, partly by forming distinct accumulations of their remains, partly by contributing their remains to be imbedded in different kinds of deposits both on land and in the sea. As examples of the first mode of chronicling their existence, we may take the growth of marsh-plants in peat-bogs, the spread of mangrove-swamps along tropical shores, and the deposition of infusorial earth on the bottom of lakes and of the sea ; the accumulation of nullipore sand into solid stone, the formation of extensive shell-banks in many seas, the wide diffusion of organic ooze over the floor of the sea, and the growth of coral reefs. As illustrations of the second method, we may cite the manner in which the remains of terrestrial plants and animals are preserved in peat-bogs, in the deltas of rivers, in the stalagmite of caverns, and in the travertine of springs ; and the way in which the hard parts of marine creatures are entombed in the sediments of the sea-floor, more especially along that belt fringing the continents and islands, where the chief deposit of sediment from the disintegration of the land takes place. Nevertheless, alike on land and sea, the proportion of organic remains thus sealed up and preserved is probably always but an insignificant part of the total population of plants and animals living at any given moment.

How the remains of plants and animals when once entombed in sediment are then hardened and petrified, so as to retain their minute structures, and to be capable of enduring for untold ages, will be treated of in Chapter XV.

CHAPTER IX

THE geological changes described in the foregoing chapters affect only the surface of the earth. A little reflection will convince us that they may all be referred to one common source of energy—the sun. It is chiefly to the daily influence of that great centre of heat and light that we must ascribe the ceaseless movements of the atmosphere, the phenomena of evaporation and condensation, the circulation of water over the land, the waves and currents of the sea, in short the whole complex system which constitutes what has been called the Life of the Earth. Could this influence be conceivably withdrawn, the planet would become cold, dark, silent, lifeless.

But besides the continual transformations of its surface due to solar energy, our globe possesses distinct energy of its own. Its movements of rotation and revolution, for example, provide a vast store of force, whereby many of the most important geological processes are initiated or modified, as in the phenomena of day and night and the seasons, with the innumerable meteorological and other effects that flow therefrom. These movements, though slowly growing feebler, bear witness to the wonderful vigour of the earlier phases of the earth's existence. Inside the globe too lies a vast magazine of planetary energy in the form of an interior of intensely hot material. The cool outer shell is but an insignificant part of the total bulk of the globe. To this cool part the name of "crust" was given at a time when the earth was believed to consist of an inner molten nucleus enclosed within an outer solid shell or crust. The term is now used merely to denote the cool solid external part of the globe, without implying any theory as to the nature of the interior.

Condition of the Earth's Interior.—It is obvious that we are not likely ever to learn by direct observation what may be the condition of the interior of our planet. The cool solid outer shell is far too thick to be pierced through by human efforts; but by various kinds of observations, more or less probable conclusions may be drawn with regard to this problem. In the first place, it has been ascertained that all over the world, wherever borings are made for water or in mining operations, the temperature increases in proportion to the depth pierced, and that the average rate of increase amounts to about one degree Fahrenheit for every 64 feet of descent. If the rise of temperature continues inward at this rate, or at any rate at all approaching it, then at a distance from the surface, which in proportion to the bulk of the whole globe is comparatively trifling, the heat must be as great as that at which the ordinary materials of the crust would melt at the surface. In the second place, thermal springs in all quarters of the globe, rising sometimes with the temperature of boiling water, and occasionally even still hotter, prove that the interior of the planet must be very much warmer than its exterior. In the third place, volcanoes widely distributed over the earth's surface throw out steam and heated vapours, red-hot stones, and streams of molten rock.

It is quite certain therefore that the interior of the globe must be intensely hot; but whether it is actually molten or solid has been the subject of prolonged discussion. Three opinions have found stout defenders. (1) The older geologists maintained that the phenomena of volcanoes and earthquakes could not be explained, except on the supposition of a crust only a few miles thick, enclosing a vast central ocean of molten material. (2) This view has been opposed by physicists who have shown that the globe, if this were actually its structure, could not resist the attraction of sun and moon, but would be drawn out of shape, as the ocean is in the phenomenon of the tides, and that the absence of any appreciable tidal deformation in the crust shows that the earth must be practically solid and as rigid as a ball of glass, or of steel. (3) A third opinion has been advanced by geologists who, while admitting that the earth behaves on the whole as a solid rigid body, yet believe that many geological phenomena can only be explained by the existence of some liquid mass beneath the crust. Accordingly they suppose that while the nucleus is retained in the solid state by the enormous superincumbent pressure under which it lies, and the crust has become solid by cooling, there is an intermediate liquid or viscous layer which has not yet cooled sufficiently to pass into the solid crust above, and does not lie under sufficient pressure to form part of the solid nucleus below. At present, the balance of evidence and argument seems to be in favour of the practical rigidity and solidity of the globe as a whole. But the materials of its interior must possess temperatures far higher than those at which they would melt at the surface. They are no doubt kept solid by the vast overlying pressure, and any change which could relieve them of this pressure would allow them to pass into the liquid form. This subject will be again alluded to in Chapter XVI. Meanwhile, let us consider how the intensely hot nucleus of the planet reacts upon its surface.

Rocks are bad conductors of heat. So slowly is the heat of the interior conducted upwards by them that the temperature of the surface of the crust is not appreciably affected by that of the intensely hot nucleus. But the fact that the surface is not warmed from this source shows that the heat of the interior must pass off into space as fast as it arrives at the surface, and proves that our planet is gradually cooling. For many millions of years the earth has been radiating heat into space, and has consequently been losing energy. Its present store of planetary vitality therefore must be regarded as greatly less than it once was.

VOLCANOES.

Of all the manifestations of this planetary vitality, by far the most impressive are those furnished by volcanoes. The general characters of these vents of communication between the hot interior and cool surface of the planet are doubtless already familiar to the reader of these chapters—the *volcano* itself, a conical hill or mountain, formed mainly or entirely of materials ejected from below, having on its truncated summit the basin-shaped *crater*, at the bottom of which lies the *vent* or funnel from which, as well as from rents on the flanks of the cone, hot vapours, cinders, ashes, and streams of molten *lava* are discharged, till they gradually pile up the volcanic cone round the vent whence they escape.

A volcanic cone, so long as it remains, bears eloquent testimony to the nature of the causes that produced it. Even many centuries after it has ceased to be active, when no vapours rise from any part of its cold, silent, and motionless surface, its conical form, its cup-shaped crater, its slopes of loose ashes, and its black bristling lava-currents remain as unimpeachable witnesses that the volcanic fires, now quenched, once blazed forth fiercely. The wonderful groups of volcanoes in Auvergne and the Eifel are as fresh as if they had not yet ceased to be active, and might break forth again at any moment ; yet they have been quiescent ever since the beginning of authentic human history.

But in the progress of the degradation which everywhere slowly changes the face of the land, it is impossible that volcanic hills should escape the waste which befalls every other kind of eminence. We can picture a time when the volcanic cones of Auvergne will have been worn away, and when the lava-streams that descend from them will be cut into ravines and isolated into separate masses by the streams that have even already deeply trenched them. Where all the ordinary and familiar signs of a volcano have been removed, how can we tell that any volcano ever existed ? What enduring record do volcanoes inscribe in geological history ?

Now, it must be obvious that among the operations of an active volcano, many of the most striking phenomena have hardly any importance as aids in recognising the traces of long-extinct volcanic action. The earthquakes and tremors that accompany volcanic outbursts, the constant and prodigious out-rushing of steam, the abundant discharge of gases and acid vapours, though singularly impressive at the time, leave little or no lasting mark of their occurrence. It is not in phenomena, so to speak, transient in their effects, that we must seek for a guide in exploring the records of ancient volcanoes, but in those which fracture or otherwise affect the rocks below ground, and pile up heaps of material above.

Keeping this aim before us, we may obtain from an examination of what takes place at an active volcano such durable proofs of volcanic energy as will enable us to recognise the former existence of volcanoes over many tracts of the globe where human eye has never witnessed an eruption, and where, indeed, all trace of what could be called a volcano has utterly vanished. A method of observation and reasoning has been established, from the use of which we learn that in some countries, Britain for example, though there is now no sign of volcanic activity, there has been a succession of volcanoes during many protracted and widely separated periods, and that probably the interval that has passed away since the last eruptions is not so vast as that which separated these from those that preceded them. A similar story has been made out in many parts of the continent of Europe, in the United States, India,

and New Zealand, and, indeed, in most countries where the subject
has been fully investigated.

A little reflection on this question will convince us that the
permanent records of volcanic action must be of two kinds : first
and most obvious are the piles of volcanic materials which have
been spread out upon the surface of the earth, not only round the
immediate vents of eruption, but often to great distances from
them ; secondly, the rents and other openings in the solid crust of
the earth caused by the volcanic explosions, and some of which
have served as channels by which the volcanic materials have
been expelled to the surface.

Volcanic Products.—We shall first consider those materials
which are erupted from volcanic vents and are heaped up on the
surface as volcanic cones or spread out as sheets. They may be
conveniently divided into two groups : 1st, Lava, and 2d, Fragmentary materials.

(1) *Lava.*—Under this name are comprised all the molten rocks
of volcanoes. These rocks present many varieties in composition

Fig. 35.—Cellular Lava with a few of the cells filled up with infiltrated
mineral matter (Amygdules).

and texture, some of the more important of which will be described
in Chapter XI. Most of them are crystalline—that is, are made up
wholly or in greater part of crystals of two or more minerals inter-
locked and felted together into a coherent mass. Some are chiefly
composed of a dark brown or black glass, while others consist of
a compact stony substance with abundant crystals imbedded in it.
Probably most of them when in completest fusion within the
earth's crust existed in the condition of thoroughly molten glass,

H

the transition from that state to a stony or lithoid one being due to a process of "devitrification" (p. 144) consequent on cooling. During this process some of the component ingredients of the glass crystallise out as separate minerals, and this crystallisation some-times proceeds so far as to use up all the glass and to transform it into a completely crystalline substance.

In many cases lavas are strikingly cellular—that is to say, they contain a large number of spherical or almond-shaped cavities somewhat like those of a sponge or of bread, formed by the expansion of the steam absorbed in the molten rock (Figs. 35 and 36 and p. 146). Lavas vary much in weight and in colour. The heavier kinds are more than three times the weight of water; or, in other words, they have a specific gravity ranging up to 3.3; and are commonly dark grey to black. The lighter varieties, on the other hand, are little more than twice the weight of water, or have a specific gravity which may be as low as 2.3, while their colours are usually paler, sometimes almost white.

When lava is poured out at the surface it issues at a white heat —that is, at a temperature sometimes above that of melting copper, or more than 2204° Fahr.; but its surface rapidly darkens, cools, and hardens into a solid crust which varies in aspect according to the liquidity of the mass. Some lavas are remarkably fluid, flow-ing along swiftly like melted iron; others move sluggishly in a stiff viscous stream. In many pasty lavas, the surface breaks up into rough cindery blocks or scoriæ like the slags of a foundry, which grind upon each other as the still molten stream underneath creeps forward (p. 146). In general, the upper part of a lava-stream is more cellular than the central portions, no doubt because the imprisoned steam can there more easily expand. The bottom, too, is often rough and slaggy, as the lava is cooled by contact with the ground, and por-tions of the chilled bottom-crust are pushed along or broken up and involved in the still fluid portion above.

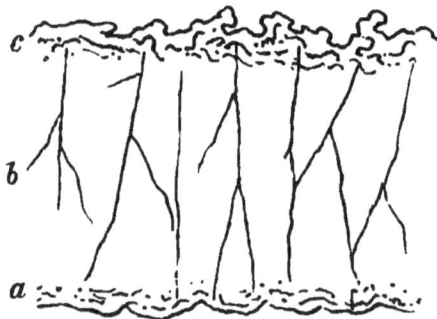

FIG. 36.—Section of a lava-current.

There are thus three more or less well-defined zones in a solidi-fied lava-current—a cellular or slaggy upper part (c in Fig. 36),

a more solid and jointed centre (*b*), embracing usually by much
the largest proportion of the whole, and a cellular or slaggy bottom
(*a*).   A rock presenting these characters tells its story of volcanic
action in quite unmistakable language.   It remains as evidence of
the existence of some neighbouring volcanic vent, now perhaps
entirely covered up, whence it flowed.   We may even be able to
detect the direction in which the lava moved.   The cells opened
by the segregation and expansion of the steam entangled in the
interstices of a mass of lava which is at rest are, on the whole,
spherical.   But if the rock is still moving, the cells will be drawn
out and flattened into almond-shaped (amygdaloidal) vesicles, with
their flat sides parallel to the surface of the lava, and their longer

FIG. 37.—Elongation of cells in direction of flow of a lava-stream.

axes ranged in one general direction, which is that of the motion
of the molten stream (Fig. 37).

At a volcanic vent, the mass of erupted lava is generally
thickest, and it thins away as its successive streams terminate on
the lower grounds surrounding the cone.   But sometimes a lava-
current may flow for 40 miles or more from its source, and may
here and there attain locally a great thickness by rolling into a
valley and filling it up, as has been witnessed among the Icelandic
eruptions.   As a rule, where ancient lava-streams are found to
thicken in a certain direction, we may reasonably infer that in that
direction lay the vent from which they flowed.

Again, sheets of lava that solidify on the slopes of a volcanic
cone are inclined ; they may congeal on declivities of as much as
30° or 40°.   If a series of ancient lavas were observed to slope
upward to a common centre, we might search there for some trace

of the funnel from which they were discharged. But, of course, in proportion to their antiquity, lava-streams, like every other kind of rock, have suffered from geological revolutions, among which those that involve upheaval and dislocation are especially important, so that the inclination of an ancient lava-bed must not be too hastily assumed as an indication of the slope of the cone of a vol-cano. It must be taken in connection with the rest of the evidence supplied by the whole district.

Where lavas reach the lower grounds beyond the foot of a volcanic cone, they may spread out in wide nearly horizontal sheets. As current succeeds current, the original features of the plain may be entirely buried under a mass of lava many feet thick. If a section could be cut through such an accumulation, it might be possible to determine the thickness of each successive lava-stream by means of the slaggy upper and lower surfaces. Here and there, too, where two eruptions were separated by an interval long enough to allow the surface of the older mass partially to crumble into soil and support some vegetation, the layer of burnt soil between the two sheets of lava would remain as a witness of this interval.

In other instances, we can understand that in the larger hollows of a lava-plain, ponds or lakes might gather, on the floor of which there might be deposited layers of fine silt full of terrestrial leaves, insect remains, land and fresh-water shells, and other organic relics of a land-surface. If, now, a lacustrine accumulation of this kind were to be buried under a new outburst of lava, it would be sealed up and might preserve its record intact for vast ages. In any section cut through such a series of lava-beds by a river or the sea, or by man, the layers of silt with their organic remains intercalated between the lava-streams would prove the eruptions to have taken place on land, and to have been separated by a long interval, during which a lake was formed on the cold and decom-posing surface of the earlier lava.

The conditions under which the volcanic outbursts occurred may be thus inferred, not so much from the nature of the volcanic materials themselves, as from that of the layers of sediment that may happen to have been preserved among them. Seams of red baked soil, with charred remains of terrestrial vegetation interposed between the upper and under sides of successive lavas would point to subaerial eruptions. Bands of hardened clay or marl, with leaves and fresh-water shells, would show that the lavas had in-vaded a lake. Beds of limestone or other rock, containing corals, sponges, marine shells, and other traces of the life of the sea, would demonstrate that the eruptions were submarine. Examples of each of these varieties of evidence occur abundantly among the old volcanic tracts of Britain.

(2) *Fragmentary Products.*—These supply some of the most striking proofs of volcanic energy. They vary in size from huge blocks of stone, weighing many tons, down to the finest dust. The coarsest materials naturally accumulate round the vent, while the finest may be borne away by wind to distances of many hundreds of miles. On the volcano itself, the stones, ashes, and dust form beds of coarse and fine texture which, on the outside of the cone, have the usual slope of the declivity. By degrees, they become more or less consolidated, and are then known by the general name of *Tuffs.* (See p. 153.)

The materials composing a tuff are generally derived from lavas. The fine dust, discharged from a large volcano in such prodigious quantities as to make the sky dark as midnight for days together, is simply lava that has been blown into this finely divided condition by the explosion of the vapours and gases which exist absorbed in it while still deep down within the earth's crust. The cinder-like fragments, and scoriæ or slags, that are ejected in such numbers and fall back into the crater and upon the outer slopes of the cone, are pieces of lava frothed up by the expansion of the imprisoned steam, torn off from the column of lava in the vent, and shot whirling up into the air. Large blocks of lava, or of the rocks through which the volcanic funnel has been opened, are often broken off by the force of the explosions and discharged with the other volcanic detritus from the vent. These materials, descending to the ground, form successive beds that vary in dimensions according to the vigour of eruption and their distance from the vent. Around the focus of activity there may be thick accumulations of blocks, bombs, and pieces of scoriæ mixed with fine ashes and sand. A notable feature is the generally cellular character of these stones—a peculiarity which marks them as made of truly volcanic materials. An examination of the finest dust likewise discloses the presence of the glass and crystals that constitute the lava from the explosion of which the dust was derived (p. 143).

From the wide area over which the fragmentary materials ejected by volcanoes are dispersed in the atmosphere before they fall to the earth, they are more likely than lavas, to be preserved among contemporaneous sedimentary accumulations. They often descend upon lakes, and must there be interstratified with the mud, marl, or other deposit in progress at the time. They are also widely diffused over the sea-floor. Recent dredgings of the ocean-basins have shown that traces of fine volcanic detritus may be detected even at remote distances from land. As active volcanoes almost always rise near the sea, as the oceans are dotted over with volcanic islands, and as, doubtless, many eruptions take place on the sea-bottom, there are obvious reasons why volcanic particles should be universally diffused over the sea-bottom. Geologists can also understand why, in the records of volcanic history in bygone ages, so large a proportion of the evidence should be of submarine eruptions.

Beds of tuff often contain traces of the plants or animals that lived on the surfaces on which the volcanic materials fell—sometimes remains of terrestrial or lacustrine vegetation and animals ; but in the great majority of instances, shells and other relics of the inhabitants of the sea.

In a series of layers of tuff round a volcanic orifice, the memorials of the earliest discharges are of course preserved in the layers at the bottom. Accordingly, in such situations, abundant fragments of the rocks of the surrounding country may be noticed. We could hardly ask for more convincing evidence of the blowing out of the vent and the ejection of the rock-fragments from it, before the volcano began to discharge only volcanic materials.

Large blocks of lava ejected obliquely from the crater may fall beyond the limits of the cone. If a block thus discharged should

FIG. 38.—Volcanic block ejected during the deposition of strata in water.

fall into a lake-basin, it would be covered up in the silt accumulating there, and might be the only remaining record of the eruption to which it belonged. In after times, were the lake-floor laid dry, the stone might be found in its place, the layers of sediment into which it fell pressed down by the force with which it landed on them and by its weight, the later layers mounting over and covering it. Examples of this kind of evidence may be gathered in many old volcanic districts. One taken from the coast of Fife is given in Fig. 38. The lowest bed there shown (1) is a brown shaly fire-clay, about 5 inches thick, which was once a vegetable soil, for abundant rootlets can be seen branching

through it.   It is overlain by a seam of coal (2), 5 or 6 inches
thick, representing the dense growth of vegetation that flourished
upon the soil; the next layer (3) is a green crumbling fire-clay
about a foot thick, covered by a dark shale with remains of plants
(4).   The feature of special interest in this section is the angular
block of lava (diabase) weighing about six or eight pounds, which
is stuck vertically in bed No. 3.   There can be no doubt that
this was ejected by an explosion, of which there is here no other
record.   It probably descended with some force, for, as shown in
the drawing, the lower layers of the fire-clay are pressed down by
it, and the coal itself is compressed.   We see that the stone fell
before the upper half of the fire-clay was formed, for the layers of
that part of the bed are heaped around the stone and finally spread
over it.   There are layers of lava and tuff above and below the
strata depicted in this section, so that there is abundant other
evidence of preceding and subsequent volcanic action here (see
Fig. 109).

As the fine volcanic dust may be transported by wind for
hundreds of miles before reaching the surface of the earth, its
presence does not necessarily show that a volcano existed in the
neighbourhood in which it fell.   The fine ashes from the Icelandic
volcanoes, for example, have been found abundantly even as far
as Sweden and the Orkney Islands.   But where the fragmentary
materials are of coarse grain, and more especially where they
contain large slags, scoriæ, and bombs, and where they are interstratified with sheets of lava, they unquestionably indicate the
proximity of some volcanic vent from which the whole proceeded.

Volcanic Vents and Fissures.—The various materials ejected
from a volcano to the surface may conceivably be in course of
time entirely swept away.   Nevertheless, though every sheet of
lava and every bed of tuff is removed, there will still remain the
filled-up *vent*, funnel, or fissure, up which these materials rose, and
out of which they were ejected.   This opening in the solid crust
of the earth must evidently be one of the most durable, as it is
certainly one of the fundamental features in a volcano.   Let us
first consider the vent of an ordinary volcano.

In many instances, there is reason to believe that volcanic
vents are opened along lines of fracture in the earth's crust.   This
seems especially to be the case where a group of volcanoes runs
along one definite line, as is represented in Fig. 39.   Along such
a fissure, either of older date or due to the energy of the volcanic
explosions themselves, there must be weaker places where the

overlying mass is less able to bear the strain of the pent-up vapours underneath. At these places, after successive shocks, openings may at length be made to the surface, whence the lavas and ashes will be emitted, and each such opening will be marked by a cone of the erupted material.

But in innumerable examples, it is found that a fissure is not necessary for the formation of a volcano. The effect of the volcanic explosions is such as to drill a pipe or funnel even through the solid unfractured crust of the earth. The volcanic energy, so far from requiring a line of fracture for its assistance, seems often to have avoided making use of such a line, even when it existed. In the volcanic plateaux of Utah, which are dotted over with little volcanic cones, and are also traversed by great

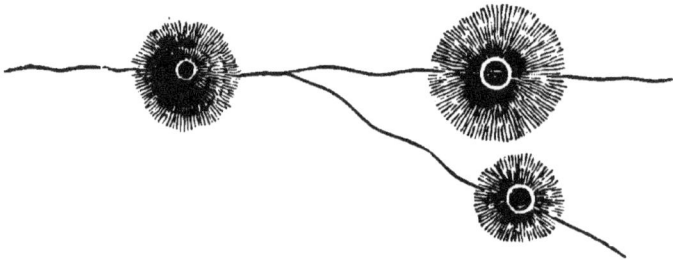

FIG. 39.—Volcanoes on lines of fissure.

dislocations, it is noticeable that the vents are not clustered along the lines of fracture. In the same region, and also along the courses of the Rhine and Moselle, volcanic vents have been opened near the brink of deep ravines rather than at the bottom. These are features of volcanic action, for which no satisfactory explanation has yet been found.

The crater of an active volcano is a hideous yawning cauldron, with rough red and black walls, at the bottom of which lie steaming pools of molten lava. Every now and then a sharp explosion tears the lava open, sending up a shower of glowing fragments and hot ashes. These pools of liquid lava lie evidently on the top of a column of melted rock which descends in the volcanic chimney to an unknown depth into the earth's interior. If the volcano were to become extinct, this lava column would cool and solidify, and even after the entire destruction and removal of the cone and crater, would remain as a stump to tell where the site of the volcano had been. Layer after layer might be stripped off the surface of the land ; hundreds or thousands of feet of rock might in this manner be removed, yet, so far as we know, the stump of the volcano would still be there. No probable amount of waste of the surface of the earth's crust could remove a vertical column of rock which descends to an unknown depth into the interior. The site of a volcanic vent can never be effaced except by being buried under masses of younger rock.

Volcanic vents, affording as they do so durable a testimony to volcanic action, deserve careful attention. At an active volcano, or even at one which, though extinct, still retains its cone of erupted materials, we cannot, of course, learn much regarding the shape and size of the funnel, for only the crater, and at most merely the upper part of the vent, are accessible. But among volcanic tracts of older date, where the cones have been destroyed, and where the filled-up funnels are laid bare, the subterranean architecture of volcanoes is revealed to us. At such places we are allowed, as it were, to descend the chimney of a volcano, and to make observations altogether impossible at a modern volcanic cone.

From observations made at such favourable localities, it has been ascertained that the funnels of volcanoes are in general rudely circular or elliptical, though liable to many modifications of outline. They vary indefinitely in diameter, according to the vigour of the volcanic outbursts that produced them. The smaller vents are not more than a few yards in width; but those of the larger volcanoes which, as in Sumatra and Java, have sometimes craters comprising an area of 40 square miles, must have enormously larger funnels.

The materials that fill up a vent are sometimes only fragments of the surrounding rocks. In such cases, we may suppose that when the volcanic explosions had spent their force and had blown out an opening to the surface of the ground, they were not succeeded by the uprise of any solid volcanic materials ; that, in short, only the first stage in the establishment of a volcano was reached, when, owing to some failure of the subterranean energy at the place, the operations came to an end. But though the upper part of the vent might remain open, surrounded with a crater formed of the fragments into which the rocks were blown by the explosions, the lower parts would undoubtedly be filled up by the fall of fragments back again into the vent. And if all the material ejected to the surface were removed, the top of this column of fragmentary materials would remain as an unmistakable evidence of the explosions that had originated it.

But in the vast majority of cases, the operations at a volcanic vent do not end with the first explosions. Clouds of ashes and stones are ejected, and streams of molten lava are poured forth. In some instances, the chimney may be finally choked with volcanic blocks, scoriæ, cinders, and ashes, in others with consolidated lava. Examples of both kinds of infilling are found, and also others where the two forms of volcanic material occur together in the same vent.

A volcanic chimney filled up in this way with volcanic materials, and exposed by the removal of the lava or ashes thrown out to the surface is known as a *Neck* (see Figs. 40, 41, 42, 111, and p. 208). As these materials are usually harder and more durable than the surrounding rocks, they project above the general surface of the ground. The stump of the volcano is left as a hill, the form and prominence of which will chiefly depend upon the nature of the material; hard tough lava will rise abruptly, as a crag or hill, above the surrounding country, while consolidated ashes, scoriæ, and other fragmentary stuff will give a smoother and less marked outline.

These features will be best understood from a series of diagrams.

FIG. 40.—Outline of a Volcanic Neck.

We may take, by way of illustration, a neck composed mainly of fragmentary ejections, but with a plug of lava reaching its summit.

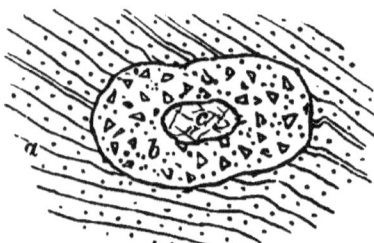

FIG. 41.—Ground-plan of the structure of the Neck shown in Fig. 40.

The usual outlines of such a neck are represented in Fig. 40. There is nothing in the general form of this hill to suggest a volcanic origin; yet, if we examine its structure and that of the ground around it, we may find them to be as represented in Fig. 41, where the surrounding rocks are supposed to consist of various sandstones, clays, limestones, and other sedimentary deposits (*a*), through which the volcanic vent (*b*, *c*) has been drilled. The

neck is represented as elliptical in cross section, composed mainly
of consolidated volcanic ashes and blocks (*b*), but with a mass of
lava (*c*) in the centre. The structure of the hill is explained in
the vertical section, Fig. 42 (see also Fig. 111). We there see
that the vent has been blown through the surrounding strata
(*a, a*), and has been filled up mainly with fragmentary materials
(*b, b*) ; but that through its centre there has risen a column or
plug of lava (*c*), which not improbably marks the last effort of the
volcano to force solid ejections to the surface. The line *s, s* indicates the present surface of the ground, after the prolonged waste
during which all the volcanic cone has been removed. But we

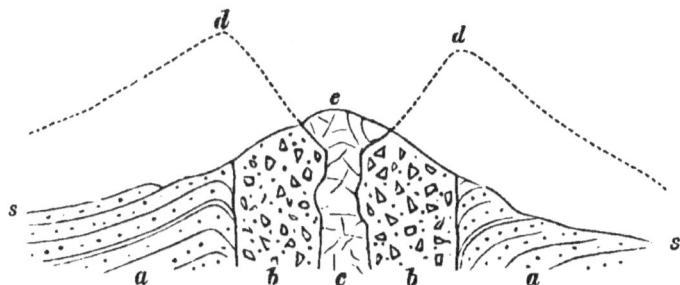

FIG. 42.—Section through the same Neck as in Figs. 40 and 41.

can in imagination restore the original surface, which may have
been somewhat as shown by the dotted lines, the position of the
crater being indicated at *e*, and its crest on either side at *d, d*.
No trace is here left of the original volcanic cone. The present
form of the ground is due to denudation, which has left the more
durable volcanic rocks projecting above the surrounding strata.
The continued progress of superficial degradation will remove still
more of the neck, but the downward continuation of the volcanic
column must always remain, and will probably always project as
a hill. A volcanic neck is thus one of the most enduring and
unmistakable evidences of the site of a volcano (see p. 208).

Besides vents or funnels, other openings are made by volcanic
explosions in the crust, which serve as receptacles of lava and
ashes, and remain as durable memorials of volcanic action. Of
these the most important are *Fissures*, which are formed in
large numbers in and around a volcanic cone, but which may
also arise at a distance from any actual volcano. During the
convulsions of an eruption, the cone and the surrounding
country are sometimes split by lines of fissure, which tend to

radiate from the centre of disturbance, somewhat as cracks do in a pane of glass through which a stone is thrown.   Sometimes the two sides of a fissure close together again, leaving no superficial trace of the dislocation.   More frequently steam and various volcanic vapours escape from the chasm, and may deposit along the walls sublimates of different minerals, such as common salt, chloride of iron, specular iron, sulphur, and sal-ammoniac.   These deposited substances may even continue to grow there until they entirely fill up the space between.   In such cases, the line of fissure is marked by a vertical or steeply inclined band of minerals interposed between the ends of the rocks that have been ruptured and separated.   But in most instances, the opening is filled up by the rise of lava from below.   At night, the vents opened on the outside of an active volcano may be traced from afar by the glow of the white-hot lava that rises in them to within a short distance from the surface.   When the lava cools and solidifies in these fissures, it forms wall-like masses, known as *Dykes* (Fig. 43). Inside many volcanic craters, the walls are traversed with dykes which, though on the whole tending to keep a vertical direction, may curve about irregularly according to the form of the vents into which the lava rose.   Like the necks above described, dykes form enduring records of volcanic action.   The superficial cones and craters may disappear, but the subterranean lava-filled fissures will still remain as records of volcanic action.

In some volcanic regions, where enormous floods of lava have been poured forth, no great central cones have existed.   Such regions extend as vast black plains of naked rock, mottled with shifting sand-hills, or as undulating tablelands carved by running water into valleys and ravines, between which the successive sheets of lava are exposed in terraced hills.   Beyond the limits over which the lava-sheets are spread, dykes of the same kinds of lava rise in abundance to the surface.   There can be no doubt that the dykes do not terminate at the edge of the lava-fields, but pass underneath them.   Indeed, as they increase in number in that direction, they are probably more abundant underneath the lava than outside of the lava-fields.   Sometimes sections are exposed showing how, after rising in a fissure, the lava has spread out on either side as a sheet.   In these vast lava-plateaux or deserts, the molten rock, instead of issuing from one main central Etna or Vesuvius, appears to have risen in thousands of fissures opened in the shattered crust, and to have welled forth from numerous vents on these fissures, spreading out sheet after sheet till, like a rising lake, it has not only overflowed the lower grounds, but even buried all the minor hills.   Such appears to have been the history of vast tracts in Western North America.   The area which has there been flooded with lava has been estimated to be larger than that of France and Great Britain together, and the depth of the total mass of lava erupted reaches in some places as much as 3700 feet.   Some rivers have cut gorges in this plain of lava, laying bare its component rocks to a depth of 700 feet or more.   Along

FIG. 43.—Volcanic dykes rising through the bedded tuff of a crater.

the walls of these ravines we see that the lava is arranged in parallel beds or sheets often not more than 10 or 20 feet thick, each of which, of course, represents a separate outpouring of molten rock.

Except where such deep sections have been cut through them by rivers, recent lava-floods can only be examined along their surface, and we are consequently left chiefly to inference regarding their probable connection with fissures and dykes underneath. But in various parts of the world, lava-plains of much older date have been so deeply eroded as to expose not only the successive sheets of lava but the floor over which they were poured, and the

abundant dykes which no doubt served as the channels wherein
the lava rose towards the surface, till it could escape at the lowest
levels, or at weaker or wider parts of the fissures.   In Western
Europe important examples of this structure occur, from the north
of Ireland through the Inner Hebrides and the Faröe Islands to
Iceland.   This volcanic belt presents a succession of lava-fields
which even yet, in spite of enormous waste, are in some places
more than 3000 feet thick.   The sheets of lava are nearly flat,
and rise in terraces one over another into green grassy hills, or
into the dark fronts of lofty sea-washed precipices.   Where this
thick cake of lava has been stripped off during the degradation of
the land, thousands of dykes are exposed, and many of these
traverse at least the lower parts of the sheets of lava.   They form,
as it were, the subterranean roots of which these sheets were the
subaerial branches ; and even where the whole of the material
that reached the surface, more than 3000 feet thick, has been worn
away, the dykes still remain as evidence of the reality and vigour
of the volcanic forces.

EARTHQUAKES.

The rise of hot springs and the explosions of volcanoes furnish
impressive testimony to the internal heat of our planet ; but they
are by no means the only proofs that the pent-up energy of the
interior of the globe reacts upon the outer surface.   By means of
delicate instruments, it can be shown that the ground beneath
our feet is subject to continual tremors which are too feeble to be
perceived by the unaided senses.   From these minuter vibrations,
movements of increasing intensity can be detected up to the
calamitous earthquake, whereby a country is shaken to its foundations, and thousands of human lives, together with much valuable
property, are destroyed.   We do not yet know by what different
causes these various disturbances are produced.   Some of the
fainter tremors may arise from such influences as changes of
temperature and atmospheric pressure, and the rise and fall of the
tides.   But the more violent must be assigned to causes working
within the earth itself.   The collapse of the roofs of underground
caverns, the sudden condensation of steam or explosion of volcanic
vapours, the snap of rocks that can no longer resist the strain to
which, by the cooling and consequent contraction of the inner hot
nucleus, they have been subjected within the earth's crust—these
and other influences may at different times come into play to

determine convulsive earthquake shocks. Without, however, entering into the difficult question of the causes of the movements, we may inquire into their effects in so far as these register their passing in the annals of geological history.

Their awful suddenness and devastation have invested earthquakes with a high importance in the popular estimate of the forces by which the surface of the globe is modified. Yet if we judge of them by their permanent effects, we must give them a comparatively subordinate place among these forces. After some of the most destructive earthquakes recorded in human history, hardly any trace of the calamity is to be seen, save in shattered and prostrate houses. But when these buildings have been repaired or rebuilt, no one visiting the ground might be able to detect any trace of the earthquake that shattered or overthrew them.

Yet severe earthquakes do not pass without their self-written chronicle which, though often evanescent on the face of nature, is at the time conspicuous enough. Landslips are caused, large masses of earth and blocks of rock being shaken down from higher to lower levels ; the ground is rent, and the fissures are sometimes subsequently widened and deepened by rain and runnels into ravines. But more important are the marked changes of level that occasionally accompany earthquake-shocks. In some cases, the ground is raised for several feet, so that along maritime tracts there is a gain of land from the sea ; in others, the ground sinks, and the sea flows in upon the land. Yet it is evident that unless these changes are actually witnessed as the accompaniments of the earthquakes, they may take place without retaining any evidence that they were produced by such a cause. The convulsion of an earthquake, notwithstanding the havoc it may bring to the human population of a country, does not always record itself in distinctive and enduring characters in geological history. Some of its most noticeable effects also are not due directly to its own action, but to the operations of the waters of the land and of the sea which, when disturbed by the shock, not infrequently acquire increased vigour in their own peculiar forms of activity. The great waves set in motion by an earthquake roll over the low lands bordering the sea, and may cause vastly more destruction than is done by the mere shock of the earthquake itself.

UPHEAVAL AND SUBSIDENCE.

It is perhaps not so much by earthquakes, as by quiet, hardly perceptible movements, that the relative positions of sea and land are undergoing change at the present time. In some parts of the world the land is gradually rising, in others it is slowly sinking. Proofs of elevation are supplied by lines of barnacles or rock-boring shells, now standing above the reach of the highest tides ; by caves that have obviously been scooped out by the sea, but now stand at a higher level than the waves can reach ; and by deposits of sand, gravel, and shells which were evidently accumulated on a beach, but which now rise above the level where similar materials are now being accumulated (*Raised Beaches*). Evidences of subsidence are furnished by traces of old land-surfaces—trees with roots *in situ*, and beds of peat, lying below the limits of the tides (*Submerged Forests*). But it must be more difficult to prove subsidence than elevation, for as the land sinks, its surface is carried below the waves, which soon efface the evidence of terrestrial characters.

The time within which man has been observing and recording the changes of the earth's surface forms but an insignificant fraction of the ages through which geological history has been in progress. We cannot suppose that during this brief period he has had experience of every kind of geological process by which the outlines of land and sea are modified. There may be great terrestrial revolutions which happen so rarely that none has occurred since man began to take note of such things. Among these revolutions, of which he has had as yet no experience, the most gigantic is the formation of a mountain-chain. That the various mountain-chains of the globe are of very different ages, and that some of the most gigantic of them are, compared with others, of recent date, are facts in the history of the globe which will be more fully referred to in later pages ; but so far as human history or tradition goes, man has never witnessed the uprise of a range of mountains. The crust of the earth has been folded and crumpled on the most colossal scale, some parts having been pushed for miles away from their original position ; it has been rent by profound fissures, on each side of which the rocks have been displaced for many thousand feet ; and it has been so broken, crushed, and sheared, that its component rocks have in some places assumed a structure entirely different from what they originally possessed. But of all these colossal mutations there is no human experience. We are driven to reason regarding them from the record of them preserved among the rocks, and from the analogies that can be suggested by experiments devised to imitate as far as possible the processes of nature. To this subject we shall return in Chapter XIII.

Summary.—The enduring records left by volcanoes, whence their former existence in almost all regions of the world may be demonstrated, are to be sought partly in the materials which they have brought up to the surface, and partly in the vents and fissures by which they have discharged these materials. Of the former kind of evidence lava furnishes a conspicuous example ; its internal crystalline or glassy structure, its steam-cavities, and the cellular slaggy upper and under parts of the sheets in which it lies, are all proofs of its former molten condition. A succession of lava-beds, piled one above another, marks a series of volcanic eruptions, and the nature of the layers of non-volcanic material intercalated between them may indicate the conditions under which the eruptions took place, whether on land, in lakes, or in the sea. The fragmentary products consolidated into beds of tuff are likewise characteristic of volcanoes ; they consist mainly of lava-dust with cindery scoriæ, slags, and blocks ; they accumulate most deeply and in coarsest material at and immediately around the volcanic vents, but their finer particles may be carried to enormous distances; they are especially liable to be intercalated with contemporaneous sedimentary deposits in lakes and on the sea-floor.

The vents through which lava and ashes are ejected to the surface form the most permanent record of volcanoes, for, being filled up with volcanic rocks to unknown depths, they cannot be destroyed by the mere denudation of the surface, and can only disappear by being buried under later accumulations. Such "necks" consist sometimes of lava, sometimes of consolidated volcanic debris, or of both kinds of material together, and remain as the stumps of volcanoes, where every other trace of volcanic action may have passed away. Not less enduring are the dykes or wall-like masses of lava which have risen and solidified in open fissures. Enormous sheets of lava appear to have flowed out from such fissures in regions where the volcanic energy never produced any great central cone.

Earthquakes do not impress their mark upon geological history so indelibly as might be supposed. In spite of the destruction which they cause to human life and property, it is by such direct

I

changes as landslips, rents of the ground, and the upheaval or depression of land, and by such indirect changes as may be produced by derangements of rivers, lakes, and the sea, that earthquakes leave their chief record behind them.

Some of the most important changes of level now going on are effected quietly and almost imperceptibly, some regions being slowly elevated, and others gradually depressed. But the time within which man has been an observer and recorder of nature is too brief to have supplied him with experience of all the ways in which the internal energy of the globe affects its surface. In particular, he has never witnessed the production of a mountain-chain, nor any of the plications, fractures, and displacements which the crust of the earth has undergone. Regarding these revolutions we can only reason from the records of them in the rocks, and from such laboratory experiments as may seem most closely to imitate the processes of nature that were concerned in their production.

PART II

ROCKS, AND HOW THEY TELL THE HISTORY OF THE EARTH

CHAPTER X

THE MORE IMPORTANT ELEMENTS AND MINERALS OF THE EARTH'S CRUST

IN the foregoing Part of this volume we have been engaged in considering the working of various processes by which the surface of the earth is modified at the present time, and some of the more striking ways in which the record of these changes is preserved. We have seen that, on the whole, it is by deposits of some kind, laid down in situations where they can escape destruction, that the story of geological revolution is chronicled. In one place it is the stalagmite of a cavern, in another the silt of a lake-bottom, in a third the sand and mud of the sea-floor, in a fourth the lava and ashes of a volcano. In these and countless other examples, materials are removed from one place and set down in another, and in their new position, while acquiring novel characters, they retain more or less distinctly the record of their source and of the conditions under which their transference was effected.

In these chapters, reference has intentionally been avoided as far as possible to details that required some knowledge of minerals and rocks, in order that the broad principles of geology, for which such knowledge is not absolutely essential, might be clearly enforced. It is obvious, however, that as minerals and rocks form the records in which the history of the earth has been preserved, this history cannot be followed into detail until some acquaintance with these materials has been made. What now lies before the reader, therefore, in order that he may be able to apply the knowledge he has gained of geological processes to the elucidation of former geological periods, is to make himself familiar with at least the more common and important minerals and rocks. This he can only do satisfactorily by handling the objects themselves, until he acquires such an acquaintance with them as to be able to recognise them where he meets with them in nature.

At first the number and variety of these objects may appear to be almost endless, and the learner may be apt to despair of ever mastering more than an insignificant portion of the wide circle of inquiry and observation which they present. But though the detailed study of this subject is more than enough to tax the whole powers of the most indefatigable student, it is not by any means an arduous labour, and assuredly a most interesting one, to acquire so much knowledge of the subject as to be able to follow intelligently the progress of geological investigation, and even to take personal part in it. This accordingly is the task to which he is invited in the present and following chapters.

SIMPLE ELEMENTS COMPOSING THE EARTH'S CRUST.

Before considering the characters presented by the various rocks that form the visible part of the earth's crust, we may find it of advantage to inquire into the general chemical composition of rocks, for by so doing we learn that though the chemist has detected more than sixty substances which he has been unable to decompose, and which therefore he calls *elements*, only a small proportion of these enter largely into the composition of the outer part of the globe. In fact, there are only about sixteen elements that play an important part as constituents of rocks; these together constitute about ninety-nine parts of the terrestrial crust. Half of them are metals; and the other half are metalloids or non-metals, as in the two subjoined lists, the most abundant being in each case placed first.

METALLOIDS OR NON-METALS.	Symbol.	Atomic Weight.	METALS.	Symbol.	Atomic Weight.
Oxygen . .	O	15.96	Aluminium .	Al	27.30
Silicon . .	Si	28.00	Calcium .	Ca	39.90
Carbon . .	C	11.97	Magnesium .	Mg	23.94
Sulphur . .	S	31.98	Potassium .	K	39.04
Hydrogen . .	H	1.00	Sodium . .	Na	22.99
Chlorine .	Cl	35.37	Iron . . .	Fe	55.90
Phosphorus .	P	30.96	Manganese .	Mn	54.80
Fluorine . .	F	19.10	Barium . .	Ba	136.80

Some of those elements occur in the free state, that is, not combined with any other element. Carbon, for instance, is found pure in the form of the diamond, and also as graphite. But in the great majority of cases, they assume various combinations. Most abundant are *oxides*, or compounds of oxygen with another element. Compounds of sulphur and a metal are known as *sulphides* ; and similar compounds with chlorine are *chlorides*. Some of the compounds form further combinations with one or more elements. Thus the acid-forming oxides unite with water to form what are called *acids*, which, combining with metallic oxides or *bases*, form with them compounds termed *salts*. Sulphur and oxygen, for example, uniting in certain proportions with water, constitute sulphuric acid (H_2SO_4) which, parting with its displaceable hydrogen and combining with the metal calcium, forms the salt known as calcium-sulphate, or sulphate of lime ($CaSO_4$).

METALLOIDS.—Of the non-metallic elements, by far the most abundant and important is **Oxygen.** In its free state, it exists as a gas which has been occasionally detected at active volcanic vents. But with this rare exception, it is always found mixed or combined with one or more elements. Thus, mixed with nitrogen, it constitutes the atmosphere, of which it forms not less than 23 per cent by weight. It takes a still larger share in the composition of water, which consists of 88.88 per cent of oxygen and 11.12 of hydrogen. There is a continual removal of oxygen from air and water in the processes of weathering described in Chapter II. Substances which can take more of this element abstract it especially from damp air or from water. A knife or any other piece of iron, for example, will remain unchanged for an indefinite length of time if kept in dry air ; but as soon as it is exposed to moisture, in which there is always some dissolved air, it begins to rust. The familiar brown rust which slowly eats into the very centre of the iron, is due to a chemical union of oxygen with the iron, forming what is called an Oxide of iron with water (see p. 127). Among the rocks of the earth's crust, a large proportion are liable to undergo a similar change, and so enormous has been the extent of this change in the past history of our globe, that somewhere about one-half of the outer and accessible part of the crust consists of oxygen, which was probably at first in the atmosphere.

Next in importance to oxygen among the metalloids is **Silicon,** which is never met with in the free state. It has been artificially obtained, however, in the form of a dull brown powder. In nature, it always occurs united with oxygen, forming the familiar substance known as Silica or Silicic Acid (SiO_2), which constitutes more than a half of all the known part of the earth's crust. Silica is indeed the fundamental compound of the crust, forming by itself entire masses of rock, and entering as a principal constituent into the majority of rocks. It occurs abundantly as the mineral Quartz, the colourless transparent forms of which are known as rock-crystal (Fig. 44), and also in combination with various metallic bases as the important family of Silicates (p. 130). It is present in solution in most natural waters, both those of the land and of the sea, whence it is secreted by plants (diatoms, grasses) and animals

FIG. 44.—Group of Quartz-crystals (Rock-crystal).

(radiolarians, sponges, p. 83). It is thus carried by percolating water into the heart of rocks, and may be deposited in their interstices and cavities. Its hardness and durability eminently fit it for the important part it plays in binding the materials of rocks together, and enabling them better to resist the decomposing effects of air and water.

Carbon, though found in a nearly pure state in the clear gem called Diamond, and also in the black opaque mineral Graphite, more usually occurs mixed with various impurities, as in the different kinds of coal. This element has a high importance in nature, because it is the fundamental substance made use of by both plants and animals to build up their structures, and because it serves as a bond of connection between the organic and the inorganic worlds. In union with oxygen, Carbon forms the widely-diffused gaseous compound known as Carbon-dioxide (CO_2), which occurs in the proportion of about four parts in every ten thousand parts of ordinary atmospheric air. From the air it is abstracted and decomposed by living plants in presence of sunshine, the oxygen being in great measure sent back into the atmosphere, while the carbon with some oxygen, nitrogen, and hydrogen is built up into the various vegetable cells and tissues. When we look at a verdant landscape or a boundless forest, it is a striking thought that all this vegetation has been chiefly constructed out of the small proportion of invisible carbon-dioxide present in the atmosphere. The vast numbers of beds of coal imbedded in the earth's crust have, in like manner, been derived from the atmosphere through the agency of former tribes of plants. Not only in beds of coal, but still more prevalently in masses of limestone, carbon enters into the composition of rocks. Carbon-dioxide, as was pointed out in Chapter II, is abstracted by rain in passing from the clouds to the earth, and is also supplied by decomposing plants and animals in the soil. It is readily dissolved in water, and forms with it carbonic acid, $CO(OH)_2$, which has been referred to as so powerful a solvent of the substance of many rocks. This acid unites with a number of alkaline and earthy bases to form the important family of Carbonates. Of these the most abundant is calcium-carbonate, or carbonate of lime $(CaCO_3)$, which consists of 44 per cent carbon-dioxide, and 56 per cent lime. This carbonate not only occurs abundantly diffused through many rocks, but in the form of limestone builds up by itself thick mountainous masses of rock many hundreds of square miles in extent. It is abstracted by plants to form calcareous tufa (Chapter V), but far more abundantly by animals, especially by the invertebrata, as exemplified by the familiar urchins, corallines, and shells of the sea-shore. The limestones of the earth's crust appear to have been mainly formed of the calcareous remains of animals. Hence we perceive that the two forms in which carbon has been most abundantly stored up in the earth's crust have been principally due to the action of organised life ; coal being chiefly carbon that has been taken out of the atmosphere by plants, and limestone consisting of carbon-dioxide, to the extent of nearly one-half, which has been secreted from water by the agency of animals.

Sulphur is found in the free state, more particularly at volcanic vents, in pale yellow crystals or in shapeless masses and grains ; but it chiefly occurs in combination. Some of its compounds are widely diffused among plants and animals. The blackening of a silver spoon by a boiled egg is an illustration of this diffusion, for it arises from the union of the sulphur in the egg with the metal. Combinations with a metal (Sulphides, p. 137) and combinations with a metal and oxygen (Sulphates, p. 136) are the conditions in which sulphur chiefly exists.

Hydrogen is a gas which has been detected in the free state at active volcanic vents ; but otherwise it occurs chiefly in combination with oxygen as the oxide water (H_2O), of which it constitutes about one-ninth, or 11.12 per cent by weight. It also enters into the composition of plant and animal substances, and forms with carbon the important group of bodies known as Hydrocarbons, of which mineral oil and coal-gas are examples. In smaller quantity, it is found united with sulphur (sulphuretted hydrogen, H_2S), with chlorine (hydrochloric acid, HCl), and a few other elements.

Chlorine is a transparent gas of a greenish-yellow colour, but except possibly at active volcanic vents it does not occur in the free state. United with the alkali metals (potassium, sodium, and magnesium), it forms the chief salts of sea-water. The most important of these salts, Sodium-chloride, or common salt (NaCl) contains 60.64 per cent of chlorine, and forms 2.64 per cent by weight of sea-water. This salt is found diffused in microscopic particles in the air, especially near the sea, and beds of it hundreds of feet thick occur in many parts of the world among the sedimentary rocks that constitute most of the dry land (pp. 137, 157).

Phosphorus does not occur free ; it has so strong an affinity for oxygen that it rapidly oxidises on exposure to the air, and even melts and takes fire. Its most frequent combination is with oxygen and calcium, as Calcium-phosphate or phosphate of lime ($Ca_3(PO_4)_2$, p. 157). Though for the most part present in minute proportions, it is widely diffused in nature. It occurs in fresh and sea water, in soil and in plants, especially in their fruits and seeds ; it is supplied by plants to animals for the formation of bones, which when burnt consist almost entirely of phosphate of lime.

Fluorine also is never met with uncombined; it never unites with oxygen, forming in this respect the sole exception among the elements. Its most frequent combination as a rock constituent is with calcium, when it forms the mineral Fluor-spar (CaF_2, p. 137). Like phosphorus, it is widely diffused in minute proportions in the waters of some springs, rivers, and the sea, and in the bones of animals.

To these metalloids we may add the colourless, tasteless gas

Nitrogen, which, though not largely present in the earth's crust, constitutes four-fifths by volume or 77 per cent by weight of the atmosphere. It does not enter into combination so readily as the other elements above enumerated, but it is always found in the composition of plants and is a constituent of many animal tissues. It is the principal ingredient of the substance called Ammonia, which is produced when moist organic matter is decomposed in the air. In many rocks composed wholly or in great part of organic remains, such, for instance, as peat and coal, nitrogen is a constant constituent.

METALS.—Though so large a proportion of the known terrestrial elements are metals, these are much less abundant in the earth's crust than the metalloids. The most frequent are Aluminium, Calcium, and Magnesium. The substances most familiar to us as metals occupy an altogether subordinate part among rocks, the most abundant of them being Iron.

Aluminium never occurs in the free state, but can be artificially separated from its compounds, when it is seen to be a white, light, malleable metal. It is almost always united with oxygen as the oxide of Alumina (Al_2O_3), which occurs crystallised as the ruby and sapphire, but is for the most part united with silica, and in this form constitutes the basis of the great family of minerals known as the Silicates of Alumina, or Aluminous Silicates. These silicates generally contain some other ingredient which is more liable to decomposition, and when they decay and their more soluble parts are removed, they pass into clay, which consists chiefly of hydrated silicate of alumina.

Calcium is not met with uncombined, but has been artificially isolated and found to be a light, yellowish metal, between gold and lead in hardness. It occurs in nature chiefly combined with carbonic acid as a carbonate (p. 134), and with sulphuric acid as a sulphate (p. 136), to both of which substances reference has already been made; it is also present in many silicates. So abundant is Calcium-carbonate or carbonate of lime in nature that it may be found in most natural waters, which dissolve it and carry it in solution into the sea. Its presence in rocks may be detected by a drop of any mineral acid, when the liberated carbon-dioxide escapes as a gas with brisk effervescence. Calcium-sulphate is likewise a common constituent of terrestrial waters, especially of those which in household management are called *hard*; it constitutes not less than 3.6 per cent of the salts in ordinary sea-water, and when sea-water is evaporated this sulphate (gypsum, p. 136), being least soluble, is the first to be precipitated in minute crystals resembling in shape those shown in Fig. 62.

Magnesium is likewise only isolated artificially, when it appears as a soft, silver-white, malleable and ductile metal. It occurs in sea-water combined with chlorine as Magnesium-chloride, which constitutes 10.8 per cent of the total proportion of salts. It unites with carbonic acid as a carbonate, which with carbonate of lime forms the widely diffused rock called magnesian limestone or Dolomite (pp. 135, 155); it also enters into the composition of the Magnesian Silicates which are only second in importance to those of alumina.

Potassium and **Sodium** (alkali metals) are only obtainable in the free state by chemical processes, when they are found to be white brittle metals that float on water, and rapidly oxidise if exposed to the air. Combined with chlorine, Sodium forms the familiar chloride known as common salt, which constitutes 77.7 per cent of the salts of sea-water, is abundantly present in salt lakes, and occurs in extensive beds among the rocks of the dry land (p. 157). Potassium-chloride likewise occurs in the sea and may be obtained from the ashes of burnt sea-weed. Enormous deposits of it, combined with chlorides of sodium and magnesium, have been met with in Germany (Stassfurt). Potassium also exists in the sea in combination with sulphuric acid as potassium-sulphate or sulphate of potash, which amounts to about 2.4 of the total salts of sea-water. Sulphates of potassium, sodium, magnesium, and calcium form thick masses of rock in the Stassfurt deposits. Potassium and sodium in combination with silica form silicates which enter largely into the composition of many rocks. They are readily attacked by water containing carbonic acid, giving rise to what are called carbonates of the alkalies, or alkaline carbonates, which are removed in solution. By this means, carbonate of potash is introduced into soil, where it is taken up by plants into their leaves and succulent parts. When wood is burnt, this carbonate in considerable quantity may be dissolved with water out of the ash.

Iron is found in the free or native state in minute grains (rarely in large blocks) in some volcanic rocks, also in granules of "cosmic dust," probably of meteoric origin, and in fragments of various size which have undoubtedly fallen upon the earth's surface from the regions of space. There is reason to believe that much of the solid interior of the globe may consist of native iron and other metals. But it is in combination that iron is chiefly of importance in the earth's crust. It has united with oxygen to form several abundant oxides (p. 128). The protoxide or ferrous oxide (FeO) contains the lowest proportion of oxygen, and being, therefore, prone to take up more, gives rise to many of the processes of decay included under the general name of Weathering (p. 14). It is readily dissolved by organic and other acids, and is then removed in solution, but on exposure rapidly oxidises and passes into the highest oxide, known as the peroxide or sesquioxide of iron or ferric oxide (Fe_2O_3), which, being the permanent insoluble form, is found abundantly among the rocks of the earth's crust. Iron is the great colouring matter of nature ; its protoxide compounds give greenish hues to many rocks, while its peroxide colours them various shades of red which, when the peroxide is combined with water, pass into many tints of brown, orange, and yellow.

Manganese is commonly associated with iron in minute proportion in many lavas and other crystalline rocks ; its oxides resemble those of iron in their modes of occurrence (p. 130).

Barium and Calcium are called metals of the alkaline earths. The former can only be obtained in a free state by artificial means, when it appears as a pale yellow very heavy metal which rapidly tarnishes. In nature it chiefly occurs as the sulphate, Barytes, or heavy spar ($BaSO_4$), a mineral of frequent occurrence in veins associated with metallic ores (p. 136).

MINERALS OF CHIEF IMPORTANCE IN THE EARTH'S CRUST

Passing now from the simple elements, we have next to note the mineral forms in which they appear as constituents of the earth's crust. A *mineral* may be defined as an inorganic substance, having theoretically a definite chemical composition, and in most cases also a certain geometrical form. It may consist of only one element, for example, the diamond, sulphur, and the native metals, gold, silver, copper, etc. But in the vast majority of cases, minerals consist of at least two, usually more, elements in definite chemical proportions. In the following short list of the more important minerals of the earth's crust they are arranged chemically, according to the predominant element in them, or the manner in which the combinations of the elements have taken place, so that their leading features of composition may be at once perceived. The two elements, Carbon and Sulphur, in their native or uncombined state, sometimes form considerable masses of rock. Some of the native metals also may be enumerated as rock-constituents when they occur in sufficient quantities to be commercially important. Gold, for example, is found in grains and strings, in veins of quartz, and in irregular pieces or nuggets dispersed through the gravel deposits of regions where gold-bearing quartz-veins traverse the solid rocks. Omitting, however, the minerals formed of a single element, we may pass on to combinations of two or more elements, and consider first those in which oxygen is combined with some other element, forming what are commonly grouped together as Oxides. Then will come the Silicates, or combinations of silica with one or more bases, followed by the Carbonates, or combinations of carbon-dioxide with some base ; the Sulphates, or compounds of sulphuric acid and a base ; the Fluorides, or compounds of fluorine and a metal ; the Chlorides, or compounds of chlorine and a metal ; and the Sulphides, or compounds of sulphur and a metal.

It cannot be too strongly impressed upon the mind of the learner that no mere description in books will suffice to make him familiar with minerals and rocks. He ought to handle actual specimens of these objects and identify for himself the several characters which he finds assigned to them in books.

One of the most obvious features in a crystal of any mineral is the regular and sharply-defined edges and corners which it presents. Take a piece of rock-crystal or quartz, for example (Fig. 44), and

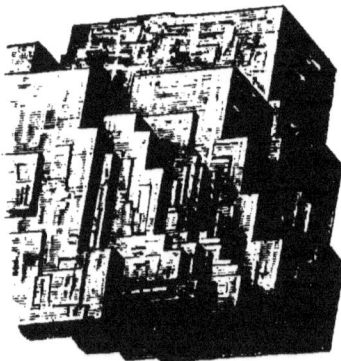

FIG. 45.—Calcite (Iceland spar), showing its characteristic rhombohedral cleavage.

you will find it to consist of six sides or faces, forming what is called a prism, and bevelled off at the end into a six-sided cone, called a pyramid. If you examine a large collection of similar crystals you may find no two of them exactly alike, yet they agree in presenting a six-sided figure. Again, procure a piece of the common mineral calcite, either a whole crystal (Figs. 59, 60), or a portion of a crystalline mass (Fig. 45) ; break it and you will find each fragment to possess the same form, that of a rhombo-hedron ; crush one of these fragments and you will observe that each little grain of the powder preserves the same shape. The rhombohedron, therefore, is called the fundamental crystalline form of the mineral. The property so strikingly shown in calcite, of breaking along definite crystalline planes, is termed *cleavage*. So perfect is the cleavage of calcite, that the crystallised mineral can hardly be broken, except along the planes that define the rhombohedron. Many minerals cleave more or less easily in one or more directions, and break irregularly in others. The cleavage affords a guide to the proper crystalline form of a mineral.

Though there are many hundreds of varieties of crystalline form, they may all be reduced to six primary types or systems. These are distinguished from each other by the number and position of their *axes*, which are mathematical straight lines, intersecting each other in the interior of a crystal, and connecting the centre of opposite flat faces of the crystal, or opposite angles or corners. The six systems, with their axes, are enumerated in the subjoined list.

I. Isometric (monometric, cubical, tesseral, regular). In this system there are three axes which are of the same length, and intersect each other at a right angle. The cube, octahedron, and dodecahedron are examples (Fig. 46). Crystals of this system are distinguished by

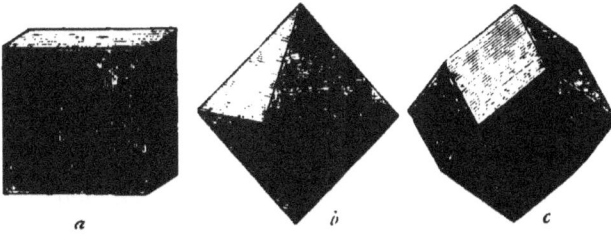

FIG. 46.—Cube (*a*), octahedron (*b*), dodecahedron (*c*).

their symmetry, their length, breadth, and thickness being equal. Common salt, fluor-spar (Fig. 64), and magnetite (Fig. 54) are illustrations.

II. Tetragonal (dimetric). The axes are three in number, and intersect each other at a right angle, but one of them, called the *vertical* axis,

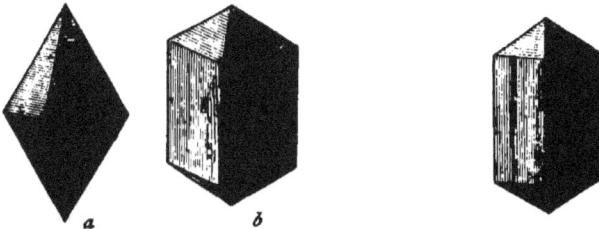

FIG. 47.—Tetragonal prism (*b*) and pyramid (*a*).   FIG. 48.—Orthorhombic prism.

is longer or shorter than the other two, which are *lateral* axes. Hence a crystal belonging to this system may either be oblong or squat (Fig. 47).

III. Orthorhombic (trimetric) has the three axes intersecting each other at a right angle, but all of unequal lengths. The rectangular and rhombic prisms and the rhombic octahedron belong to this system (Fig. 48).

IV. Hexagonal. This is the only system with four axes (Fig. 49). The lateral axes are all equal, intersect at right angles the vertical axis

FIG. 49.—Hexagonal prism (*a*), rhombohedron (*b*), and scalenohedron (*c*).

(which is longer or shorter than they are), and form with each other angles of 60°. Water, for instance, crystallises in this system, and the six-rayed star of a snow-flake is an illustration of the way in which the lateral axes are placed. Quartz is an example (Fig. 44), also calcite (Figs. 59, 60).

V. Monoclinic, with all the axes of unequal length. One of the lateral axes cuts the vertical axis at a right angle, the other intersects the vertical axis obliquely. Augite (Fig. 50), Hornblende (Fig. 57), and Gypsum (Fig. 62) are examples.

VI. Triclinic, the most unsymmetrical of all the systems, all the axes being unequal and placed obliquely to each other (Fig. 51).

FIG. 50.—Monoclinic prism.
Crystal of Augite.

FIG. 51.—Triclinic prism.
Crystal of Albite felspar.

Every mineral that takes a crystalline form belongs to one or other of these six systems, and through all its varieties of external form the fundamental relations of the axes remain unchanged.

Some minerals have crystallised out of solutions in water. How this may take place can be profitably studied by dissolving salt, sugar, or alum in water, and watching how the crystals of these substances gradually shape themselves out of the concentrated solution, each according to its own crystalline pattern. Other minerals have crystallised from hot vapours (sublimation), as may be observed at the fissures of an active volcano (p. 108). Others have crystallised out of molten solutions, as in the case of lava. Thoroughly fused lava is a glassy or vitreous solution of all the mineral substances that enter into the composition of the rock (p. 97), and when it cools, the various minerals crystallise out of it, those that are least fusible taking form first, the most fusible appearing last ; but a residue of non-crystalline glass sometimes remaining even when the rock has solidified (p. 144).

It is evident that minerals can only form perfect crystals where they have room and time to crystallise. But where they are crowded together, and where the solution in which they are dissolved dries or cools too rapidly, their regular and symmetrical growth is arrested. They then form only imperfect crystals, but their internal structure is crystalline, and if examined carefully will be found to show that in the attempt to form definite crystals each mineral has followed its own crystalline type. These characters are of much importance in the study of rocks, for rocks are only large aggregates of minerals, wherein definite crystals are exceptional, though the structure of the whole mass may still be quite crystalline.

But minerals also occur in various indefinite or non-crystalline shapes. Sometimes they are *fibrous* or disposed in minute fibre-like threads (Fig. 56); or *concretionary* when they have been aggregated into various irregular concretions of globular, kidney-shaped, grape-like, or other imitative shapes (Figs. 61, 64, 65, 75); or *stalactitic* (Fig. 20) when they have been deposited in pendent forms like stalactites ; or *amorphous* when they have no definite shape of any kind, as, for instance, in massive ironstone.

OXIDES occur abundantly as minerals. The most important are those of Silicon (Quartz) and Iron (Hæmatite, Limonite, Magnetite, Titanic Iron).

Quartz (Silica, Silicic Acid, SiO_2), already alluded to, is the most abundant mineral in the earth's crust. It occurs crystallised, also in various crystalline and non-crystalline varieties. In the crystallised form as common quartz it is, when pure, clear and glassy, but is often coloured yellow, red, green, brown, or black, from various impurities. It crystallises in the six-sided prisms and pyramids above referred to, the clear colourless varieties being rock-crystal (Fig. 44). When purple it is called *amethyst* ;

yellow and smoke-coloured varieties, found among the Grampian Mountains, are popularly known as Cairngorm stones. In many places, silica has been deposited as *chalcedony*, in translucent masses with a waxy lustre, and pale grey, blue, brown, red, or black colours. Deposits of this kind are not infrequent among the cavities of rocks. The common pebbles and agates with concentric bands of different colours are examples of chalcedony, and show how the successive layers have been deposited from the walls of the cavity inwards to the centre which is often filled with

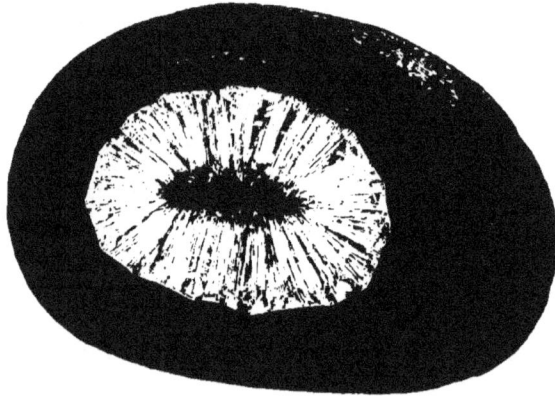

FIG. 52.—Section of a pebble of chalcedony. The outer banded layers are chalcedony, the interior being nearly filled up with crystalline quartz.

crystalline quartz (Fig. 52). The dark opaque varieties are called *jasper*.

Quartz can be usually recognised by its vitreous lustre and hardness ; it cannot be scratched with a knife, but easily scratches glass, and it is not soluble in the ordinary acids. It is an essential constituent of many rocks, such as granite and sandstone. Silica being dissolved by natural waters, especially where organic acids or alkaline carbonates are present, is introduced by permeating water into the heart of even the most solid rocks. Hence it is found abundantly in strings and veins traversing rocks, also in cavities and replacing the forms of plants and animals imbedded in sedimentary deposits. Soluble silica is abstracted by some plants and animals and built up into their organic structures (diatoms, radiolarians, sponges).

Four minerals composed of Oxides of Iron occur abundantly among rocks. The peroxide is found in two frequent forms, one without water (Hæmatite), the other with water (Limonite). The peroxide and protoxide combine to form Magnetite, and a mixture of the peroxide with the peroxide of the metal titanium gives Titanic Iron.

Hæmatite or Specular Iron ($Fe_2O_3 = Fe_7O_3O$) occurs in rhombohedral crystals that can with difficulty be scratched with a knife ; but is more usually found in a massive condition with a compact, fibrous, or granular texture, and dark steel-grey or iron-black colour, which becomes bright red when the mineral is scratched or powdered. The earthy kinds are red in colour, and it is in this earthy form that hæmatite plays so important a part as a colouring material in nature. Red sandstone, for example, owes its red colour to a deposit of earthy peroxide

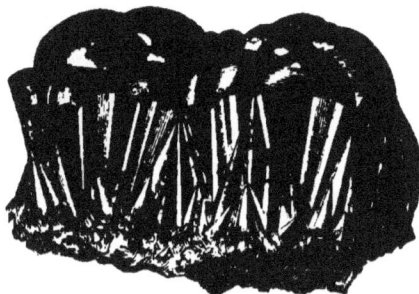

FIG. 53.—Piece of hæmatite, showing the nodular external form and the internal crystalline structure.

of iron round the grains of sand. Hæmatite occurs crystallised in fissures of lavas as a product of the hot vapours that escape at these places ; but is more abundant in beds and concretionary masses (Fig. 53) among various rocks.

Limonite or Brown Iron-ore differs from Hæmatite in being rather softer, in containing more than 14 per cent of water which is combined with the iron to form the hydrated peroxide, in being usually massive or earthy, in presenting a dark brown to yellow colour (ochre), and in giving a yellowish-brown to dull yellow powder when scratched or bruised. It may be seen in the course of being deposited at the present time through the action of vegetation in bogs and lakes (p. 47), hence its name of Bog-iron-ore, likewise in springs and streams where the water carries much sulphate of iron. The common yellow and brown colours of sandstones and many other rocks are generally due to the presence of this mineral.

Magnetite (Fe_3O_4) occurs crystallised in octahedrons and dodecahedrons of an iron-black colour, giving a black powder when scratched. It is found abundantly in many rocks (schists, lavas, etc.), sometimes in large crystals (Fig. 54), sometimes in such minute form as can only be detected with the microscope.

K

It also forms extensive beds of a massive structure.   Its presence in rocks may be detected by its influence on a magnetised needle. By pounding basalt and some other rocks down to powder, minute crystals and grains of magnetite may be extracted with a magnet.

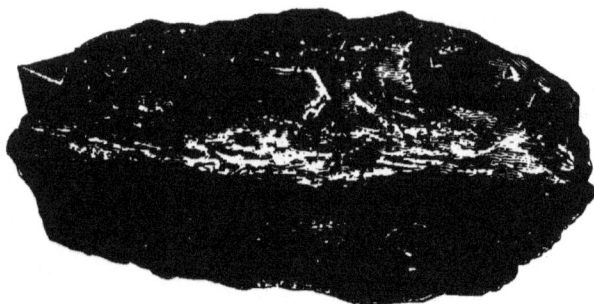

FIG. 54.—Octahedral crystals of magnetite in chlorite schist.

Titanic Iron $(FeTi)_2O_3)$ occurs in iron-black crystals like those of hæmatite, from which they may be distinguished by the dark colour and metallic lustre of its surface when scratched. Though it occurs in beds and veins in certain kinds of rock (schists, serpentine, syenite), its most generally diffused condition is in minute crystals and grains scattered through many crystalline rocks (basalt, diabase, etc.)

Manganese Oxides are commonly associated with those of iron in rocks.   They are liable to be deposited in the form of bog-manganese, under conditions similar to those in which bog-iron is thrown down.   Earthy manganese oxide (wad) not infrequently appears between the joints of fine-grained rocks in arborescent forms that look so like plants as to have been often mistaken for vegetable remains.   These plant-like deposits are called *Dendrites* or dendritic markings (Fig. 55).

SILICATES.—Compounds of Silica with various bases form by far the most numerous and abundant series of minerals in the earth's crust.   They may be grouped according to the chief metallic base in their composition, the most important are the Silicates of Alumina, and the Silicates of Magnesia.   Of the aluminous silicates we need consider here only the Felspars, Zeolites, and Mica.   Among the magnesian silicates it will be enough to note the leading characters of Hornblende, Augite, Olivine, Talc, Chlorite, and Serpentine.   When the learner has made himself so familiar with these as to be able readily to recognise them, he may proceed to the examination of others, of which he will find descriptions in treatises on Mineralogy and in more advanced text-books of Geology.

Felspars.—This family of minerals plays an important part in the construction of the earth's crust, for it constitutes the largest part of the crystalline rocks which, like lava, have been erupted from below; is found abundantly in the great series of schists; and by decomposition has given rise to the clays, out of which so many sedimentary rocks have been formed. The felspars are divided into two series, according to crystalline form.

Orthoclase or potash-felspar contains about 16.89 per cent of potash, crystal-lises in monoclinic or oblique rhombic prisms, but also occurs massive; is white, grey, or pink in colour; has a glassy lustre; can with difficulty be scratched with a knife, but easily with quartz. Associated with quartz, it is an abundant ingredient of many ancient

Fig. 55.—Dendritic markings due to arborescent deposit of earthy manganese oxide.

crystalline rocks (granite, felsite, gneiss, etc.) In the form of *sanidine* it is an essential constituent of many modern volcanic rocks.

Plagioclase.—Under this name are grouped several species of felspar which, differing much from each other in chemical composition, agree in crystallising in the same type or system, which is that of a triclinic or oblique rhomboidal prism. As abundant ingredients of rocks they commonly appear as clear, colourless, or white glassy strips, on the flat faces of which a fine minute parallel ruling may be detected with the naked eye, or with a lens. This striation or lamellation is a distinctive character, which proves the crystals in which it occurs not to be

orthoclase. The plagioclase felspars occur as essential constituents
of many volcanic rocks, and also among ancient eruptive masses
and schists.   Among them are *Microcline* (a potash-felspar), with
15 per cent of potash ; *Albite* or Soda-felspar, containing nearly 12
per cent of soda (Fig. 51) ; *Anorthite* or Lime-felspar, with 20.10
per cent of lime ; *Soda-lime felspar, Lime-soda felspar*—a group
of felspars containing variable proportions of soda, lime, and
sometimes potash ; the chief varieties are *Oligoclase* (Silica, 62-
65 per cent), *Andesine* (Silica, 58-61 per cent), *Labradorite*
(Silica, 50-56 per cent).

Zeolites, a characteristic family of minerals, composed essentially of silicate of alumina and some alkali with water : often

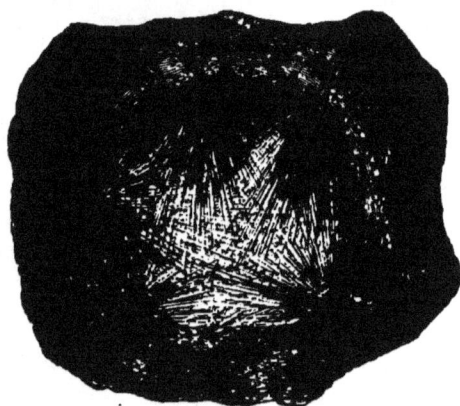

FIG. 56.—Cavity in a lava, filled with zeolite which has crystallised in long
slender needles.

marked by a peculiar pearly lustre, especially on certain planes
of cleavage ; usually found filling up cavities in rocks where they
have been deposited from solution in water.   Some of the species
commonly crystallise in fine needles or silky tufts.   The zeolites
have obviously been formed from the decomposition of other
minerals, particularly felspars.   They are especially abundant in
the steam-cells of old lavas in which plagioclase felspars prevail,
either lining the walls of the cavities, and shooting out in crystals
or fibres towards the centre (Fig. 56), or filling the cavities up
entirely.

Mica, a group of minerals (monoclinic) specially distinguished
by their ready cleavage into thin, parallel, usually elastic silvery
laminæ.   They are aluminous silicates with potash (soda), or with

magnesia and ferrous oxide, and always with water.   They occur
as essential constituents of granite, gneiss, and many other eruptive
and schistose rocks, also in worn spangles in many sedimentary
strata (micaceous sandstone).   Among their varieties the two
most important are *Muscovite* (white mica, potash-mica), and
Black mica (magnesia-mica, Biotite).

Hornblende or Amphibole, a silicate of magnesia, with lime,
iron-oxides, and sometimes alumina, occurs in monoclinic (oblique
rhombic) prisms, also columnar, fibrous, and
massive.   It is divisible into (1) a group of
pale-coloured varieties, containing little or no
alumina, white or pale green in colour, often
fibrous (*Tremolite*, *Actinolite*, *Asbestus*), found
more particularly among gneisses, marbles, and
associated rocks, and (2) a dark group containing 5 to 18 per cent of alumina, which replaces
the other bases ; dark green to black in colour,
in stout, dumpy prisms (Fig. 57), and in columnar
or bladed aggregates (*Common hornblende*).

FIG. 57.—Horn-
blende crystal.

Abundant in many eruptive rocks, and also forming almost entire
beds of rock among the crystalline schists.

Augite (Pyroxene), in composition resembles hornblende ;
indeed, they are only different forms of the same substance, differing slightly in crystalline form, hornblende being the result of slow
and augite of rapid crystallisation.   Many rocks in which the
dark silicate was originally augite have that mineral now replaced
by hornblende, as the result of a gradual internal
alteration.   Like hornblende, augite occurs in
two groups : (1) pale non-aluminous, found more
especially among gneisses, marbles, and associated rocks ; and (2) dark green or black
(Fig. 50), occurring abundantly in many eruptive
rocks, such as black heavy lavas (basalts, etc.)

FIG. 58.—Olivine crys-
tal ; the light portions
represent the unde-
composed mineral, the
shaded parts show the
conversion of the oli-
vine into serpentine.

Olivine (Peridot) (SiO_2 41.01, MgO 49.16,
FeO 9.83) occurs in small orthorhombic prisms
and glassy grains in basalts and other lavas ; of
a pale yellowish-green or olive-green colour,
whence its name.   These grains can often be
readily detected on the black ground of the
rock, through which they are abundantly dis-
persed.   Olivine is liable to alteration, and especially to conversion
into serpentine by the influence of percolating water (Fig. 58).

Chlorite (SiO_2 25-28, Al_2O_3 19-23, FeO 15-29, MgO 13-25, H_2O 9-12) is a dark olive-green hydrated magnesian silicate. It is so soft as to be easily scratched with the nail, and occurs in small six-sided tables, also in various scaly and tufted aggregations diffused through certain rocks. It appears generally to be the result of the alteration of some previous anhydrous magnesian silicate, such as hornblende.

Serpentine ($Mg_3Si_2O_7 + 2H_2O$) is another hydrated magnesian silicate, containing a little protoxide of iron and alumina, usually massive, dark green but often mottled with red. It occurs in thick beds among schists, is often associated with limestones, and may be looked for in all rocks that contain olivine, of the alteration of which it is often the result. In many serpentines, traces of the original olivine crystals can be detected.

CARBONATES.—Though these are abundant in nature, only three of them require notice here as important constituents of the earth's crust, those of lime, magnesia and lime, and iron.

Calcite (calcium-carbonate, carbonate of lime, $CaCO_3$) crystallises in the hexagonal system, and has for its fundamental crystal-

FIG. 59.—Calcite in the form of "nail-head spar."

line form the rhombohedron, as already mentioned (p. 124). When quite pure it is transparent (Iceland spar, Fig. 45), with the lustre of glass ; but more usually is translucent or opaque and white. Its crystals, where the chief axis is shorter than the others, sometimes take the form of flat rhombohedrons (nail-head spar, Fig. 59) ; where, on the other hand, that axis is elongated, they present pointed pyramids (scalenohedrons, dog-tooth spar, Fig. 60). The mineral occurs also in fibrous, granular, and compact forms. The decomposition of silicates containing lime by permeating water gives rise to calcium-carbonate, which is removed in solution.

Being readily soluble in water containing carbonic acid, this carbonate is found in almost all natural waters, by which it is introduced into the cavities of rocks. Some plants and many animals secrete large quantities of carbonate of lime, and their remains are aggregated into beds of limestone, which is a massive and more or less impure form of calcite (pp. 154, 158). Calcite

FIG. 60.—Calcite in the form of dog-tooth spar.

is easily scratched with a knife, and is characterised by its abundant effervescence when acid is dropped upon it.

A less frequent and stable form of calcium - carbonate is **Aragonite** which crystallises in orthorhombic forms, but is more usually found in globular, dendritic, coral-like, or other irregular shapes, and is rather harder and heavier than calcite.

Dolomite assumes a rhombohedral crystallisation, and is a compound of 54.4 of magnesium-carbonate, with 45.6 of calcium-carbonate. It is rather harder than calcite, and does not effervesce so freely with acid. It occurs in strings and veins like calcite, but also in massive beds having a prevalent pale yellow or brown colour (owing to hydrated peroxide of iron), a granular and often cavernous texture, and a tendency to crumble down on exposure (p. 155).

Siderite (chalybite, spathic iron, ferrous carbonate, $FeCO_3$), another rhombohedral carbonate, contains 62 per cent of ferrous oxide or protoxide of iron. In its crystalline form it is gray or brown, becoming much darker on exposure as the protoxide passes into peroxide. It also occurs mixed with clay in concretions and beds, frequently associated with remains of plants and animals (*Sphærosiderite, Clay-ironstone*, Figs. 61, 65).

SULPHATES.—Two sulphates deserve notice for their importance among rock-masses—those of lime and baryta.

Gypsum (hydrous calcium-sulphate, $CaSO_4 + 2HO_2$) occurs in
monoclinic crystals, commonly with the form of right rhomboidal
prisms (Fig. 62, *a*), which not infrequently appear as *macles* or
twin-crystals (Fig. 62, *b*). When pure it is clear and colourless,
with a peculiar pearly lustre
(*Selenite*) ; it is found fibrous with
a silky sheen (*Satin-spar*), also
white and granular (*Alabaster*).
It is so soft as to be easily cut
with a knife or even scratched
with the finger-nails. It is readily
distinguished from calcite by its
crystalline form, softness, and non-
effervescence with acid. When
burnt it becomes an opaque white
powder (plaster of Paris). Gypsum occurs in beds associated
with sheets of rock-salt and dolomite (pp. 47, 156) ; it is soluble
in water, and is found in many
springs and rivers, as well as in
the sea. One thousand parts of
water at 32° Fahr. dissolve 2.05
parts of sulphate of lime ; but
the solubility of the substance is increased in the presence of
common salt, a thousand parts of a saturated solution of common
salt taking up as much as 8.2 parts of the sulphate.

FIG. 61.—Sphærosiderite or Clay-ironstone
concretion enclosing portion of a fern.

Anhydrous calcium-sulphate or **Anhydrite** is harder and
heavier than gypsum, and is found extensively in beds associated
with rock-salt deposits. By absorbing water, it increases in bulk
and passes into gypsum.

Barytes (Heavy spar, barium-sulphate, $BaSO_4$), the usual
form in which the metal barium is distributed over the globe,
crystallises in orthorhombic prisms which are generally tabular ;
but most frequently it occurs in various massive forms. The
purer varieties are transparent or translucent, but in general the
mineral is dull yellowish or pinkish white, with a vitreous lustre,
and is readily recognisable from other similar substances by its great
weight ; it does not effervesce with acids. Barytes is usually met with
in veins traversing rocks, especially in association with metallic ores.

PHOSPHATES.—Only one of these requires to be enumerated in the present list of minerals—the phosphate of lime or Apatite.

Apatite (tricalcic phosphate, phosphate of lime) crystallises in hexagonal prisms which, as minute colourless needles, are abundant in many crystalline rocks ; it also occurs in large crystals and in amorphous beds associated with gneiss. It is soluble in water containing carbonic acid, ammoniacal salts, common salt, and other salts. Hence its introduction into the soil, and its absorption by plants, as already mentioned (p. 120).

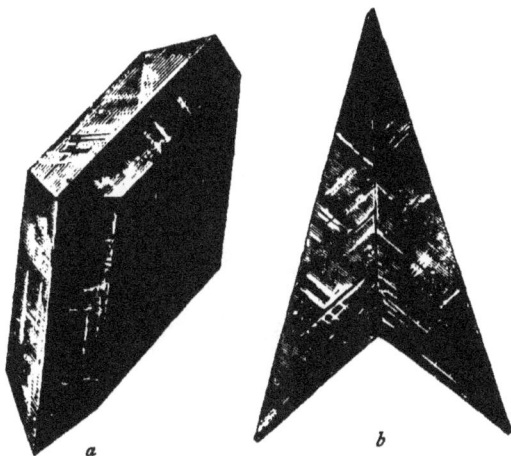

FIG. 62.—Gypsum crystals.

FLUORIDES.—The only member of this family occurring conspicuously in the mineral kingdom is calcium fluoride or **Fluor-Spar** (Fluorite, CaF_2), which, in the form of colourless, but more commonly light green, purple, or yellow cubes, is found in mineral veins not infrequently accompanying lead-ores (Fig. 63).

CHLORIDES.—Reference has already been made to the only chloride which occurs plentifully as a rock-mass, the chloride of sodium, known as **Halite** or Rock-salt (NaCl, chlorine 60.64, sodium 39.36). It crystallises in cubical forms, and is also found massive in beds that mark the evaporation of former salt-lakes or inland seas (p. 157).

SULPHIDES.—Many combinations of sulphur with the metals occur, some of them of great commercial value ; but the only one that need be mentioned here for its wide diffusion as a rock-constituent is the iron-disulphide (FeS_2), in which the elements are combined in the proportion of 46.7 iron and 53.3 sulphur. This substance assumes two crystalline forms : (1) **Pyrite** which occurs in cubes and other forms of the first or monometric system, of a bronze-yellow colour and metallic lustre, so hard as to strike fire with steel, and giving a brownish-black powder when scratched. This mineral is abundantly diffused in minute grains, strings, veins, concretions (Fig. 64, *c*), and crystals in many different kinds of rocks ; it is usually recognisable by its colour,

FIG. 63.—Group of fluor-spar crystals.

lustre, and hardness ; (2) **Marcasite** (white pyrite) crystallises in the tetragonal system, has a paler colour than ordinary pyrite, and is much more liable to decomposition. This form, rather than pyrite, is usually associated with the remains of plants and animals imbedded among rocks. The sulphide has no doubt often been precipitated round decaying organisms by their effect in reducing sulphate of iron. By its ready decomposition, marcasite gives rise to the production of sulphuric acid and the consequent formation of sulphates. One of the most frequent indications of this decomposition is the rise of chalybeate springs (p. 59).

CHAPTER XI

FROM the distribution of the more important elements in the earth's crust and the mineral forms which they assume, we have now to advance a stage farther and inquire how the minerals are combined and distributed so as to build up the crust. As a rule, simple minerals do not occur alone in large masses ; more usually they are combined in various proportions to form what are known as Rocks. A rock may be defined as a mass of inorganic matter, composed of one or more minerals, having for the most part a variable chemical composition, with no necessarily symmetrical external form, and ranging in cohesion from loose or feebly aggregated debris up to the most solid stone. Blown sand, peat, coal, sandstone, limestone, lava, granite, though so unlike each other, are all included under the general name of Rocks.

In entering upon the study of rocks, or the division of geology known as Petrography, it is desirable to be provided with such helps as are needed for determining leading external characters ; in particular, a hammer to detach fresh splinters of rock, a pocket-knife for trying the hardness of minerals, a small phial of dilute hydrochloric acid for detecting carbonate of lime, and a pocket lens. The learner, however, must bear in mind that the thorough investigation of rocks is a laborious pursuit, requiring qualifications in chemistry and mineralogy. He must not expect to be able to recognise rocks from description until he has made good progress in the study. As already stated on a previous page, he must examine the objects themselves, and for this purpose he will find much advantage in procuring a set of named specimens, and making himself familiar with such of their characters as he can himself readily observe.

Great light has in recent years been thrown upon the structure

and history of rocks by examining them with the microscope.
For this purpose, a thin chip or slice of the rock to be studied is
ground smooth with emery and water, and after being polished
with flour-emery upon plate-glass, the polished side is cemented
with Canada balsam to a piece of glass, and the other side is then
ground down until the specimen is so thin as to be transparent.

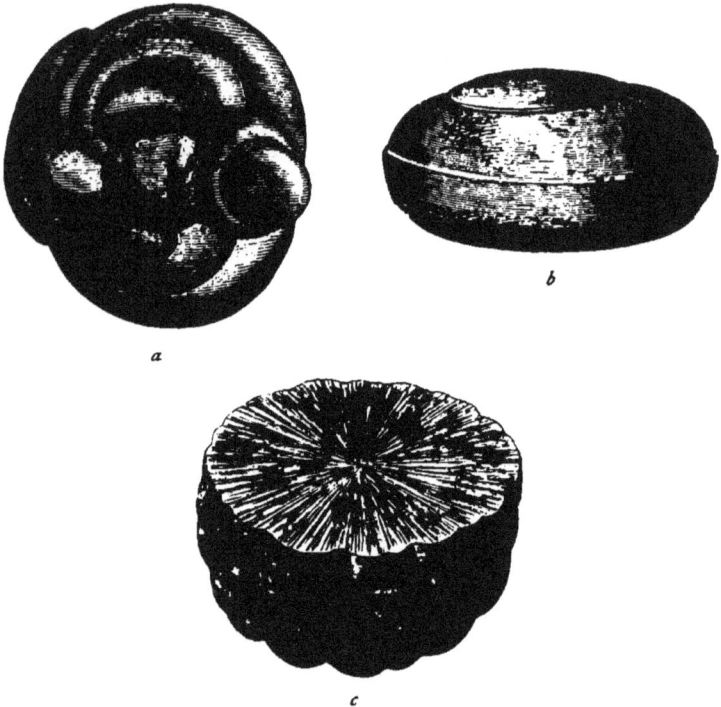

FIG. 64.—Concretions.
a, b, " Fairy stones ;" c, Pyrite, showing internal radiated structure.

Thin sections of rock thus prepared (which can now be obtained
from any good mineral-dealer) reveal under the microscope the
minutest kinds of rock-structure.   Not only can the component
minerals be detected, but it is often possible to tell the order in
which they have appeared, and what has been the probable
origin and history of the rock.   Some illustrations of this method
of investigation will be given in a later part of the present chapter.
It will be of advantage to begin by taking note of some of the
more important characters of rocks, and of the names which
geologists apply to them.

Sedimentary—composed of sediment which may be either a mechanically suspended detritus, such as mud, sand, shells, or gravel; or a chemical precipitate, as rock-salt and calcareous tufa. The various deposits which are accumulated on the floors of lakes, in river-courses, and on the bed of the sea, are examples of sedimentary rocks.

Fragmental, *Clastic*—composed of fragments derived from some previous rock. All ordinary detritus is of this nature.

Concretionary—composed of mineral matter which has been aggregated round some centre so as to form rounded or irregularly-shaped lumps. Some minerals, particularly pyrite (Fig. 64 *c*), marcasite, siderite, and calcite, are frequently found in concretionary forms, especially round some organic relic, such as a shell or plant (Figs. 61, 65). In alluvial clay, calcareous concretions which often take curious imitative shapes, are known as "fairy stones" (Fig. 64, *a*, *b*; see p. 177).

When nodules of limestone, ironstone, or cement-stone are marked internally by cracks which radiate towards, but do not

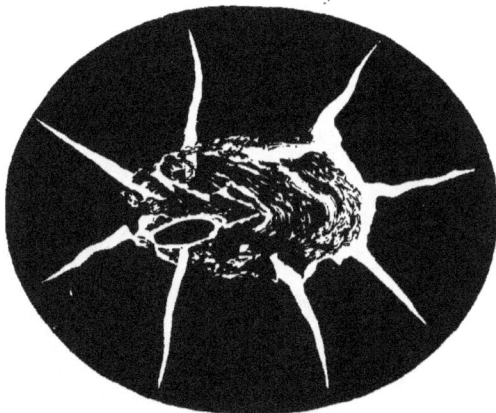

FIG. 65.—Section of a septarian nodule, with coprolite of a fish as a nucleus.

reach, the outside, and are filled up with calcite or other mineral, they are known as *Septaria* or septarian nodules (Fig. 65, and layer 13 in Fig. 80).

Oolitic—made up of spherical grains, each of which has been formed by the deposition of successive coatings of mineral matter round some grain of sand, fragment of shell, or other foreign particle (Fig. 66).   A rock with this structure looks like fish-roe,

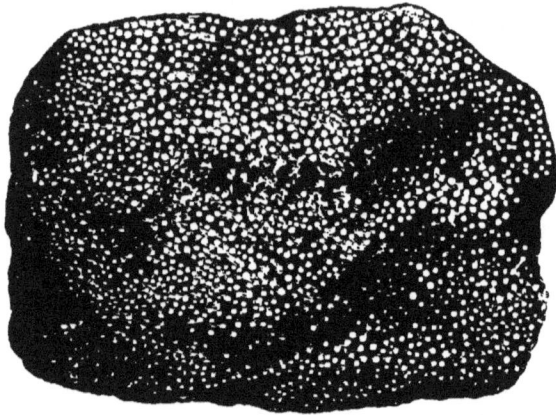

Fig. 66.—Piece of oolite.

hence the name oolite or roe-stone ; but when the granules are like peas, the rock becomes *pisolitic* (pea-stone, Fig. 67).   This peculiar structure is produced in water (springs, lakes, or enclosed

Fig. 67.—Piece of pisolite.

parts of the sea), wherein dissolved mineral matter (usually carbonate of lime) is so abundant as to be deposited in thin pellicles round the grains of sediment that are kept in motion by the current (p. 88).

Stratified, Bedded—arranged in layers, strata, or beds lying generally parallel to each other, as in ordinary sedimentary deposits (Fig. 79, p. 172).

Aqueous—laid down in water, comprising nearly the whole of the sedimentary and stratified rocks.

Unstratified, Massive — having no arrangement in definite layers or strata. Lavas and the other eruptive rocks are examples (Chapter XIV).

Eruptive, Igneous—forced upwards in a molten or plastic condition into or through the earth's crust. All lavas are Eruptive or Igneous rocks, also called *Volcanic* because erupted to the surface by volcanoes. In the same division must be classed granite and allied masses, which have been thrust through rocks at some depth within the earth's crust and may not have been directly connected with any volcanic eruption ; such rocks are sometimes called *Plutonic* or *Hypogene*.

Crystalline—consisting wholly or chiefly of crystals or crystalline grains. Rocks of this nature may have arisen from (*a*) igneous fusion, as in the case of lavas, where the minerals have separated out of a molten glass, or what is called a *Magma*; (*b*) aqueous solution, as where crystalline calcite forms stalactite and stalagmite in a cavern ; (*c*) sublimation, where the materials have crystallised out of hot vapours, as in the vents and clefts of volcanoes.

By the aid of the microscope many rocks which to the naked eye show no definite structure can be shown to be wholly or partially crystalline. Moreover, it can often be ascertained that the crystals or crystalline grains in a rock, as they were crystallising out of their solution, have en-closed various foreign bodies. Among the objects thus taken up are minute globules of gas, which are prodigi-ously abundant in certain minerals in some lavas ; liquids, usually water, enclosed in cavities of the crystals, but not quite filling them, and leaving a minute freely-moving bubble (Fig. 68) ; glass, filling globular spaces, probably part of the original glassy magma of the rock ; crystals and crystallites (rudimentary crystalline

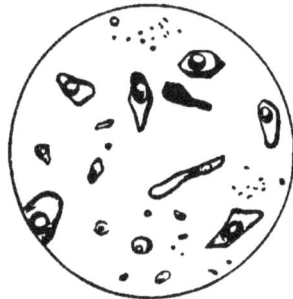

Fig. 68.—Cavities in quartz con-taining liquids (magnified).

forms, Fig. 69) of other minerals. Thus a crystal, which to the eye may appear quite free from impurities, may be found to be full of various kinds of enclosures. Obviously the study of these enclosures cannot but throw light on the conditions under which the rocks enclosing them were produced.

There are various types of crystalline structure which can best be examined under the microscope, as *Holocrystalline*, composed entirely of crystalline elements without any interstitial glass —one of the most characteristic types of this structure is found in

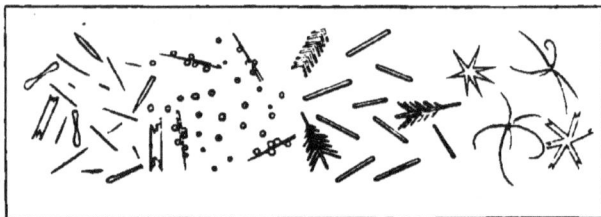

FIG. 69.—Various forms of crystallites (highly magnified).

granite, hence it is sometimes termed the granitic or granitoid structure ; *Semi-crystalline*, consisting partly of crystals, but with a ground mass or base which may be partly glassy or variously devitrified ; *Felsitic* or *microfelsitic*—composed of indefinite half-effaced granules and filaments (p. 163).

Glassy, Vitreous—having a structure and aspect like that of artificial glass. Some lavas, obsidian for example, have solidified as natural glasses, and look not unlike masses of dark bottle-glass. In almost all cases, however, they contain dispersed crystals, crystallites, or other enclosures. These substances have generally multiplied to such an extent in most lavas as to leave only small interstitial portions of the original glass, while in many cases the glass has entirely disappeared. When a glass has thus been converted into a dull, opaque, stony, or lithoid mass, or into a completely crystalline substance, it is said to be *devitrified*. The microscope enables us to prove many crystalline eruptive rocks to have been once molten glass which by a process of devitrification have been brought into their present more or less crystalline condition (p. 97).

Porphyritic—composed of a compact or crystalline base or matrix, through which are scattered conspicuous crystals much larger than those of the base, and generally of some felspar. Many eruptive rocks have this structure and are sometimes spoken of as " porphyries." The large crystals existed in the rock while still in a mobile state within the earth's crust, while the minuter crystals of the base were developed by a later process of crystallisation during the consolidation of the rock.   In the successive zones of growth which porphyritic crystals often present, we may note by the enclosed minerals some of the successive stages of consolidation.

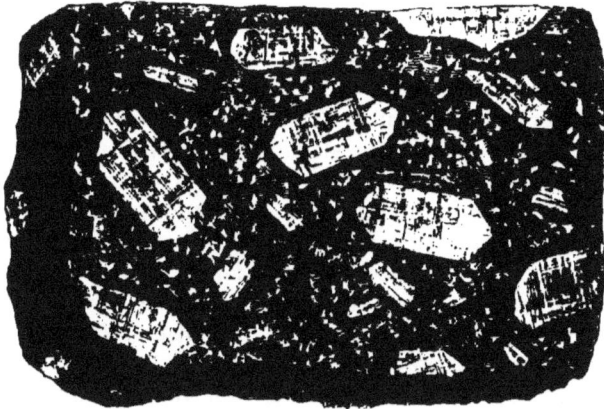

FIG. 70.—Porphyritic structure.

Spherulitic—composed of or containing small pea-like globular bodies (*Spherulites*) which show a minutely fibrous internal structure radiating from the centre (Fig. 71, A).   This structure is

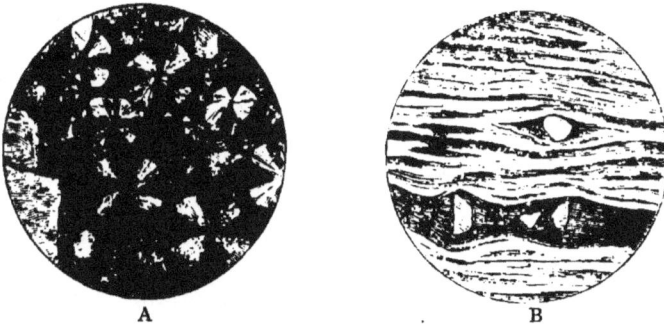

A . B

FIG. 71.—Spherulites and fluxion structure.   A, Spherulites, as seen under the microscope (with polarised light).   B, Fluxion-structure of Obsidian, as seen under the microscope.

particularly observable in vitreous rocks, where it appears to be one of the stages of devitrification (p. 144).

Perlitic.—Many vitreous rocks show a minute fissured structure as one of the accompaniments of devitrification.   In the structure termed perlitic the original glass has had a series of reticulated and globular or spiral cracks developed in it, sometimes giving rise to globules composed of successive thin shells.

L

Vesicular, Cellular—containing spheroidal or irregularly shaped cavities. In many eruptive rocks (as in modern lavas) the expansion of interstitial steam, while the mass was still in a molten condition, has produced this cellular structure (Fig. 35), the vesicles have usually remarkably smooth walls; they may form a comparatively small part of the whole mass, or they may so increase as to make pieces of the rock capable of floating on water. Where the vesicular structure is conjoined with more solid parts, as in the irregular slags of an iron furnace, it may be called *slaggy*. Where, as in the scoriæ of a volcano, the cellular and solid parts are in about equal proportions, and the vesicles vary greatly in numbers and size within short distances, the structure may be termed *scoriaceous*. The lighter and more froth-like varieties that can float on water are said to be *pumiceous*, or to have the characters of pumice (p. 162). Exposed to the influence of percolating water, vesicular rocks have had their vesicles filled up by the deposition of various minerals from solution, especially quartz, calcite, and zeolites. These substances first begin to encrust the walls of the cells, and as layer succeeds layer they gradually fill the cells up (Fig. 52); as the cells have not infrequently been elongated in one direction by the motion of the rock before consolidation was completed (Fig. 37), the mineral deposits in them, taking their exact moulds, appear as oval or almond-shaped bodies. Hence rocks which have been treated in this way are called *Amygdaloids*, and the kernels filling up the cells are known as *Amygdules* (Fig. 35). An amygdaloidal rock, therefore, was originally a molten lava, rendered cellular by the expansion of its absorbed steam and gases, its vesicles having been subsequently filled up by the deposit in them of mineral matter, often derived out of the surrounding rock by the decomposing and rearranging action of percolating water.

Flow-structure, Fluxion-structure — an arrangement of the crystallites, crystals, or particles of a rock in streaky lines, the minuter forms being grouped round the larger, indicative of the internal movement of the mass previous to its consolidation. The lines are those in which the particles flowed past each other, the larger crystals giving rise to obstructions and eddies in the movement of the smaller objects past them. This structure is characteristic of many once molten rocks; it is well seen in obsidian (Fig. 71, B). But it is also found in rocks which, by enormous stresses within the earth's crust, have been crushed and made to undergo an interstitial movement like that of the flow of liquids. The most solid gneisses and granites have in this way been so sheared and squeezed that their component minerals have been crushed into a fine compact mass, through which the streaking lines of flow are sometimes displayed with singular clearness.

Mylonitic—a name sometimes applied to rocks which by terrestrial movement have had their original structure entirely obliterated, and which now present only a dull, crushed felsitic mass, sometimes partially or completely recrystallised.

Schistose, Foliated—consisting of minerals that have crystallised in approximately parallel, wavy, and irregular laminæ, layers,

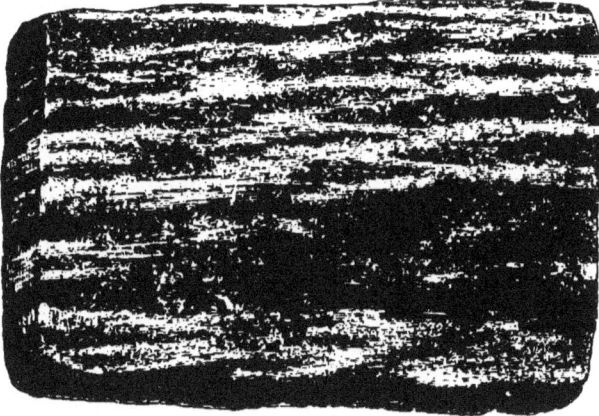

FIG. 72.—Schistose structure.

or folia (Fig. 72). Such rocks are called generally schists. They have, in large measure, been formed by the alteration or metamorphism of other rocks of various kinds by the vast terrestrial movements referred to in the foregoing paragraphs (see Chapter XIII).

CLASSIFICATION OF ROCKS

Various schemes of classification of rocks are in use among geologists, some based on mode of origin, others on mineral composition or structure. For the purpose of the learner, perhaps the most instructive and useful arrangement is one which as far as possible combines the advantages of both these systems. Accordingly, in the following account of the more important rocks which enter into the structure of the earth's crust, a threefold subdivision will be adopted into : (i) sedimentary rocks ; (ii) eruptive rocks ; (iii) schistose rocks.

I. SEDIMENTARY ROCKS.

This division includes the largest number, and to the geologist the most important of the rocks accessible to our notice.   It comprises the various deposits that arise from the decay of the surface of the land and are laid down on the land or over the bed of the sea, together with all those directly or indirectly due to the growth of plants and animals.   It thus embraces those which constitute the main mass of the earth's crust so far as known to us, and which contain the evidence whence the geological history of the earth is chiefly worked out.   It is, therefore, worthy of the earliest and closest attention of the student.

Sedimentary rocks, being due to the deposition of some kind of sediment or detritus, are obviously not original or primitive rocks.   They have all been derived from some source, the nature of which, if not its actual site, can usually be determined.   In no case, therefore, can sedimentary rocks carry us back to the beginning of things ; they are themselves derivative and pre-suppose the existence of some older rock or material from which they could be derived.

One of their most obvious characters is that, as a rule, they are *stratified*.   They have been deposited, usually in water, some-times in air, layer above layer, and bed above bed, each of these strata marking a particular interval in the progress of deposition (Chapter XII).   As regards their mode of origin, they may be subdivided into three great sections : (1) fragmental or clastic, composed of fragments of pre-existing rocks ; (2) chemically pre-cipitated, as in the deposits from mineral springs ; and (3) formed of the remains of organisms, as in peat and coral-rock.

(1) *Fragmental or Clastic Rocks.*

These are masses of mechanically-formed sediment, derived from the destruction of older rocks ; they vary in coherence from loose sand or mud up to the most compact sandstone or con-glomerate ; they are accumulating abundantly at the present time in the beds of rivers and lakes, and on the floor of the sea, and they have been formed in a similar way all over the globe from the earliest periods of known geological history.   Some of the more frequent kinds are the following : —

Cliff-Debris—coarse angular rubbish, including large blocks of stone, disengaged by the weather from cliffs and other bare faces of rock. This kind of detritus is formed abundantly in rugged and mountainous regions, especially where the action of frost is severe ; it slides down the slopes and accumulates at their foot, unless washed away by torrents. In glacier-valleys it descends to the ice, where, gathering into moraines (Chapter VI), it is transported to lower levels. The perched blocks of such valleys are some of the larger fragments of this cliff-debris left stranded by the ice, and from around which the smaller detritus has been washed away (Fig. 23).

Soil, Subsoil, described in Chapter II, represent the result of the subaerial decomposition of the surface of the land.

Breccia—a rock composed of angular fragments. Such a rock shows that its materials have not travelled far ; otherwise, they

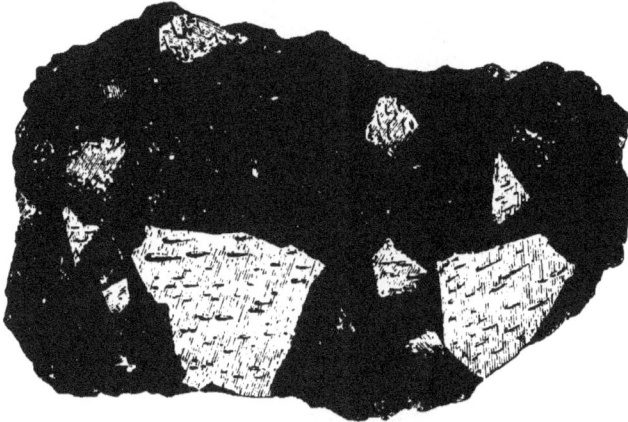

FIG. 73.—Brecciated structure—volcanic breccia, a rock composed of angular fragments of lava, in a paste of finer volcanic debris.

would have lost their edges, and would have been more or less rounded. Ordinary cliff-debris may consolidate into a breccia, more especially where it falls into water and is allowed to gather on the bottom. The angular fragments shot out of a volcano often accumulate into volcanic breccia (Fig. 73). A rock with abundant angular fragments is said to be *brecciated*.

Gravel—loose rounded water-worn detritus, in which the pebbles range in average size between that of a small pea and that of a walnut ; where they are larger they form Shingle. They may consist of fragments of any kind of rock, though having resulted from more or less violent water-action, as a rule, pieces of only the more durable stones are found in them.   Quartz and other siliceous materials, from their great hardness, are better able to withstand the grinding to which the detritus on an exposed sea-shore, or in the bed of a rapid stream, is subjected.   Hence quartzose and siliceous pebbles are the most frequent constituents of gravel and shingle.

Conglomerate—a name given to gravel and shingle when they have been consolidated into stone, the pebbles being bound to-gether by some kind of paste or cementing material, which may

FIG. 74.—Conglomerate.

be fine hardened sand, clay, or some calcareous, siliceous, or ferruginous cement (Fig. 74).   As above remarked with regard to gravel, the component materials of conglomerate may have been derived from any kind of rock, but siliceous pebbles are of most common occurrence.   Different names are given to conglomerates, according to the nature of the pebbles, as quartz-conglomerate, flint-conglomerate, limestone-conglomerate.

Sand—a name given to fine kinds of detritus, the grains of which may vary from the size of a small pea down to minute particles that can only be detected with a lens.   In general, for the reason already assigned in the case of gravel, the component grains of sand are of quartz or of some other durable material. Examined with a good magnifying glass, they are seen to be usually rounded, water-worn, but sometimes angular, unworn particles of indefinite shapes which, except in their smaller size, resemble those of gravel-stones. Sand may be formed by the disintegration of the surface of rocks exposed to the weather, more especially in dry climates, where there is a great difference between the temperature of day and night (p. 13). The loosened particles are blown away by the wind, and may be heaped up into great sand-wastes, as in the tracts known as Deserts. On a sea-coast, where a sandy beach is liable to be laid bare and exposed to be dried between tides by breezes blowing from the sea, the upper particles of sand are lifted up by the wind and borne away landward, to be piled up into dunes (p. 20). In some places the materials are derived mainly from the remains of calcareous sea-weeds, shells, corallines, and other calcareous organisms exposed to the pounding action of the surf. A sand composed of such materials speedily hardens into a more or less coherent and even compact limestone, for rain falling on it dissolves some carbonate of lime which, being immediately deposited again, as the moisture evaporates, coats the grains of sand and cements them together. At Bermuda, as already stated, all the rock above sea-level has been formed in this way, and some of it is hard enough to make a good building stone (p. 84). Ordinary siliceous or quartzose sand remains loose, unless its grains are made to cohere by some kind of cement, when it becomes sandstone.

Sandstone—consolidated sand. The grains are chiefly quartz, but may include particles of any other mineral or rock ; they are bound together by some kind of cement which has either been laid down with them at the time of their deposition, or has subsequently been introduced by water permeating the sand. The cementing material may be *argillaceous*—that is, some kind of clay ; or *calcareous*, consisting of carbonate of lime ; or *ferruginous*, composed mainly of peroxide of iron ; or *siliceous*, where silica has been deposited in the interstices of the mass. The colours of sandstone vary chiefly with the nature of this cementing material. The hydrous peroxide of iron colours them shades of yellow and brown ; the anhydrous peroxide of iron gives them different hues of red ; the mineral glauconite tints them a greenish hue. Some varieties of sandstone are named after a conspicuous component or structure ; thus *micaceous* sandstone is distinguished by abundant spangles of mica deposited along the bedding planes, whereby the rock can be split up into thin layers ; *freestone*—a thick-bedded sandstone that does not tend to split up in any one direction, and can therefore be cut into blocks of any size and form ; *glauconitic* sandstone (green sand), containing green grains and kernels of glauconite ; *quartzose* sandstone, conspicuously composed of quartz-grains ; *grit*—a sandstone formed of coarse or sharp, somewhat angular grains of quartz.

Greywacke—a greyish, compact, granular rock, composed of rounded or subangular grains of quartz and other minerals or rocks, cemented together in a compact paste ; it differs from sandstone chiefly in its darker colour, in the proportion of other grains than those of quartz, and in the presence of a tough cement.

The rocks above enumerated represent the coarser and more durable kinds of detritus derived from the weathering of the surface of the land ; but during the progress of the decomposition from which these materials are derived some of the component ingredients of the rocks decay into clay, or what is called *argillaceous* sediment. This more particularly occurs in the case of felspars and other aluminous silicates, the decomposition of which produces minute particles capable of being lifted up and carried a great distance by running water. Hence argillaceous sediment, being commonly finer in grain, travels farther, on the whole, than quartzose sediment ; and beds of clay denote, generally, deeper and stiller water than beds of sand.

Clay—a fine-grained argillaceous substance, derived from the decay and hydration of aluminous silicates, white when pure, but usually mixed with impurities, which impart to it various shades of grey, green, brown, red, purple, or blue ; it usually contains interstitial water, and when wet can be kneaded between the fingers ; when dry it is soft and friable, and adheres to the tongue. Shaken with water it becomes *Mud* ; even a small quantity will make a glass of water turbid, so fine are the particles of which it is composed.

Kaolin—the name given to the white purer forms of clay, resulting from the decomposition of the felspars of granite or similar rocks ; it is sometimes called *China-clay*, from its use in the manufacture of porcelain.

Fire-clay—a white, grey, yellow, or black clay, nearly free from alkalies and iron, and capable of standing a great heat without fusing ; it is abundantly found underneath coal-seams, where it represents the ancient soil on which the plants grew that have been converted into coal.

Brick-clay—a name commonly applied to any clay, loam, or earth from which bricks can be made. Such deposits are always more or less sandy and impure clays ; in the south of England they have largely arisen from the prolonged subaerial waste of the Cretaceous and Tertiary formations.

Mudstone—a compact solidified clay or clay-rock, having little or no tendency to split into thin laminæ.

Shale—clay that has become hard and splits into thin laminæ which lie parallel with the planes of deposit (p. 172). A thoroughly fissile shale can be subdivided into leaves as thin as fine cardboard. This is the common form which the clays of the older geological formations have assumed. Gradations can be traced from shale into other sedimentary rocks ; thus, by additions of sand into fissile sandstones, of calcareous cement into limestone, of carbonate of iron into ironstone, of carbonaceous matter into coal. These passages are interesting as indications of the conditions under which the rocks were formed. Where, for example, shale shades off into coral-limestone, we see that mud gathered over one part of the sea-floor, while not far off, probably in clearer water, corals flourished and built up a limestone out of their remains (see p. 179).

Loess—a pale somewhat calcareous and sandy clay, found in regions where it has probably been accumulated by the drifting action of the wind. It is sufficiently coherent to be capable of excavation into tunnels and passages, and in China is even dug out into houses and subterranean villages. It occupies parts of the valleys of the Rhine, Danube, Mississippi, and other large rivers, but also crosses watersheds (p. 363).

Fragmental rocks of volcanic origin may be enumerated here. They consist partly of materials ejected in fragmentary form from volcanic vents, and partly of the detritus derived from the disintegration of volcanic rocks already erupted to the surface. They are comprised under the general name of *Tuff* (p. 101).

Bombs—round elliptical or discoidal pieces of lava which have been ejected in a molten state from an active vent, and have acquired their form from rapid rotation in the air during ascent and descent. They are often very cellular or even quite empty inside. Where the large ejected stones are of irregular forms, and appear to have been thrown out in an already solidified condition, as from the consolidated crust of the lava-plug, or from the sides of the funnel or crater, they are called *Volcanic Blocks* (p. 101).

Lapilli—ejected pieces of lava, usually vesicular or porous, from the size of a pea to a walnut (Fig. 73).

Volcanic Ash—the fine dust produced by the explosion of the

superheated steam absorbed in molten lava. Under the microscope, it is often found to consist of minute grains of glass, and
in such cases, shows that the lava from which it was derived, rose
from below in the condition of a liquid glassy magma. In other
instances, it is made up of the crystallites and crystals that arose
during the devitrification of the glass. It consolidates into a more
or less coherent mass, which is known as *Tuff*, and which may
receive some distinctive name according to the nature of the lava
that has supplied it, as *Basalt-tuff* and *Trachyte-tuff*. Most tuffs
contain angular and vesicular pieces of lava, and sometimes pass
into coarse breccias (*Volcanic Breccia*). In many cases, they
enclose the remains of plants and animals which, if of terrestrial
kinds, indicate that the eruptions took place on land ; if of marine
species, that the volcanoes were probably submarine (pp. 101-103).

Agglomerate—a coarse, usually unstratified accumulation of
blocks of lava and other rocks, not infrequently filling up the
chimney or neck of a volcanic vent.

(2) *Rocks formed by Chemical Precipitation.*

In Chapter V it was pointed out that all natural waters contain
in solution invisible mineral matter which they have dissolved out
of the rocks of the earth's crust, and that the quantity of this
material is sometimes so great that it is precipitated into visible
form as the water evaporates. The substance most abundantly
dissolved and deposited is Carbonate of lime. Others of frequent occurrence are Sulphate of lime, Chloride of sodium, Silica,
Carbonate of magnesia, and various salts of iron. Among the
rocks of the earth's crust, considerable masses of these substances
have been piled up by chemical precipitation.

Limestone—compact or crystalline calcium-carbonate (carbonate of lime) which may be nearly pure, or may contain sand, clay,
or other impurity, and may consequently pass into sandstone,
shale, or other sedimentary rock. Probably the great majority
of the limestones in the earth's crust have been formed by the
agency of animals, as more particularly referred to at p. 158. We
are here concerned only with those which have been deposited
from chemical solution. The most familiar example of this kind
of limestone is afforded by stalactites and stalagmite, which have
already been described (Chapter V and Fig. 20). Large masses
of it have been deposited by calcareous springs and streams (p. 57).

At first, it is a fine white milky precipitate, but gradually crystals of calcite shape themselves and grow out of it, with their vertical axes usually at right angles to the surface of deposit.   In a vertical stalactite, consequently, the prisms radiate horizontally from the centre outwards ; on a horizontal surface of stalagmite they diverge perpendicular to the floor.   A mass of limestone, not originally crystalline, may thus acquire a thoroughly crystalline internal structure by the action of infiltrating water in dissolving the carbonate of lime and redepositing it in a crystalline condition.

Limestones vary greatly in texture and purity.   Some are snow-white and distinctly crystalline ; others are grey, blue, yellow, or brown, dull and compact, and full of various impurities. They may usually be detected by the ease with which they can be scratched, and their copious effervescence when a drop of weak acid is put on the scratched surface.   Pure limestone dissolves entirely in hydrochloric acid, so that the amount of residue is an indication of the proportion of insoluble impurity. Among the varieties of limestone the following may be named :— *Oolite*, a limestone composed of minute spherical grains like the roe of a fish, each grain being composed of concentrically deposited layers or shells of calcite (Fig. 66) ; *Pisolite*, a similar rock, where the grains are as large as peas (Fig. 67) ; *Travertine* or calcareous tufa, a white porous crumbling rock which, by infiltration of carbonate of lime, may acquire a compact texture, and become suitable for building stone (p. 57) ; *Hydraulic limestone*, containing 10 to 30 per cent of fine sand or clay, and having the property, after being burnt, of hardening under water into a firm compact mortar.

Dolomite, Magnesian Limestone—this substance has been already referred to as a mineral (p. 135) ; but it also occurs in large masses as a white or yellowish crystalline or compact rock. The white varieties look like marble.   The yellow and brown kinds contain various impurities, and are coloured by iron-oxide. Dolomite differs from limestone in its greater hardness and feebler solubility in acid, in its frequently cellular or cavernous texture, in its tendency to assume spherical, grape-shaped, or other irregular concretionary forms (Fig. 75), and in its proneness to crumble down into loose crystals.   It occurs in beds, not uncommonly associated with gypsum and rock-salt, and in such conditions it may have been deposited first as limestone which, by the chemical action of the magnesian salts in the saline water, had its carbonate of lime partially replaced by carbonate of

magnesia.   It is also found in irregular bands traversing limestone which, probably by the influence of percolating water
containing carbonate of magnesia in solution, has been changed
into dolomite.

Gypsum is not only a mineral (p. 136) but also a rock, white,
grey, brown, or reddish in colour, granular to compact, some-

FIG. 75.— Concretionary forms assumed by Dolomite, Magnesian Limestone, Durham.

times fibrous or coarsely crystalline in texture.   It consists of
sulphate of lime, is easily scratched with the nail, and is not
affected by acids, being thus readily distinguishable from limestone.   It is found in beds or veins, especially associated with
layers of red clay and rock-salt, and in these cases has evidently
resulted from the evaporation of water containing it in solution,
such as that of the sea.   The lime-sulphate being less soluble
than the other constituents is precipitated first.   Hence in a
thick series of alternations of beds of gypsum (or anhydrite) and
rock-salt, each layer of sulphate of lime indicates a new supply of
water into the natural reservoirs where the evaporation took place.
The overlying bed of salt, usually much thicker than the gypsum,
points to the condensation of the water into a strong brine, from

which the salt was ultimately precipitated. And the next sheet of sulphate of lime tells how, by the breaking down of the barrier, renewed supplies of salt water were poured into the basin (pp. 47, 136).

Rock-salt occurs in beds or layers, from less than an inch to hundreds or even thousands of feet in thickness. One mass of salt in Galicia is more than 4600 feet thick, and a still thicker mass occurs near Berlin. When quite pure, rock-salt is clear and colourless, but it is usually more or less mixed with impurities, particularly with red clay, and in association with beds of gypsum, as above remarked. It has been formed in inland salt lakes or basins by the evaporation and concentration of the saline water. It is being deposited at the present time in the Dead Sea, the Great Salt Lake, and the salt lakes so frequent in the desert regions of continents, where the drainage does not flow outwards to the sea (p. 47).

Ironstone.—Various minerals are included under this name as large rock-masses. One of the most important of them is *Hæmatite* (p. 129), which occurs in large beds and veins, as well as filling up caverns in limestone. *Limonite* or bog-iron ore is formed in lakes and marshy places (p. 47), and occurs in beds among other sedimentary accumulations. *Magnetite* (p. 129) is found in beds and huge wedge-shaped masses among various crystalline rocks, as in Scandinavia, where it sometimes forms an entire mountain. *Carbonate of iron* (Siderite, Sphærosiderite, Clay-ironstone) occurs in concretions and beds among argillaceous deposits (Figs. 61, 65, and p. 135). In the Coal-measures, for example, it is largely developed, much of the iron of Britain being obtained from this source. As many ironstones are largely due to the influence of plants and animals, the rock is alluded to again on p. 160.

Siliceous Sinter—a white powdery to compact and flinty deposit from the hot water of springs in volcanic districts, consisting of 84 to 91 per cent of silica, with small proportions of alumina, peroxide of iron, lime, magnesia, and alkali, and from 5 to 8 per cent of water. It accumulates in basin-shaped cavities round the mouths of hot springs and geysers, and sometimes forms extensive terraces and mounds, as at the geyser regions of Iceland, Wyoming, and New Zealand.

Vein-quartz—a massive form of quartz, which occurs in thin veins and in broad dyke-like reefs, traversing especially the older rocks.

(3) Rocks formed of the Remains of Plants or Animals.

In Chapter VIII an account was given of the manner in which extensive accumulations are now being formed of the remains of plants and animals. Similar deposits have constantly been accumulated from an early period in the history of the earth. Regarding them with reference to their mode of origin, we observe that in some cases they have been piled up by the unremitting growth and decay of organisms upon the same site. In a thick coral-reef, for example, the living corals now building on the surface are the descendants of those whose skeletons form the coral-rock underneath (p. 87). In other cases, the remains of the organisms are broken up and carried along by moving water, which deposits them elsewhere as a sediment. Strictly speaking, these last deposits are fragmental, and might be classed with those described at p. 148 ; they pass into ordinary sand, sandstone, clay, or shale. But it will be more convenient to class together all the rocks which consist mainly of organic remains, whether they have been directly built up by the organisms, or have only been formed out of their detrital remains.

Limestone.—As carbonate of lime is so largely secreted by animals in their hard parts which are more or less durable, it is naturally the most common substance among rocks of organic origin. The limestones that form so large a proportion of the stratified rocks of the earth's crust have been, for the most part, formed out of the remains of marine animals. The following are some of the more important or interesting varieties of this rock :— *Shell-marl,* a soft white earthy crumbling deposit formed chiefly of fresh-water shells (pp. 4, 46); by subsequent infiltration it may be hardened into a compact stone, when it is known as fresh-water limestone ; *Calcareous sand*—a mass of broken-up shells, calcareous algæ, and other calcareous organisms (p. 84), often cemented by percolating water into solid stone ; *Coral rock*—a limestone formed by the continuous growth of corals and cemented into a solid compact and even crystalline rock by the washing of calcareous mud into its interstices and the permeation of sea-water and rain-water through it, whereby crystalline calcite is deposited within it (p. 87) ; *Chalk*—a soft, white rock, soiling the fingers, formed of a fine calcareous powder of remains of foraminifera, shells, etc. (see *Ooze,* p. 86) ; *Crinoidal limestone*—composed chiefly of the calcareous joints of the marine creatures known as crinoids, with foraminifera, shells, corals, and other

organisms.   A limestone composed in great part of organic
remains may show little trace of its origin on a fresh fracture of
the stone ; but a weathered surface will often reveal its true
nature, the fossils being better able to withstand the action of the
atmosphere than the surrounding matrix which is accordingly
removed, leaving them standing out in relief (Fig. 76).

FIG. 76.—Weathered surface of crinoidal limestone.

Peat—a yellow, brown, or black fibrous mass of compressed
and somewhat altered vegetation.   It occurs in boggy places in
temperate latitudes where it largely consists of bog-mosses and
other marshy plants (p. 82).   Its upper parts are loose and full
of the roots of living plants, while the bottom portions may be
compact and black like clay, and with little trace of vegetable
structure.

Lignite or Brown Coal is a more compressed and chemically
changed condition of vegetation.   It varies in colour from yellow
to deep brown or black, and may be regarded as an intermediate
stage between peat and coal.   It occurs in beds intercalated
between layers of shale, clay, and sandstone.

Coal—a compact, brittle, black, or dark brown stone, formed
of mineralised vegetation, and found in beds or seams usually
resting on clay, and covered with sandstone, shale, etc. (see Figs.
79 and 140).   There are many varieties of coal, differing from
each other in the relative proportions of their constituents.
Caking-coal, such as is ordinarily used in England, contains from
75 to 80 per cent of carbon, 5 or 6 per cent of hydrogen, and 10

or 12 per cent of oxygen, with some sulphur and other impurities. *Anthracite*, the most thoroughly mineralised condition of vegetation, is a hard, brittle, lustrous substance, from which the hydrogen and oxygen have been in great measure driven away, leaving 90 per cent or more of carbon.

Ironstone.—Reference was made at pp. 47, 59, to ironstone precipitated from chemical solution. This precipitation is often caused through the medium of decomposing organic matter. Organic acids, produced by the decay of plants in marshy places and shallow lakes, attack the salts of iron contained in the rocks or detritus of the bottom, and remove the iron in solution. On exposure, the iron oxidises and is thrown down as a yellow or brown precipitate of *limonite* or bog-iron-ore (p. 129), which is found in layers and concretions. *Clay-ironstone*, composed of a mixture of carbonate of iron, with clay and carbonaceous matter, occurs abundantly both as nodules and in layers, with remains of plants, shells, fishes, etc., in the Coal-measures (Figs. 61, 65, and bed 13 in Fig. 80), and has, no doubt, been also formed through the agency of organic acids which, passing into carbonic acid, have given rise to the solution and subsequent deposit of the iron as carbonate mingled with mud and with entombed plants and animals.

Flint.—Some siliceous deposits, due to organic agency, have been already referred to at p. 84. Besides these, mention may be made of *Flint*, which occurs as dark lumps and irregular nodular sheets in chalk and other limestones, frequently enclosing urchins, shells, and other organisms, which are sometimes converted into flint. Its mode of origin is not yet thoroughly understood, but there is reason to regard it as due to the abstraction of silica from sea-water, either directly, by such animals as sponges, or indirectly, by the decomposition of animal remains. *Chert* is a more impure siliceous aggregate found under similar conditions, especially among the older limestones.

Guano—a brown, light, powdery deposit, formed of the droppings of sea-birds in rainless tracts of the west coasts of South America and Africa. Containing much phosphate of lime as well as ammoniacal salts, it has great commercial value as an important manure.

Bone-beds—deposits composed of fragmentary or entire bones of fish, reptiles, or higher animals, as in the well-known bone-bed of the Rhætic series (p. 298). The floors of some caverns are covered with stalagmite, so full of pieces of the bones of cave-bears, hyænas, and other extinct and living species, as to be called *Bone-breccia.* Layers of stone, full of the *coprolites* (fossil excrement) or of the rolled bones of various vertebrate animals, have, in recent years, been largely worked as sources of phosphate of lime for the manufacture of artificial manures.

II. ERUPTIVE ROCKS.

Under this division are grouped all the massive rocks which have been erupted from underneath into the crust or to the surface of the earth. They are composed chiefly of silicates of alumina, magnesia, lime, potash, and soda, with different proportions of free silica, magnetic or other oxide of iron, and phosphate of lime. The principal silicate is generally some felspar, the number of eruptive rocks without felspar being comparatively small. The felspar is, in different rocks, conjoined with mica, hornblende, augite, magnetite, or other minerals.

No perfectly satisfactory classification of the eruptive rocks has yet been devised ; they have been grouped according to their presumed mode of origin, some being classed as *plutonic* or *hypogene*, from their supposed origin, deep within the earth's crust, others as *volcanic*, from having been ejected by volcanoes. They have likewise been arranged according to their chemical composition, and also with reference to their internal structure. In the following enumeration of some of the more abundant and important varieties, it may be enough to adopt an arrangement in three sections, according to the nature of the predominant silicate : viz. (1) Orthoclase rocks; (2) Plagioclase rocks; and (3) Olivine and Serpentine rocks. It has already been pointed out that the original condition of many lavas and other eruptive rocks has been that of molten glass, their present stony structure being due to the more or less complete devitrification and disappearance of the glass by the development of crystals and crystallites out of it during the process of cooling and consolidation (p. 144). Though there is no evidence that all crystalline eruptive rocks have once been in the state of molten glass, it may be useful to begin with the vitreous varieties, which we know to represent the earliest forms of many that are now quite crystalline.

(1) *Orthoclase Rocks.*

In this section the prevalent silicate is Orthoclase, either in its common dull, white, or pink form, or in the glassy condition

M

(sanidine). In many of the rocks, free quartz occurs either in irregular crystalline blebs or in definite crystals, which frequently take the form of double pyramids. Among other minerals, hornblende, white and black mica, and apatite are of common occurrence. The rocks of this division are the most acid of the eruptive series—that is, they contain the largest proportion of silica or silicic acid, sometimes more than 75 per cent. Some of them (granite) are only found as masses that have consolidated deep beneath the surface ; others (trachyte, rhyolite, obsidian) are abundant as superficial volcanic products.

Obsidian—a black, brown, or greenish (sometimes yellow, blue, or red) glass, breaking with a shell-like or *conchoidal* fracture and into sharp splinters, which are translucent at the edges. Examined in a thin section under the microscope, the rock is found to owe its usual blackness to the presence of minute opaque crystallites (Fig. 69) which are crowded through it, not infrequently drawn out into streaky lines and curving round any larger crystal that may be embedded in the mass (Fig. 71 B). These arrangements, called flow-structure (p. 146), have evidently been caused by the movement of the rock while still in a fused state, the crystallites and other objects being borne onward by the currents of molten glass. In some obsidians, little spherulites of a dull grey enamel-like substance have made their appearance as stages in the devitrification of the rock (Fig. 71) ; but the mass has consolidated before the stony condition could be completed. In other instances, the whole rock has passed into a stony enamel-like mass with perlitic structure (*pearlstone*, p. 145). Where a still molten obsidian has been frothed up by the expansion of steam or gas through it, so as to become a spongy cellular substance which will float on water, it is called *pumice*. Obsidian occurs in many volcanic regions, sometimes as streams of lava which have been poured forth at the surface, sometimes in dykes and veins, and often in fragments ejected with the other detritus that now forms tuffs.

Trachyte—a compact porphyritic rock, consisting mainly of orthoclase (sanidine), with some plagioclase and usually with some hornblende, or with augite, mica, magnetite, or other minerals ; having a peculiar matrix which, under the microscope, is found to consist mainly of minute felspar-crystallites. Large crystals of orthoclase (sanidine) are frequent, and also scales of dark mica. This rock is found abundantly among some of the younger volcanic regions of the world, where it occurs in lava-streams and also in intrusive sheets and dykes. **Quartz-trachyte** (*Liparite*, *Rhyolite*) is a rock composed of a compact, often rough and somewhat porous base, through which are scattered crystals of felspar and blebs of quartz, often also with hornblende and mica.

Felsite—an exceedingly close-grained rock, composed of an intimate mixture of quartz and orthoclase. The felspar often occurs as large disseminated crystals, giving the porphyritic structure. Where the quartz appears as distinct blebs or crystals (sometimes double pyramids) the rock becomes **Quartz-porphyry**. The felsites and quartz-porphyries play an important part among the eruptive rocks of older geological time, occurring both in the form of lavas erupted to the surface and of intrusive masses that have consolidated below ground. Many of them can be proved to have been originally in the condition of molten glass which has been devitrified. Rocks which show the characteristic closeness of grain characteristic of the felsites are said to be *felsitic* or to have a felsitic ground mass (p. 144).

Syenite—a thoroughly crystalline rock, consisting essentially of orthoclase and hornblende, and distinguished from granite chiefly by the absence or small amount of quartz. It occurs in bosses and veins which have been erupted into older rocks.

Granite—a thoroughly crystalline (holo-crystalline) compound of felspar, quartz, and mica, the individual minerals being large enough to be distinctly recognised by the naked eye. Sometimes large crystals of felspar are porphyritically scattered through the rock. Granite occurs in large eruptive masses which have been intruded into many different kinds of rocks, also in smaller bosses and veins. Round the outside of a mass of granite there frequently diverge from it dykes and veins (p. 203) which, where of great width, may show the usual granitic structure ; but which, when of small dimensions, are apt to appear as felsite or quartz-porphyry. There can be no doubt that such fine-grained veins are actually portions of the same mass of rock as the granite, so that granite and felsite or quartz-porphyry are only different conditions of the same substance, the differences being probably due to variations in the circumstances under which the cooling and consolidation took place. In the crystalline-granular structure so distinctive of granite (*granitic* or *granitoid*, p. 144) the constituent minerals have not had room to assume perfect crystallised shapes, but occasionally they have been able to shoot out in perfect crystals where cavities occur. Fig. 77, for example, shows a group of the ordinary crystals of this rock which have crystallised in a cavity of the granite of the Mourne Mountains, Ireland.   It is in such cavities also that the rarer minerals of this rock, such as topaz and beryl, may be looked for.

Fig. 77.—Group of crystals of felspar, quartz, and mica, from a cavity in the Mourne Mountain granite.

(2) *Plagioclase Rocks.*

In this section the felspar is some variety of plagioclase, and the other most frequent silicate is either augite or hornblende. Though free quartz occurs in some of the rocks, they contain generally so much less silica than the orthoclase rocks that instead of being *acid* they are commonly *basic* compounds.   A range of texture can be observed in them similar to that characteristic of the orthoclase series, from a true glass up to a thoroughly crys-talline granitoid rock.   Some of them, more especially the coarsely crystalline varieties, are probably of deep-seated origin ; others (and these include the great majority) are truly volcanic ejections which have risen in volcanic pipes and fissures, and have been poured forth at the surface as actual lava-streams.

Basalt-Rocks—a group of rocks consisting of plagioclase, augite, olivine, and magnetite or titaniferous iron, to which apatite and other minerals may be added.   These rocks range in texture from a black glass up to a coarsely crystalline mass wherein the component minerals are distinctly visible to the naked eye. Different names are employed to distinguish these varieties.

Basalt-glass (*Tachylyte, Hyalomelan*) is a general epithet to denote the vitreous varieties. These are particularly to be observed along the edges of dykes and other intrusive masses, where they represent the outer surface of the basalt that was suddenly chilled and consolidated by coming in contact with the cold walls of the vent or fissure into which it was injected, and where they no doubt show what was the original state of the whole basalt before devitrification converted the rock into its present crystalline structure (see pp. 97, 143). *Basalt*—a black, compact, heavy, homogeneous rock, breaking with a conchoidal fracture, showing sometimes large porphyritic crystals of plagioclase, olivine, or augite, but too fine-grained for the component minerals of the base to be determined except with the microscope. The coarser varieties, where the minerals can be recognised with the naked, eye, are known as *Dolerite.* The basalt-rocks are pre-eminently volcanic lavas, occurring both as intrusive masses that consolidated underground, and as sheets that were poured out in successive streams at the surface. The black, compact kinds (true basalt) are particularly prone to assume columnar forms (Fig. 78), whence columnar rocks are sometimes spoken of as *basaltic.* In some varieties of basalt the mineral leucite takes the part of the plagioclase ; and in others this is done by another mineral, nepheline.

Diabase—a name given to some ancient basalt-rocks in which, owing to alteration of their augite or olivine, a greenish chloritic discoloration has often taken place. The lavas of early geological time are to a large extent diabase.

Andesite is closely allied to basalt ; but contains no olivine. It sometimes includes free quartz, and hornblende may be substituted in it for augite. *Hornblende-andesite* and *Augite-andesite* are lavas which have been extensively erupted in later geological time.

Diorite—a crystalline aggregate of plagioclase and hornblende, usually with magnetite and apatite, sometimes with augite and mica. The hornblende is black or dark green and often more or less decomposed, giving rise to a greenish chloritic discoloration of the felspar. From its prevalent green colour, the rock was formerly known as "greenstone." It occurs in intrusive masses, and seems generally if not always to have consolidated below ground instead of being poured out at the surface.

Gabbro, Diallage-rock—a thoroughly crystalline granitoid aggregate of plagioclase and the variety of augite known as diallage, which appears in distinct brown or greenish crystals, with

FIG. 78.—Columnar basalts of the Isle of Staffa, resting upon tuff (to the right is Fingal's Cave).

a peculiar metalloidal or pearly lustre ; it is found in bosses associated with granite, gneiss, etc., and also sometimes with volcanic rocks in centres of eruption.

(3) *Olivine and Serpentine Rocks.*

In this group may be included a comparatively small number of rocks which consist principally of olivine, and which by gradual alteration pass into serpentine (Fig. 58). **Olivine-rocks** (Peridotites) are liable to remarkably rapid changes of texture and composition. In some places they are mainly made up of olivine, augite, or hornblende, magnetite, and brown mica, but some of these minerals may disappear and some felspar may take their place. They are intrusive masses which appear to have been generally injected into the crust in connection with volcanic eruptions, rather than to have been poured out at the surface in true lava-streams.

Serpentine—a compact, dull, or faintly glimmering rock, with a general dark dirty green colour, variously mottled, greasy to the touch, easily scratched, and giving a white powder which does not effervesce with acids. It is a massive form of the mineral serpentine described on p. 134 ; frequently containing disseminated crystals of the minerals bronzite, enstatite, and chromic iron, and veins of a delicately fibrous silky variety of serpentine known as chrysotile. Many serpentines were originally olivine-rocks which, by hydration and alteration of their magnesian silicates, have assumed their present characters. Serpentine occurs in bosses, dykes, and veins, which were evidently of eruptive origin and were at first probably olivine-rocks ; it is also found in thick beds associated with limestones and crystalline schists, where it may be a metamorphosed sedimentary rock.

III. The Schists and their Accompaniments.

This section includes a remarkable series of rocks of which the leading character is the possession of a schistose or foliated character (Fig. 72). They are, in their more typical varieties, distinctly crystalline. Some of them shade off into ordinary fragmental rocks, such as shale and sandstone ; others agree in chemical and mineral composition with some of the eruptive rocks already enumerated, into which they may often be traced by imperceptible gradations.

In the schists, therefore, we see an assemblage of rocks which, though possessing distinct characters of their own, may yet be observed to shade off into fragmental rocks on the one side, and into eruptive rocks on the other. In Chapter XIII some further account of them will be given, with special reference to their probable origin, and to the grounds on which they have been regarded as *metamorphic* or altered rocks. For the present, in taking notice of their composition and structure, it will be enough to state that in many cases they can be shown to be more or less altered and crystalline transformations of what were originally sedimentary rocks ; and that in other instances they represent original crystalline eruptive masses, which have been subjected to such enormous pressure and shearing, that a foliated structure and recrystallisation of minerals have been superinduced in them. The essential feature which unites masses of such different origin is the possession of that common schistose structure which they have derived from having all been alike subjected to the same kind of intense terrestrial movements.

Clay-slate—a hard fissile clay-rock, through which minute scales of mica and crystals or crystallites of other minerals have been developed ; generally bluish-grey to purple or green, and splitting into thin parallel leaves. As this rock often contains remains of marine animals and plants, and is interstratified with bands of sandstone, grit, conglomerate, and limestone, it was undoubtedly at first in the condition of soft mud on the sea-bottom. Sometimes the organic remains in it are so curiously elongated or distorted in one general direction as to show that the rock has been drawn out by intense pressure and shearing (Figs. 98, 103, 104). The planes along which clay-slate splits are generally independent of the original surfaces of deposit, sometimes cross these at a right angle, and have been superinduced by mechanical movements (Cleavage), as explained in Chapter XIII. Different varieties of clay-slate have received special names. *Roofing slate* is the fine compact durable kind, employed for roofing purposes and also for the manufacture of cisterns, chimneypieces, writing-slates ; *Alum-slate*—dark, carbonaceous, and pyritous, the iron-disulphide oxidising into sulphuric acid, and giving rise to an efflorescence of alum ; *Whet-slate honestone*—exceedingly hard, fine-grained, and suitable for making hones ; sometimes owing its hardness to the presence of microscopic crystals of garnet ; *Chiastolite-slate*—containing disseminated crystals of chiastolite, and found especially around eruptive bosses of granite. By increase of its mica-flakes a clay-slate passes into a *Phyllite*, which has a more silvery sheen, and represents a farther stage of metamor-phism. Phyllite, by increase of the mica, becomes *Mica-slate*, so that a transition may be traced from sedimentary fossiliferous rocks through clay-slate and phyllite into thoroughly crystalline schist. Clay-slate occurs extensively among the older geological formations in all parts of the world.

Amphibolites—rocks composed mainly of hornblende, but with quartz, orthoclase, and other minerals in minor proportions ; sometimes they are massive and granular (*Hornblende-rock*), and in this condition doubtless represent eruptive rocks. Grada-tions can be followed from such rocks (originally diorite, diabase, etc.) into perfect schist (*Hornblende-schist*), so that the development of the schistose structure can be traced from rocks that were at first as structureless as any amorphous eruptive mass can be. Amphibolites occur among the crystalline schists in most parts of the world as occasional bands or bosses, which probably mark zones of basic igneous rock, either intruded into the accompanying masses, or contemporaneously erupted with them.

Chlorite-schist—a scaly, schistose aggregate of greenish chlorite with quartz, and often with felspar, mica, and octahedra of magnetite (Fig. 54) ; it occurs in beds associated with gneiss and other schists. Some chloritic schists may represent old lavas or other erupted rocks which have been crushed down and be-come schistose ; others, especially where they contain pebbles of quartz, etc., and are banded with quartzites and schistose con-glomerates, not improbably mark where fine volcanic ashes fell over a sea-bottom, and were then mingled and interstratified with the ordinary sediment that happened to be accumulating at the time.

Mica-schist (Mica-slate)—a schistose aggregate of quartz and mica, the two minerals being arranged in irregular but nearly parallel wavy folia. The rock splits along the laminæ of mica, so that its flat surfaces have a bright silvery sheen, and the quartz is not well seen except on the cross fracture, where only the thin edges of the mica-plates present themselves. Mica-schist is often remarkably crumpled or puckered—a structure bearing witness to the intense compression it has undergone (Fig. 114). It abounds in most regions where schists are extensively developed (Chapter XVI). Some mica-schists contain fossil shells and corals (Bergen), and must thus represent what were originally sediment-ary deposits ; others may be highly deformed eruptive rocks.

Gneiss—a schistose aggregate of orthoclase, quartz, and mica, varying in texture from a fine-grained rock up to a coarse crystal-line mass which, in hand specimens, may not be distinguishable from granite.   There is no difference indeed as regards composi-tion between gneiss and granite ; gneiss may be called a foliated granite.   There is good reason to believe that some, if not all, true gneisses have been made out of granite or allied rocks by the process of shearing above referred to.   Gneiss occurs abundantly among the oldest known rocks of the earth's crust, and may be found in most large regions of crystalline schists (Chapter XVI).

A few rocks which are found associated with the schists, or with evidence of metamorphism, may be noticed here—marble, quartzite, and schistose conglomerate.

Marble—a crystalline granular aggregate of calcite, white when pure, and having the texture of loaf-sugar, but passing into various colours according to the nature of the impurities.   It occurs in beds among the schists, and is no doubt a limestone, formed either by chemical precipitation or by organic agency, which has been metamorphosed by heat and pressure into its present thoroughly crystalline character.   Some of the fossiliferous limestones through which the Christiania granite rises have been changed into crystal-line marble, but their original corals and shells have not been wholly effaced (see Chapter XIV).

Quartzite—a hard, compact, granular rock, composed of adherent quartz-grains, and breaking with a characteristic lustrous fracture.   It occurs in beds and thick masses, not infrequently associated with slates, mica-schists, and limestones ; it sometimes contains organic remains; and is evidently an indurated siliceous sand.

Schistose Grit and Conglomerate.—Interstratified with clay-slates and mica-schists there are sometimes found beds of grit and conglomerate, the grains and pebbles of which consist of quartz or other durable material, imbedded in slate or schist. The original fragmental character of such rocks admits of no doubt ; they were obviously at one time sheets of fine and coarse gravel mixed with sandy mud ; and their presence among schistose rocks furnishes additional corroborative evidence of the original sedimentary character of some of these rocks.   The clay or mud which formed the matrix has been metamorphosed into a more or less thoroughly crystalline micaceous substance, while in many cases the pebbles have been flattened and pulled out of shape. Hence these rocks afford important evidence as to the nature of the processes whereby the schists have been produced.

PART III

THE STRUCTURE OF THE CRUST OF THE EARTH

CHAPTER XII

SEDIMENTARY ROCKS—THEIR ORIGINAL STRUCTURES

HAVING in the two foregoing chapters considered the more important elementary substances of which the earth's crust is composed and their combinations in minerals and rocks, we have to inquire how these minerals and rocks have been put together so as to build up the crust. A very little examination will suffice to show us that the upper or outer parts of the solid globe consist chiefly of sedimentary rocks. All over the plains and low grounds of the earth's surface, which cover so large a proportion of the whole area of the land, some kind of sediment underlies the soil—clay, sand, gravel, limestone. It is for the most part only in hilly or mountainous regions that anything has been pushed up from below, so as to indicate the nature of the materials underneath. But everywhere we encounter proofs that the sedimentary rocks do not remain as they were deposited. In the first place, most of them were laid down on the sea-floor, and they have been upraised into land. In the next place, not only have they been upheaved, they have not infrequently been bent, broken, and crushed, until sometimes their original condition can no longer be determined. Moreover they have been invaded by masses of lava and other eruptive rocks, which have been thrust in among them and have often burst through them to form volcanoes at the surface. We must now endeavour to form as clear a conception as possible of what, after all these changes, the present structure of the crust actually is.   In this chapter, therefore, we may examine some of the leading characters of sedimentary rocks in the architecture of the crust, more particularly those which have been determined by the conditions under which the rocks were formed.   In the next chapter we shall consider some of the more important characters which have been superinduced upon the rocks since their formation.

Stratification.—It has been shown (p. 148) that one of the most distinctive features in sedimentary rocks is that they are stratified—that is, are arranged in layers one above another. As those at the bottom must have been deposited before those at the top, a succession of layers of stratified rocks forms a record of deposition, in which the early stages are chronicled by the lower, and the later stages by the upper layers. An illustration of this kind of record has already been given in the introductory chapter.   As a further example, the accompanying section (Fig. 79) may be taken.   At the bottom lies a bed (*a*) of dark shale or clay with fragments of crinoids, corals, shells, and other marine organisms.   Such a bed unmistakably points to a former muddy sea-floor, on which the creatures lived whose remains have been preserved in the hardened mud or shale.   The next bed (*b*) is one of limestone full of similar organic remains ; it shows that the supply of mud, which had previously made the water turbid and had been slowly gathering in successive layers on the bottom, now ceased.   The water became clear and much better fitted for the life of the crinoids, corals, and shells.   These creatures accordingly flourished abundantly, living and dying on the spot generation after generation, until their accumulated remains had built up a solid sheet of limestone several feet thick.   But once more muddy currents spread over the place, and from the cloud of suspended mud there slowly settled down the layer of blue clay (*c*) which overlies the limestone.   As hardly any remains of organisms are to be seen in it, we may infer that the inroad of mud killed them off.   Next, owing to some new shifting of the currents, a quantity

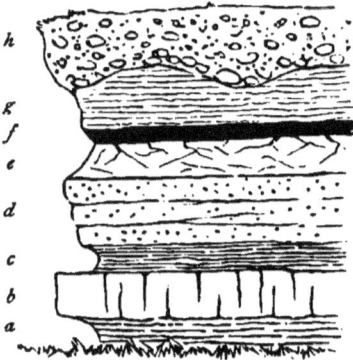

FIG. 79.—Section of stratified rocks.

of sand was brought in and spread out over the mud, forming the sandstone beds (*d*). The sea in which these various strata were deposited was probably shallow; or its floor may have been gradually rising. At all events, the last layers of sand could have been only slightly below the surface of the water, for they are immediately covered by a hardened silt or fire-clay (*e*) which, from the abundant roots and rootlets that run through it in all directions, was clearly once a soil whereon plants grew. It was probably part of a mud-flat, on which vegetation spread seaward from the land where the water shallowed, as happens at the present day among the tropical mangrove-swamps (p. 83). The plants that grew on this soil have formed the coal-seam (*f*), no doubt representing the growth of a long period of time. But the existence of the coal-jungle came to an end probably by a sinking of the ground beneath the water. Mud, once more carried hither from the neighbouring land, settled down upon the submerged vegetation and formed the clay (*g*). But that land plants still abounded in the immediate neighbourhood, is shown by their numerous remains in this clay. We notice too that the salts of iron dissolved in the water were eliminated by the decaying plants and animals and were precipitated in the form of carbonate, so as to form concretions round occasional dead shells, fishes, fern-fronds, and seed-cones. What were the immediately succeeding events in this ancient history we cannot tell; the layer next in order is a coarse conglomerate (*h*), originally gravel, which must have been swept along by a swift current that tore away the upper part of the clay-beds (*g*) and any strata which may once have overlain them.

The whole stratified part of the earth's crust is composed of materials which in this way may be made to tell their story. In forcing them to yield up their records of the ancient changes of which they are memorials, scope is afforded for the most accurate and laborious investigation and for the closest reasoning from the facts collected. At the same time, it is obvious that the pursuit is one which constantly exercises the imagination, and that, indeed, it cannot be adequately followed unless, by the proper use of the imagination, the former conditions of the earth's surface are vividly realised.

The thinnest layers of a stratified rock form *laminæ*, such as the thin paper-like leaves into which shale can be split. A number of laminæ may be united in a *stratum* or *bed* which may vary from less than an inch to several feet or yards in thickness. It is only

the finer kinds of sedimentary rock that, as a rule, are *laminated*.
In other cases a stratum or bed is the thinnest subdivision ; it can usually be separated easily from those above and below it, and it may generally be regarded as marking one continued phase of deposit, while the break between it and the next bed above or below probably denotes an interruption of the deposit. The study of the relations of strata to each other is called *Stratigraphy*.

Layers of deposit usually lie parallel with each other, their flat surfaces marking the general floor of the water at the time of their formation (Figs. 79, 80). But sometimes a series of layers may be found inclined at various angles to what was obviously the original general plane of deposition. In Fig. 81, for example, a series of strata is presented, which are distinguished by a diagonal lamination. This is known as *False bedding* or *Current-bedding*. As explained in Chapter III (p. 37), it has been caused by the pushing of layers of sediment over the advancing front of a stratum, and may be compared to the oblique bedding often to be seen in an earthwork, such as a railway embankment, the upper surface of which may be in a general sense parallel with the flat bottom of the valley, while the successive layers of which the mound is made are inclined at angles of 30° or more. False bedding is interesting as affording some indication of the nature and direction of the currents by which sediment has been transported.

Proofs of former Shores.—Along the margin of the sea, of lakes, and of rivers, several interesting kinds of markings may be seen impressed on surfaces of sand or mud from which the water has retired. Every one who has walked on a tidal sea-beach is familiar

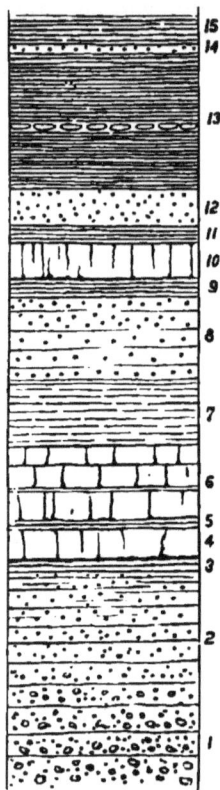

FIG. 80.—Section showing alternation of beds.

15. Shale.   14. Seam of sandstone.   13. Shale with septarian nodules. 12. Sandstone.   11. Mudstone.   10. Limestone. 9. Clay.   8. Sandstones. 7. Sandy clays.   6. Limestone with parting of shale.   5. Shale.   4. Limestone.   3. Shale with cement-stone passing down into sandstone (2), which graduates into fine conglomerate (1).

with the *Ripple-marks* left by the retreating tide upon the
bare sands. They are produced by the oscillation of the water
driven into movement by wind playing over its surface. They are
usually effaced by the next advancing tide ; hence, out of the
same sand new sets of ripple-marks are made by each tide. But
we can understand that now and then, under peculiarly favourable
conditions, the markings may not be destroyed. If, for instance,
they were made in a kind of muddy sand, which, in the interval
between two tides and under a strong sun, could become hard and
coherent on the surface, and if the next tide advanced so quietly
as not to disturb them, but to lay down upon them a fresh layer

FIG. 81.—False-bedded sandstone.

of sand or mud, they might be covered up and preserved. They
would then remain as a memorial of the shallow rippling water
and bare sandy shore where they had been formed.

Now evidence of this kind regarding the conditions of deposition occur abundantly among sedimentary rocks (Fig. 82).
Ripple-marked surfaces may be traced one over another for many
hundred feet in a thick series of sandstones. They bring clearly
to the mind that the strata on which they lie were accumulated in
shallow water, or along beaches that were often laid dry.

Land-Surfaces.—Other traces of exposure to the air may be
noticed, where ripple-mark is abundant, in what are termed *Sun-
cracks*, *Foot-prints*, and *Rain-prints*. Those who have observed
what takes place in muddy places during dry weather will

remember that, as the mud dries and contracts it splits up into a network of cracks ; and that, on its hardened surface, it retains impressions of the feet of birds or of insects that may have walked over it while still soft.   The geological history recorded at such places cannot be mistaken ; first, the rainy period, with the rush of muddy water down the slopes and the formation of pools in which the mud is allowed to settle ; then the season of warm weather when the pools gradually dry up and birds seek their edges to drink.   If by any means a layer of sediment could be

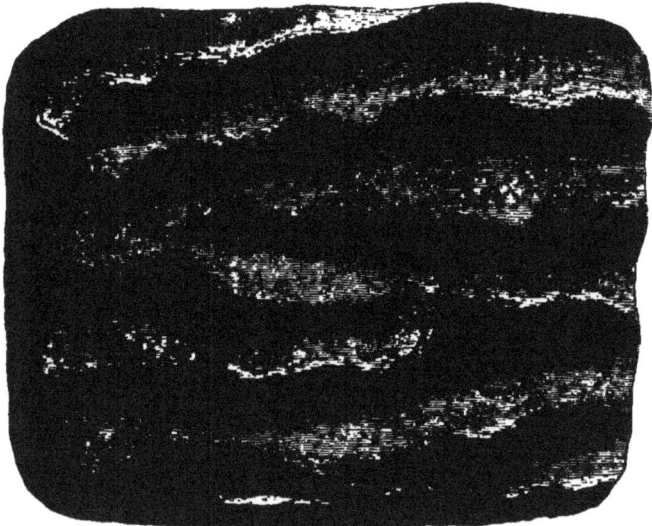

FIG. 82.—Ripple-marked surface of sandstone.

laid down upon one of these desiccated basins so gently as not to efface its peculiar markings, the cracked surface of mud, with its footprints, would contain a perfectly intelligible record of the changes which it had witnessed (Fig. 83).

Now surfaces of this kind abound among the sedimentary rocks of the earth's crust.   They are found upon strata which, from the presence of marine organic remains in them, were certainly deposited under the sea.   But these strata cannot have accumulated in deep water ; they must have been formed along flat shores, where the sheets of sand and mud were liable from time to time to be laid bare to the sun and wind, where animals of various kinds left their footmarks or trails on the still soft sediment, where the evaporation and desiccation were so rapid as to cause the exposed mud to harden on the surface and to crack up into irregular polygonal cakes, and where the next succeeding layers of sediment were deposited so gently as to cover up and preserve the sun-cracked surfaces.

One further piece of evidence to indicate land-surfaces, or, at least, shore-surfaces, in a series of aqueous sedimentary strata, is that furnished by Rain-prints.   A brief shower of rain leaves upon a smooth surface of fine sand or mud a series of small pits,

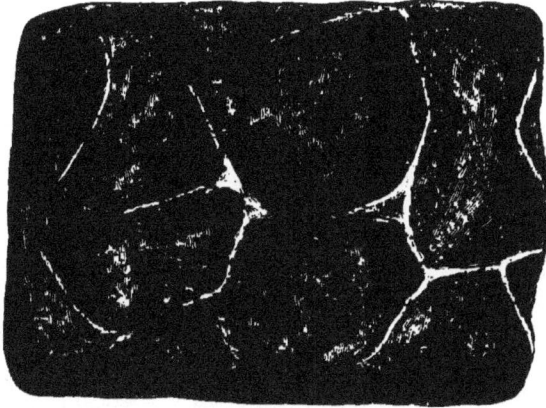

FIG. 83.—Cast of a sun-cracked surface preserved in the next succeeding layer of sediment.

each of which is the imprint of a descending raindrop (Fig. 84). Where this takes place along the edge of a muddy pool which is rapidly being dried up, the prints of the drops may remain quite distinct on the hardened surface of mud.   And here, again, we can suppose that if another layer of mud were gently deposited above this surface the rain-prints would be sealed up and preserved.   We might even be able to tell from what quarter the wind blew that brought the rain-cloud.   If, for example, the rain-prints were ridged up on one side in one general direction this would show that the shower fell aslant and with some force, and that the side on which the mud round the imprints was forced up was that towards which the rain was driven.   Such indications of ancient weather may here and there be detected among stratified rocks.

Concretions.—Another original characteristic of many sedimentary rocks is a concretionary structure, particularly observable in clays, limestones, and ironstones. In many cases, the con-cretions have gathered round some fragment of a plant or an animal. Clay-ironstone and impure limestone have been aggregated into spherical or elliptical forms (septaria), which are of frequent occurrence in clay or shale (Figs. 61, 65). Flint has also gathered round some organic nucleus, which it has often entirely replaced. But many concretions may be found where no organic fragment as a starting-point can be detected. Some of the most curious are the so-called Fairy-stones (Fig. 64), found in alluvial clays, with so many imitative shapes, which have been

FIG. 84.—Rain-prints on fine mud.

popularly supposed to be works of human or even preternatural construction. They have probably been produced by the irregular cementing of clay, owing to the spread of carbonate of lime through it, carried down by permeating water. Some of the most extraordinary concretionary masses are to be seen in certain magnesian limestones, which appear to be built up of petrified lumps of coral, bunches of grapes, cannon-balls, and other objects (Fig. 75). In reality, all these diversified figures are due to the irregularly varied way in which a concretionary structure has been developed in the limestone.

Association and Alternation of Strata.—Certain kinds of sedimentary rocks are apt to occur together to the exclusion of others. This association depends on the circumstances of

deposition.   Ironstone concretions, for example, are much more
frequent among clays or shales than in any other strata, because
it was during the deposit of fine mud with abundant decomposing
organic matter that the most favourable conditions were supplied
for the precipitation of carbonate of iron.   Clays and limestones
frequently alternate, as also do sandstones and conglomerates,
because the circumstances of deposition were somewhat alike
(see Fig. 80).   But we need not expect to encounter a bed of
coarse conglomerate in a group of fine clays, for the current
that was strong enough to sweep along the stones of the
conglomerate was too powerful to allow the fine silt to lie undisturbed.   For a similar reason, we should be surprised to meet
with a layer of well-stratified shale in a mass of conglomerate.
The agitated water in which these coarse materials were heaped
up would have swept away any fine sediment and prevented it
from being deposited.   In all cases, the manner in which the
different kinds of sediment are associated with each other leads
us back directly to the original conditions of deposit, and is only
intelligible in proportion as these conditions are clearly realised.

Relative Areas of Stratified Rocks.—Moreover, some kinds
of sedimentary material must obviously spread over wider areas
than others.   The coarse gravel and shingle of the present beach
do not extend far seawards ; they are confined to the margin of
the land.   Sand covers the sea-floor over a wider area ; and
beyond the limits of the sand, in the deeper and stiller water,
mud is allowed to accumulate.   Roughly speaking, therefore, the
area of the distribution of sediment is in inverse proportion to
the coarseness of the materials.   The same law has regulated the
accumulation of detritus from early geological time.   Coarse
conglomerates, which represent ancient shingles and gravels,
thicken and thin out rapidly, and do not usually cover a large
area, though they may sometimes be traced for long distances
in the direction probably of the original coast or line of heaping
up of the shingle.   They pass laterally and vertically into grits
and sandstones which have a much wider distribution, and these
again shade off into clays and shales that range also over large
areas.

Chronological Value of Strata.—No clue has yet been found
to determine the length of time required for the accumulation of a
stratum or group of strata ; but some indications are afforded of
relative lapse of time: Here and there, for instance, vertical
trunks of trees are met with standing in their positions of growth,

but imbedded in solid sandstone (Fig 85). These stems, sometimes 20 feet or more in height, prove that a mass of sand of that depth must have been accumulated around them before they had time to decay. We know little about the durability of the submerged trees; but they probably could not have lasted long unless covered up by sediment; so that the mass of strata in which they are enclosed may be supposed to have been accumulated within a few years. The nature of the material composing

FIG. 85.—Vertical trees (*Sigillaria*) in sandstone, Swansea (Logan).

sedimentary rocks may likewise furnish indications of relative rate of deposition. Thus finely laminated clays were evidently deposited with extreme slowness. Beds of limestone, composed of the crowded remains of successive generations of marine creatures, must also have required prolonged periods of time for their growth. On the other hand, thick beds of sandstone presenting great uniformity of characters may not improbably have been laid down with comparative rapidity.

No reliable inference can be drawn from the mere thicknesses of strata as to the lapse of time which they represent. A mass of sandstone 20 feet thick may have accumulated round a sub-merged tree in a few years.   On the other hand, a corresponding depth of fine laminated clay may have required tenfold more time for its deposition.   But the same thickness of rock composed of alternations of shale and limestone might represent a still longer period.   For it is obvious that the change from one kind of sediment to another must often have been brought about by an extremely gradual modification of the geography of the region from which the supply of sediment was derived.   Hence the interval between two beds or groups of beds, differing much from each other in mineral composition, may have been considerably longer than the time required for the actual deposition of the strata of either or both beds or groups of beds.

On any probable estimate, the deposition of sedimentary rocks to a depth of many thousand feet and over areas many thousands of square miles in extent, must have demanded enormous periods of time.   Side by side with the growth of mechanical sediments, there must have been a corresponding wasting of land.   Every bed of conglomerate, sand, or mud represents at least an equivalent amount of rock worn away from the land and trans-ported as sediment to the floor of the sea.   During such prolonged ages as these changes required, there was ample time for the outburst of many successive volcanoes, for the passage of many earthquake-shocks, and for the subsidence or upheaval of many parts of the earth's crust.

Proofs of Subsidence.—A mass of sedimentary material of

Fig. 86.—Hills formed out of horizontal sedimentary rocks.

great thickness which, from the remains of sun-cracks and other evidence, was obviously deposited in shallow water near land can only have been accumulated on an area that was gradually sink ing. Suppose, for instance, that a hill formed out of such strata rises a thousand feet above the valley at its foot (Fig. 86), and that proofs of deposition in shallow water can be detected from the lowest beds all the way up to the highest. The lowest beds having once been close to the surface, as shown by the sun-cracks and other evidence, could only be covered with hundreds of feet of similar strata by a gradual sinking of the ground, during which fresh sediment was poured in, so that, although the original bottom sank a thousand feet, the water may never have become sensibly deeper, the rate of deposit of sediment having, on the whole, kept pace with that of the subsidence.

Overlap.—During such tranquil movements, as the area of land lessens and that of the sea increases, the later sedimentary accumulations must needs extend beyond the limits of the older ones. Suppose, for instance, that such a sloping land-surface as that

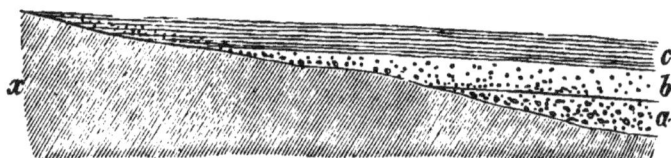

FIG. 87.—Section of overlap.

represented in the section (x, Fig 87) were slowly to subside beneath the sea, the first-formed strata (a) will be covered and overlapped by the next series (b), and these in turn, as the sea-floor sinks, will be similarly concealed by the following group (c). This structure, termed *Overlap*, may usually be regarded as evidence of a gentle subsidence of the area of deposit.

Conformability, Unconformability.—When stratified deposits are laid down regularly and continuously upon each other, with no interruption of their generally level position, they are said to be *conformable*. In the section Fig. 80, for instance, the series of sediments there represented has evidently been deposited under the same general conditions. The nature of the sediment has of course varied from time to time; limestones, shales, and sandstones have alternated with each other; but there has been no marked interruption or disturbance in their sequence. Suppose, however, that owing to subterranean movements, a series of rocks (a in Fig. 88) is shifted from its original position, and after being uplifted, is exposed to the wearing action of the sea, rivers, air, rain, frosts, and the other agents concerned in the degradation of the surface of the land.   If a new series of deposits (*b*) is laid down upon the denuded edges of these rocks, the bedding of the whole will not be continuous.   The younger strata will rest successively upon different parts of the older group, or, in other words, will be *unconformable.*   Such a relation or *unconformability* (unconformity) implies a terrestrial disturbance, and usually also the lapse of a long interval of time between the respective periods of the older and younger rocks, during which denudation of the older strata took place.   It serves to mark one of the breaks or gaps in geological history.   Unconformabilities differ much from each other in regard to the length of interval which they denote. In some cases, the blank may be of comparatively slight moment ; in others, it is so vast as to include the greater part of the time represented by the stratified rocks of the earth's crust.

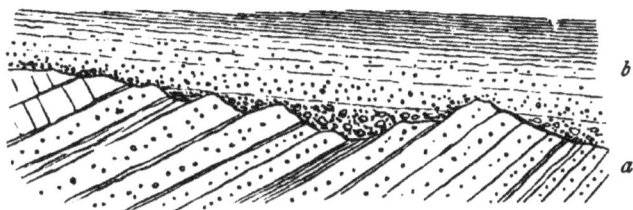

Fig. 88.—Unconformability.

By means of unconformabilities the different ages of mountain-chains are determined.   If, for example, a mountain showed the structure represented in Fig. 88, its upheaval must obviously have taken place between the deposition of the two series of rocks. Suppose the series *a* to represent Lower Silurian, and *b* Carboniferous rocks, the date of the mountain would be between the Lower Silurian and Carboniferous periods.   If, in another mountain, series *b* were unconformably overlain by a younger series, say of Jurassic age, this mountain would thereby be shown to have undergone a subsequent uplift in the long interval between the Carboniferous and the Jurassic periods.

Summary.—In this Lesson some of the more characteristic original features of sedimentary rocks have been considered.   Of these features, one of the most distinctive is the arrangement into layers of beds, each of which is the record of a portion of geological history, the oldest being below and the youngest above.   The smallest subdivision of these records is a lamina or thin leaf, such as those into which shales may be split.   A stratum or bed, which may contain many laminæ or none, is a thicker layer separable with more or less ease from those below and above it. Though strata lie on the whole parallel with each other, they often show oblique current-bedding, especially in sandstones. Traces of shore-lines and of surfaces laid bare by the retirement of the water in which they were deposited, are found in sun-cracks, rain-pittings, and footprints. Not infrequently, instead of being evenly spread out in layers, the sedimentary material has been aggregated into variously-shaped concretions. Certain kinds of sedimentary rocks are apt to occur together, such as clays and limestones, clay-ironstones and shales, coals and fire-clays ; because the conditions under which they were respectively deposited were on the whole similar. As a rule, the finer the detritus, the wider the area over which it is spread ; hence clays generally cover wider tracts than conglomerates. No inference can safely be drawn from the relative thickness of strata as to the length of time which they respectively represent ; they must vary widely in this respect, and it is quite conceivable that, in many cases, the interval of time between the deposition of two successive beds of very different character and composition may have been actually longer than the period required for the deposition of the two beds. A thick series of sedimentary deposits usually indicates that the sea-bottom on which it was laid down was slowly sinking. In subsiding, the later deposits spread beyond the limits of the earlier ones, and thus present what is called an overlap. Where they have been laid down continuously one upon another they are said to be conformable ; where one group has been deposited on the disturbed and worn edges of an older series the two are unconformable to each other.

CHAPTER XIII

SEDIMENTARY ROCKS—STRUCTURES SUPERINDUCED IN THEM AFTER THEIR FORMATION

AFTER their deposition sedimentary materials have undergone various changes before assuming the aspect which they now wear.

Consolidation.—The most obvious of these changes is that, instead of consisting of loose materials, gravel, sand, mud, and so on, they are now hard stone. This consolidation has sometimes been the result of mere pressure. As bed was piled over bed, those at the bottom would gradually be more and more compressed by the increasing weight of those that were laid down upon them, the water would be squeezed out, and any tendency which the particles might have to cohere would promote the consolidation of the mass. Mud, for example, might in this way be converted into clay, and clay in turn might be pressed into mudstone or shale. But besides cohesion from the pressure of overlying masses, sedimentary matter has often been bound together by some kind of cement, either originally deposited with it or subsequently introduced by permeating water. Among natural cements, the most common are silica, carbonate of lime, and peroxide of iron. In a red sandstone, for example, the quartz-grains may be observed to be coated over with earthy iron peroxide, which serves to unite them together into a more or less coherent stone. The effect of weathering is not infrequently to remove the binding cement, and thereby to allow the stone to return to its original condition of loose sediment.

Joints.—Next to their consolidation into stone, the most common change which has affected sedimentary rocks is the production in them of a series of divisional planes or fractures termed *Joints*. Except in loose incoherent materials, this structure is hardly ever absent. In any ordinary quarry of sandstone, limestone, or other

sedimentary rock, or along a natural cliff of the same materials, a
little attentive observation will show that the bare wall of rock
forming the back of the quarry or the face of the cliff has been
determined by one or more natural fissures in the stone,
and that there are other fissures running parallel with it through
every outstanding buttress of rock.   Moreover, we may observe that these vertical or highly inclined lines of fissure are
cut across by others, more or less nearly at a right angle, and
that the sides of the buttresses have been defined by these trans-
verse lines, just as the main face of rock has been formed by the

Fig. 89.—Joints in a stratified rock.

first set.   Such lines of division are Joints.   In close-grained
stone they may be imperceptible until it is quarried or broken,
when they reveal themselves as sharply defined, nearly vertical
fractures, along which the stone splits.   There are usually at
least two series of joints crossing each other at right angles or
obliquely, whereby a rock is divided into quadrangular blocks.
In the accompanying diagram (Fig. 89) a group of stratified rocks
is seen to be traversed by two sets of joints, one of which (*dip-
joints*, "cutters" of the quarrymen) defines the faces that are in
shadow, the other (*strike-joints*, "backs" of the quarrymen) those
that are in light.   By help of these divisional planes, it is possible
to obtain large blocks of stone for building purposes.   The art of
the quarryman, indeed, largely consists in taking advantage of
these natural lines of fracture, so as to obtain his materials with
the least expenditure of time and labour, and in large masses.

In nature also the existence of joints is a fact of the highest importance. Reference has already been made to the way in which they afford a passage for the descent of water from the surface. It is in great measure along joints that the underground circulation of water is conducted. At the surface, too, where rocks yield to the decomposing influence of the weather, it is by their joints that they are chiefly split up. Along these convenient planes of division, rain-water trickles and freezes ; the walls of the joints are separated, and the space between them is slowly widened, until in the end it opens into yawning rents, and portions of a cliff are overbalanced and fall, while detached pinnacles are here and there isolated. The picturesqueness of the scenery of stratified rock is, in great measure, dependent upon the influence of joints in promoting their dislocation and disintegration by air, rain, and frost.

In many cases, joints may be due to contraction. A mass of sand or mud, as it loses water and as its particles are more firmly united to each other, gradually occupies less room than at first. In consequence of the contraction strains are set up in the stone, and relief from these is eventually found in a system of cracks or fissures. In other instances, joints have been produced by the compression or torsion to which large masses of rock have been exposed during movements of the earth's crust.

Original Horizontality.—As laid down upon the margin or floor of the sea, on the bottoms of lakes, and on the beds or alluvial plains of rivers, sedimentary accumulations are in general nearly flat. They slope gently, indeed, seawards from a shelving shore, and they gather at steeper angles on slopes of debris at the foot of cliffs, or down the sides of mountains. But, taken as a whole, and over wide areas, their original position is not far removed from the horizontal. If we turn, however, to the sedimentary rocks that form so large a part of the earth's crust, and so much of the dry land, we find that although originally deposited for the most part over the sea-bottom, they are now inclined at all angles, and even sometimes stand on end. Such situations, in which their deposition could never have taken place, show that they have been disturbed. Not only have they been upraised into land, but they have been tilted unequally, some parts rising or sinking much more than others.

Dip.—The inclination of bedded rocks from the horizon is called their *Dip*. The amount of dip is reckoned from the plane of the horizon. A face of rock standing up vertically above that plane is said to be at 90°, while midway between that position and horizontally it lies at an inclination of 45°.   The angle of dip is accurately measured with an instrument called a *Clinometer*, of which there are various forms.   One of the simplest kinds is a brass half-circle graduated into 90° on each side of the vertical, on which a pendulum is hung as in Fig. 91.   The instrument is held between the eye and the angle to be measured, and the upper

FIG. 90.—Dip and Strike.   The arrow shows the direction of dip ; the line *s s* marks the strike.

edge is made to coincide with the line of the inclined rock.   The pendulum, remaining vertical, points to the angle of inclination from the horizon.   A little practice, however, enables an observer to estimate the amount of dip by the eye with sufficient accuracy for most purposes.   The direction of dip is the point of the compass toward which a stratum is inclined (shown by the arrow in Fig. 90), and is best ascertained with a magnetic compass.   But

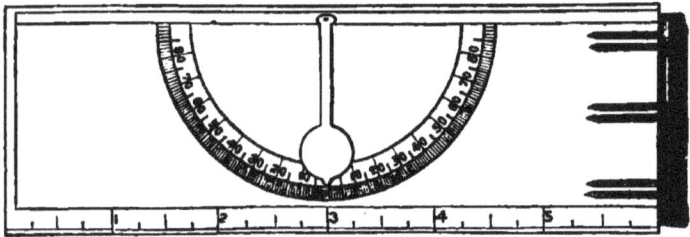

FIG. 91.—Clinometer.

here again a little experience in judging of the quarters of the sky without an instrument will usually enable us to tell the direction of dip with as much precision as may be required.

Strike.—A mathematical line running at a right angle to the direction of dip is called the *Strike* (*s s* in Figs. 90, 92).   Where a series of strata dips due north or due south the strike is east and west ; but the direction of strike changes with that of the dip.

Suppose, for example, that certain strata dip due east, then veer round by south-east to south, and so on by west and north, back to east again. The strike following this change would describe a circle. In fact, the beds would be included in a basin-shaped or dome-shaped arrangement and the strike would be the lip of the basin or rim of the truncated dome. Though the dip may slightly vary from place to place, still, if it remains in the same general direction along the line of certain strata, their strike is on the whole uniform.

Outcrop.—The actual edge presented by a stratum at the surface of the ground is called its *Outcrop.* On a perfectly level

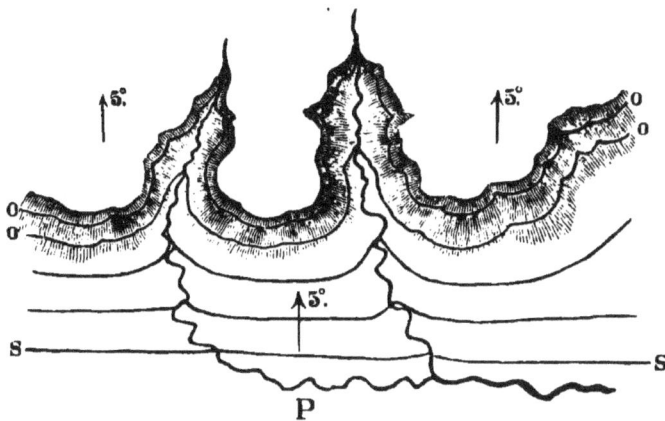

Fig. 92.—Dip, Strike, and Outcrop.

surface, strike and outcrop must coincide ; but as ground is seldom quite level they usually diverge from each other, and do so the more in proportion to the lowness of angle of dip and the inequalities of the ground. This may be illustrated by a diagram such as that given in Fig. 92, which represents a portion of the edge of a table-land, deeply trenched by two valleys that discharge their waters into the plain below (P). The arrows point out that the strata dip due N. at 5°. On the level plain, the outcrop and the strike (s s) of the beds are coincident and run due E. and W. But as the surface rises towards the high ground and the deep valleys, the outcrop (o o) is observed to depart more and more from the strike till in some places the two lines are at right angles ; yet, as the dip remains the same, the strike is likewise unchanged, the sinuosities of the outcrop being entirely due to the irregularities of the surface of the ground.

Curvature.—It requires no long observation to perceive that in being tilted from their original more or less level positions, stratified rocks have been thrown into curves. Suppose, for instance, that in walking along a mile of coast-line, where all the successive strata of a thick series are exposed to view, we should

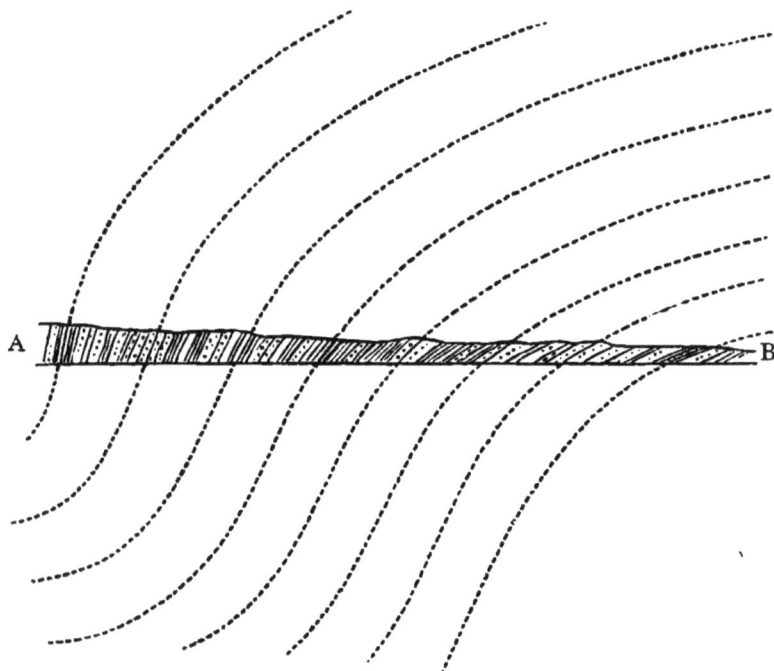

FIG. 93.—Inclined strata shown to be parts of curves.

observe such a section as is drawn in Fig. 93. Beginning at A, we find the beds tilted up at angles of 70° which gradually lessen, till at B they have sunk to 15°. As there is no break in the series, it is evident that the lines of bedding must be prolonged downward, and must once have been continued upward in some such way as is expressed by the dotted lines. The visible portion which is here shaded must thus form part of a great curvature of the rocks.

But the actual curvature may often be seen on coast-cliffs, ravines, or hillsides. In Fig. 94, for example, a simple arch is

shown from the Berwickshire coast, wherein hard beds of greywacke and shale have been folded.   Again, in Fig. 95, the reverse
structure is exhibited, beds of grit and slate being there curved
into a trough.   Where rocks dip away from a central line of axis
the structure is known as an *Anticline* ; where, on the other hand,
they dip towards an axis it is called a *Syncline*.   In Figs. 94 and
95 these two structures are presented on so small a scale as to be
visible in a single section.   More usually, however, it is only by

Fig. 94.—Curved strata (anticlinal fold), near St. Abb's Head.

observing the upturned edges of strata that anticlines and synclines
can be detected.   The dark part of Fig. 96 represents all that can
be actually seen ; but the angles and direction of dip leave no
doubt that if we could restore the amount of rock which has here
been worn away from the surface of the land, the present truncated
ends of the strata would be prolonged upward in some such way
as is indicated by the dotted lines.   By observations of this
truncation of strata some of the most interesting and important
evidence is obtained of the enormous extent to which the land has
been reduced by the removal of solid material from its surface.

Plication, Shearing.—From such simple curvatures as those depicted in the foregoing diagrams, we may advance to more complex foldings, wherein the solid strata have been doubled up and crumpled together, as if they had been mere layers of carpet. So far is this plication sometimes carried, that the lowest rocks are brought up and thrown over the highest, the more yielding materials being squeezed into the most intricate frillings and puckerings. It is in mountainous regions, where the crust of the earth has been subjected to the most intense corrugation, that

FIG. 95.—Curved strata (synclinal fold), near Banff.

these structures are best seen. We can form some idea of the gigantic energy of the earth-movements that produced them, when we see a whole mountain-range made up of solid limestones or sandstones which have been bent, twisted, crumpled, and inverted, as we might crush up sheets of paper (Fig. 97).

So enormous has been the compression produced by important movements of the earth's crust, that the solid rocks have actually been squeezed out of shape or have undergone a process of shearing. The amount of distortion may sometimes be measured by the extent to which shells or other organic remains are pulled out in the direction of movement. In Fig. 98 the proper shape of a

trilobite (*Angelina Sedgwickii*) is given, and alongside of it is a
view of the same organism which has been elongated by the distortion of the mass of rock in which it lies.   Further results of

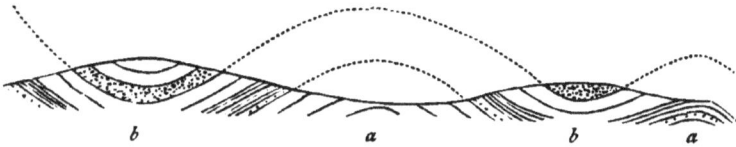

FIG. 96.—Anticlines (*a a*) and Synclines (*b b*).

shearing will be immediately referred to in connection with the
cleavage and metamorphism of rocks.

Cleavage.—One of the most important structures developed by
the great compression to which the rocks of the earth's crust have
been exposed is known as *Cleavage.*   The minute particles of
rocks, being usually of irregular shapes, have been compelled to
arrange themselves with their long axes perpendicular to the
direction of pressure during the interstitial movements consequent

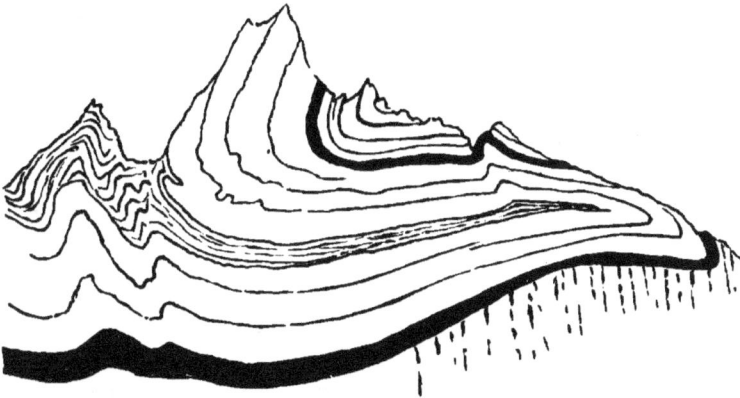

FIG. 97.—Section of folded and crumpled strata forming the Grosse Windgälle (10,482
feet), Canton Uri, Switzerland, showing crumpled and inverted strata (after Heim).

upon intense subterranean compression.   Hence, a fissile tendency
has been imparted to a rock, which will now split into leaves
along the planes of rearrangement of the particles.   This superinduced tendency to split into parallel leaves, irrespective of what
may have been the original structure of the rock, constitutes
cleavage.   It is well developed in ordinary roofing-slate   Though

O

the leaves or plates into which a slate splits resemble those in a
shale, they have no necessary relation to the layers of deposition
but may cross them at any angle.   In Fig. 99, for instance, the
original bedding is quite distinct and shows that the strata have
been folded by a force acting from the right and left of the section ;
the parallel highly inclined lines traversing the folds of the bedding
represent the planes of cleavage.   Where the material is of exceedingly fine grain, such as fine consolidated mud, the original
bedding may be entirely effaced by the cleavage, and the rock
will only split along the cleavage-planes.   Indeed, the finer the
grain of a rock, the more perfect may be its cleavage, so that
where alternations of coarser and finer sediment have been sub-

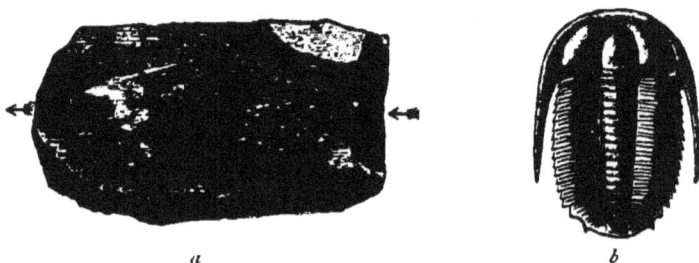

FIG. 98.—Distortion of fossils by the shearing of rocks ; (a) a Trilobite (*Angelina Sedgwickii*) distorted by shearing, the direction of movement indicated by the arrows ; (b)
the same fossil in its natural form.

jected to the same amount of compression, cleavage may be perfect
in the one and rudely developed in the other, as is indicated
in Fig. 99.

Cleavage may be regarded as one of the first stages in the
mechanical deformation of a rock, and the production of schistose
metamorphism (p. 167).   Besides being compressed and having
its component particles rearranged in definite planes, the rock
may likewise reveal under the microscope that new minerals, such
for example as crystallites or minute flakes of some mica, have been
developed out of the general matrix, as may be seen in common
roofing-slate.   By increasing stages of crystallisation we trace
gradations into phyllites and mica-schists.

Dislocation.—Another important structure produced in rocks
after their formation is *Dislocation.*   Not only have they been
folded by the great movements to which the crust of the earth has
been subjected, but the strain upon them has often been so great
that they have snapped across.   Such ruptures of continuity pre-

sent an infinite variety in the position of the rocks on the two sides.
Sometimes a mere fissure has been caused, the rocks being simply
cracked across, but remaining otherwise unchanged in their relative

FIG. 99.—Curved and cleaved rocks. Coast of Wigtonshire. The fine parallel oblique
lines indicate the cleavage, which is finer in the dark shales and coarser in the thicker
sandy beds.

situations.   But, in the great majority of instances, one or both of
the walls of a fissure have moved, producing what is termed a
Fault. Where the displacement has been small, a fault may
appear as if the strata had been sharply sliced through, shifted,

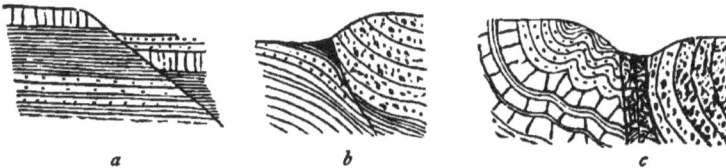

FIG. 100.—Examples of normal Faults.

and firmly pressed together again (*a* in Fig. 100).   Usually, however, they have not only been cut, but bent or crushed on one or
both sides (*b*); while not infrequently the line of fracture is represented by a band of broken and crushed material (*Fault-rock*, *c*).
The fracture is seldom quite vertical; almost always it is inclined

at angles varying up to 70° or more from the vertical.   In by far the largest number of faults, the inclination of the plane of the fissure, or what is called the *Hade* of the fault, is away from the side which has risen or toward that which has sunk.   In the examples given in Fig. 100, *a*, *b*, this relation is expressed ; but in nature it often happens that the beds on two sides of a fault are entirely different (Fig. 100, *c*), and consequently that the side of upthrow or downthrow cannot be determined by the identification of the two severed positions of the same bed.   But if the hade of the fault can be seen, we may usually be confident that the strata on the upper or hanging side belong to a higher part of the series than those on the lower side.   Faults that follow this rule (*normal* faults) are by far the most frequent.   They occur universally, and are probably for the most part caused by subsidence in the earth s crust.   In adjusting themselves to the new position into which a downward movement brings them, rocks must often be subject to such strains that their limit of elasticity is reached, and they break across, one portion settling down farther than the part next to it. In a normal fault, the same bed can never be cut twice by a vertical line.

In mountainous districts, however, and generally where the rocks of the earth's crust have been disrupted and pushed over

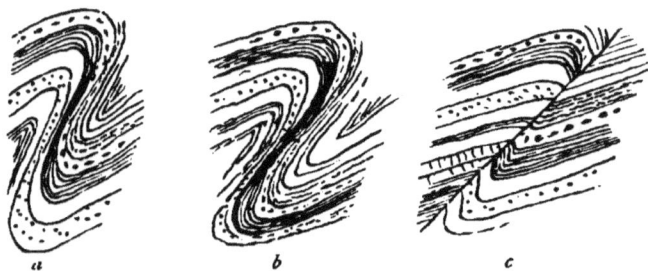

a   *b*   *c*

FIG. 101.—Sections to show the relations of Plications (*a, b*) to reversed Faults (*c*).

each other, what are termed *reversed* faults occur.   In these, the hade slopes in the direction of upthrow, and a vertical line may cut the same beds twice on opposite sides of the fracture (Fig. 101). Such faults may be observed more particularly where strata have been much folded.   A fold may be seen to have snapped asunder, the whole being pushed over, and the upper side being driven forward over the lower.

The amount of vertical displacement between the two fractured ends of a bed is called the *Throw* of a fault. In Fig. 102, for example, where bed *a* has been shifted from *b* to *d*, a vertical line dropped from the end of the bed at *b* to the level of the corresponding part of the bed at *e* will give the amount of the subsidence of *d*, which is the throw. Faults may be seen with a throw of less than an inch—mere local cracks and trifling subsidences in a mass of rock ; in others the throw may be many thousand feet. Large faults often bring rocks of entirely different characters together, as, for instance, shales against limestones or sandstones, or sedimentary against eruptive rocks. Consequently they are not infrequently marked at the surface by the difference between the form of ground characteristic of the two kinds of rock. One side, perhaps, rises into a hilly or undulating region, while the

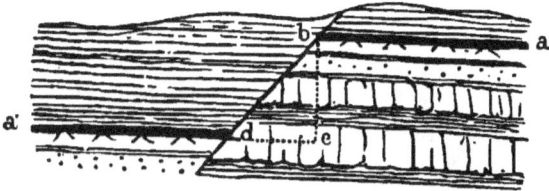

FIG. 102.—Throw of a Fault.

other side may be a plain. Comparatively seldom does a fault make itself visible as a line of ravine or valley. On the contrary, most faults cut across valleys or only coincide with them here and there. They run in straight or wavy lines which, where the amount of displacement is great, may be traced for many miles. The Scottish Highlands, for example, are bounded along their southern margin by a great fault which places a thick series of sandstones and conglomerates on end against the flanks of the mountains. This fault may be traced across the island from sea to sea—a distance of fully 120 miles, and by bringing two distinct kinds of rocks next each other along a nearly straight line it has given rise to the boundary between Highland and Lowland scenery which, in some places, is so singularly abrupt.

In regions of the most intense terrestrial disturbance, tracts of rock many square miles in area and hundreds or thousands of feet in thickness, have been torn away and pushed upward and forward until they have come to rest on rocks originally much higher in geological position. Such displaced cakes or slices of the earth's crust sometimes rest upon an almost horizontal or gently inclined platform of undisturbed materials. Vertical or

contorted strata are thus placed above others which may be flat
or but little inclined.   The plane of separation between the moved
and unmoved masses is really a dislocation, but to distinguish it
from faults, which are generally placed at steep angles, it is called
a *Thrust-plane.*   Structures of this kind on a colossal scale are
traceable for about 100 miles in the north-west of Scotland.

Metamorphism.—The last structure, which will be mentioned in
this chapter as having been superinduced upon rocks, is connected
with the movements to which plication, cleavage, and reversed
faults are due.   So enormous has been the energy with which
these movements have been carried on, that not only have the
rocks been crumpled, ruptured, and pushed over each other, but

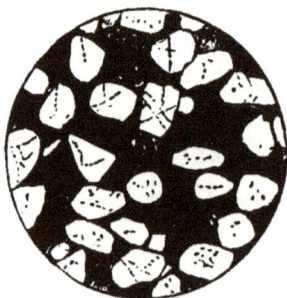

Fig. 103.—Ordinary unaltered red
sandstone, Keeshorn, Ross-shire
(magnified).

Fig. 104.—Sheared red sandstone
forming now a micaceous schist,
Keeshorn, Ross-shire (magnified).

they have undergone such intense shearing that, as was pointed
out at p. 168, their original structure has been partially or wholly
effaced.   They have been so crushed that their component particles have been reduced, as it were, to powder, and have assumed
new crystalline arrangements along the shearing-planes or surfaces
of movement.   A sandstone, for example, which in its ordinary
state shows, when magnified, such a structure as is represented in
Fig. 103, when it has come within the influence of this crushing
process has its grains of quartz, felspar, and other materials
flattened and squeezed against each other in one general direction
as in cleavage, while out of the crushed debris a good deal of new
mica has been developed.   This change may be intensified until
the component grains are hardly recognisable, and the proportion
of new mica has so increased that the rock has become a mica-
schist.   Other new minerals, such as garnet, may likewise make

their appearance, until the rock assumes an entirely crystalline structure. Such an alteration of the internal structure of a rock is known as *Metamorphism*. Where the change arises from mechanical movements combined with chemical rearrangement, it usually affects a wide district, and is then spoken of as *regional metamorphism*.

There are wide regions of the earth's surface where schists of various kinds form the prevailing rock. Whether they have all been produced by the shearing and alteration of previously-formed rocks has not yet been determined. But that a large number of schists are truly altered or *metamorphosed* rocks admits of no doubt. Sandstones, shales, limestones, quartzites, diorites, syenites, granites, in short, any old form of rock that has come within the crushing and shearing movements here referred to has been converted into schist. The gradation between the unaltered and the metamorphic condition can often be clearly traced. Granite, by crushing, passes into gneiss, diorite into hornblende-schist, sandstone into quartz-schist or mica-schist, and so on. Even where it is no longer possible to tell what the original nature of the metamorphosed material may have been, there is usually abundant evidence that the rock has undergone great compression (see pp. 167-170).

Summary.—In this Lesson attention has been directed to new structures produced in sedimentary rocks after their formation. Beginning with the simplest and most universal of these, we find that sediments have been consolidated into stone, partly by pressure, and partly by some kind of cement, such as silica or carbonate of lime. In the process of consolidation and contraction, they have been traversed by systems of joints, or have had these subsequently produced by the torsion accompanying movements of the crust. Though at first nearly flat, they have by these movements been thrown into various inclined positions, and more especially into undulating folds, or more complicated plication and puckering. So great has been the compression under which they have been moved, that a cleavage has been developed in them. They have also been everywhere more or less fractured, the dislocations being due either to their gradual subsidence or to excessive plication. Their most complete alteration is seen in metamorphism, where, under the influence of intense shearing, their original structure has been more or less completely effaced, and a new crystalline rearrangement has been developed in them, converting them into schists.

CHAPTER XIV

ERUPTIVE ROCKS AND MINERAL VEINS IN THE ARCHITECTURE
OF THE EARTH'S CRUST

NOT only have sedimentary formations since their deposition been hardened, plicated, fractured, and sometimes even turned into crystalline schists, but into the rents opened in them new masses of mineral matter have been introduced which, in many regions, have entirely changed the structure of the crust below and the appearance of the surface above. Broadly speaking, there are two ways in which these new masses have been wedged into their places. First of all, eruptive material in a molten, or at least in a viscous or plastic condition, has been thrust upward into the cool and consolidated crust of the earth ; and in the next place, various ores and minerals have been deposited from solution in cracks and fissures, which they have entirely filled up. To each of these two kinds of later rocks attention will be given in this chapter.

Eruptive Rocks.

The rise of eruptive matter thrust upwards from lower depths within the planet is one of the causes by which the structure of the crust has been most seriously affected. In Chapter IX reference was made to some of the features connected with the protrusion of molten rocks in the production of volcanoes, and more particularly to those subterranean changes which, when all the outer and ordinary tokens of a volcano have been swept away, remain as evidence of former volcanic action, even in districts where every symptom of volcanic activity has long vanished. We have now to inquire, generally, in what forms eruptive matter has been built into the earth's crust, and what changes it has produced there, apart from those superficial manifestations which are the visible signs of volcanic action.

When a mass of lava is forced upwards from the heated interior of the earth towards the surface, the form which it finally takes and in which it cools and solidifies must depend upon the shape of the rent or cavity into which it has been thrust. We may compare such a mass to a quantity of melted iron escaping from a blast-furnace. The shape taken by the iron will, of course, be fixed by that of the mould into which it is allowed to run. The crust of the earth, as was pointed out in the previous chapter, has undergone extensive movements, whereby its rocks have been crumpled and broken. It consequently presents in different parts very various degrees of resistance to any force acting upon it from below. The eruptive materials have sometimes risen in the fissures, sometimes have forced their way between the beds and joints of the strata. According to the form of the mould in which they have solidified, we may classify the eruptive rocks of the crust into (1) bosses ; (2) sheets ; (3) veins and dykes ; and (4) necks.

Bosses.—These are circular, elliptical, or irregularly shaped masses of rock which, while still in a liquid or viscous state, have been ejected into irregular rents of the earth's crust and have solidified there. They consist of various crystalline rocks, more especially granite, syenite, quartz-porphyry, diorite, diabase, and basalt-rocks, and vary in width from a few yards to several miles. Being generally harder than the surrounding rocks, they commonly stand up as prominent knobs, hills, or ridges. Their presence at the surface, however, is due, not to their original protrusion there, as in a volcanic cone, but to the removal of the overlying part of the original crust under which they cooled and consolidated. Every boss is thus a witness of the extensive wearing away of the surface of the land (Fig. 105).

In some large bosses, there may have been a complex system of fissures in which the eruptive material rose. Forced upwards into these, the molten rock would no doubt envelope separated masses of the crust, and might bear them along with it in its ascent. We may even conceive it to have melted down such enveloped masses. Pushing the rocks aside and thrusting itself into every available crack in them, the eruptive mass would work its way across the crust. Where it succeeded in opening a passage to the surface, ordinary volcanic phenomena would take place, such as disruption of the ground, ejection of stones and ashes, and outflow of lava. But, no doubt, in a vast number of cases, no such communication was ever effected. The eruptive material paused in its upward passage and consolidated below ground.

No rock affords more interesting bosses than granite. Two features are especially well displayed by it—the marginal veins or dykes and the surrounding ring of metamorphism produced in the rocks through which granite has risen. Granite has invaded many different kinds of rocks, and has effected various kinds of change in them. Round its margin, large numbers of veins or dykes of granite or quartz-porphyry often strike out from it into

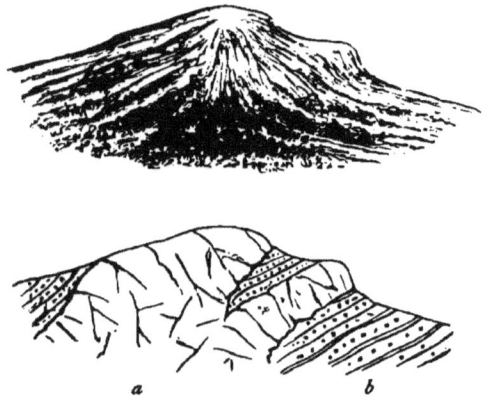

FIG. 105.—Outline and section of a Boss (*a*) traversing stratified rocks (*b b*).

the surrounding rocks. There can be no doubt that these are portions of the granite material, squeezed into cracks that opened in the crust around it during its ascent. More important is the change that can be observed to have taken place in the rocks immediately surrounding the boss. The granite at the time of its protrusion was probably in a molten or pasty condition and impregnated with hot water or steam and vapours. For a distance varying from a few feet up to two or three miles, according chiefly to the size of the granite mass, the rocks next to it have undergone alteration, the nature and amount of which appear to have been in great measure dependent on the chemical and mineralogical composition of the rocks themselves. This metamorphism consists partly in mere induration, but still more in the development of new minerals, or a new crystalline structure, even out of non-crystalline sedimentary materials. The very same rock, which is elsewhere a dark limestone full of shells, corals, or other organic remains, becomes a white crystalline marble next the granite, with no trace of any organisms, and so unlike its usual condition that no one would readily believe it to be the same rock.   Again, a dark shaly sandstone or greywacke traced towards the granite begins to show an increasing amount of mica.   New minerals likewise make their appearance, particularly garnets, until the rock entirely loses its sedimentary structure and becomes a garnetiferous mica-schist.   Shales and slates, as they approach the granite, likewise present a remarkable development of fine mica-plates, and pass into argillaceous schists or phyllites, with crystals of

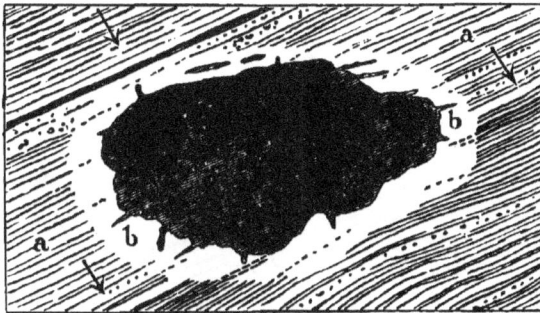

FIG. 106.—Ground-plan of Granite-boss with ring of Contact-Metamorphism ; (a) sandstones, shales, etc., dipping at high angles in the direction of the arrows ; (b) zone or ring within which these rocks are metamorphosed ; (c) granite sending out veins into b.

chiastolite or other minerals developed in them.   The alteration of rocks round eruptive masses is called *Contact-metamorphism.*

What the cause may be of this remarkable alteration has not yet been satisfactorily made out.   In some bosses, the mere heat of the eruptive material was probably sufficient to produce change. There must often have been also a copious discharge of hot vapours and water, which would powerfully affect the adjacent rocks.   Silica and other substances might then be introduced, leading to induration and new chemical rearrangements of the constituents.   The protrusion of enormous masses of granite may also have given rise to mechanical movements in the earth's crust, like those which have produced the shearing and schistose structure, seen in regional metamorphism (p. 199).

Sheets.—Sometimes the easiest passage for the erupted material from below has lain between the bedding of strata.   The molten rock, after ascending some fissure or pipe, has found its farther

progress barred, and has escaped by forcing up the overlying beds
and thrusting itself in below them.   On cooling and consolidating,
it appears as a sheet or bed intercalated between older rocks.
This structure is represented in Fig. 107.   Any one examining
such a section on the ground, might naturally regard the sheet *s*
as a bed of lava erupted at the surface after the formation of the
strata *a* and before that of *b*.   But various features characteristic
of intrusive or subsequently injected sheets enable us to distinguish
them from those which have been poured out during the deposition
of the strata among which they lie.   For example, intrusive sheets
break across the strata (as at *d* in Fig. 107) and send veins into
them.   They are commonly most close-grained along their edges,
where they have been most rapidly chilled by contact with the cool

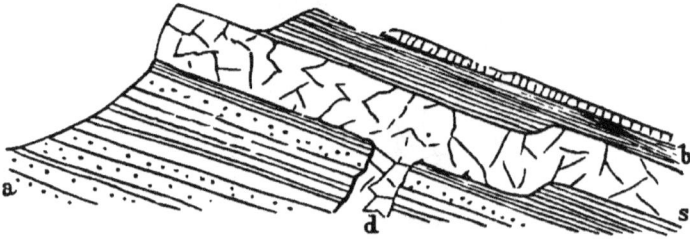

FIG. 107.—Intrusive Sheet.

rocks ; while, on the contrary, true lava-streams erupted at the
surface are generally most slaggy and scoriform on their upper and
under surfaces.   Lastly, they have generally hardened and otherwise altered the rocks above and below them, sometimes baking
or even fusing them.   Where these characters are present, we
may confidently infer that, though a sheet of crystalline rock, so
far as visible at the surface, may seem to be regularly interstratified
between sedimentary beds, as if it had been contemporaneously
poured forth among them, it has nevertheless been thrust in
between them and may be of much younger date.

On the other hand, a truly contemporaneous sheet or group of
sheets marking the actual outpouring of lava-streams at the surface,
during the deposition of the strata among which they now lie,
may be recognised by equally distinctive characters.   Thus they
do not break across nor send veins into the overlying or underlying beds, while their upper and under surfaces are usually their
most open cellular portions, though they are often more or less
vesicular or amygdaloidal throughout.   In Fig. 108 the beds

marked 1, 2, 3, and 4 are sheets of different lavas interstratified contemporaneously in the series of sandstones, shales, limestones, and other strata among which they lie. Fragments of them are not infrequently to be detected in the overlying strata, which are thus shown to be of later origin, and bands of tuff are commonly

FIG. 108.—Interstratified or contemporaneous Sheets.

associated with them, just as showers of ashes accompany the lava-streams of living volcanoes. As an illustration of the way in which the evidence of ancient volcanic action may be gathered, the section given in Fig. 109 may be taken supplementary to the data given already in Chapter IX. At the bottom of the section we stand on the slaggy upper surface of a lava-stream (1) which was poured out under water, for directly above it comes a seam of dark shale (2) representing fine mud that was deposited from suspension in water. That volcanic explosions still continued after the outflow of the lava, is indicated by the abundant bits of slaggy lava and volcanic detritus scattered through the shale, and that the scene of these operations was the sea-floor is conclusively proved by the numerous shells, crinoids, and other marine remains that lie in some bands of the shale. The bottom must then have been muddy and not so well

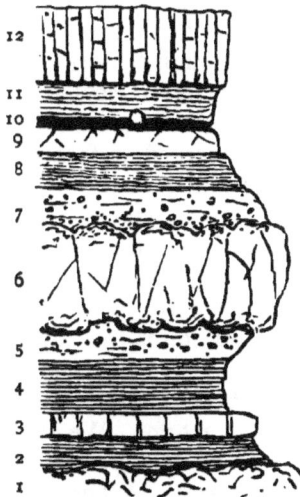

FIG. 109.—Section to illustrate evidence of contemporaneous volcanic action.

suited as it afterwards became for the support of life. Above the shale we find two feet of limestone (3) which is entirely made up of fragments of marine organisms. These creatures, when the water had cleared, continued to flourish abundantly until their congregated remains formed a bed of solid limestone. But from some change in the geography of the region, currents bearing dark mud once more invaded this part of the sea and threw down the material that now forms the band of shale (4). The absence of organic remains in this band probably shows the inroad of mud to have destroyed the life that had previously been prolific. When this condition of things had been brought about, renewed volcanic explosions took place in the neighbourhood. First came showers of dust, ashes, and stones, which fell over the sea, and are now represented by the band of tuff (5). Then followed the outpouring of a stream of lava (6), with its characteristic cellular structure. But this did not quite exhaust the vigour of the volcano, for the band of tuff (7) points to successive showers of dust and stones. When the explosion ceased, the deposition of dark mud, which had been interrupted by the volcanic episode, was resumed, and the band of shale (8) was laid down. From the fragments of ferns and other plants in this shale it is clear that land was not far off. The sea had evidently been gradually shallowing by the infilling of sediment and volcanic materials, and at last, on the muddy flat, represented by the layer of fire-clay (9), marshy vegetation sprang up into a thick jungle like the mangrove-swamps of tropical shores at the present day. But after growing long enough to form the bed of matted vegetable matter now represented by the coal-seam (10), the verdant jungle was invaded by the sea, and sank under the muddy water that threw down upon its submerged surface the grey shale (11). In this shale we detect interesting traces of the renewal of volcanic activity, more especially in occasional large blocks of lava, which have evidently been ejected by some volcanic explosion, as in the example already cited on p. 102 (Fig. 38). A more vigorous volcanic outburst poured out the stream of columnar lava (12) which buried the whole and forms the top of the section.

Veins and **Dykes.**—These have already been referred to in Chapter IX as part of the evidence for volcanic action. We have here to consider how they occur in connection with the protrusion of eruptive material within the crust of the earth. Where the material so erupted has solidified in a vertical or nearly vertical fissure so as to form a wall-like mass, it is called a *dyke* (Fig. 43 and *d* in Fig. 107). Otherwise the portions of erupted rock that have consolidated in irregular rents are known as *veins.*

Veins are of common occurrence round bosses of granite, where they can be traced into the parent mass from which they have proceeded (Fig. 106). They may likewise be observed in connection with intrusive sheets and bosses of basalt and diorite, from which they ramify outwards into the surrounding rocks. Their occurrence there is one of the proofs of the intrusive character and subsequent date of such sheets (p. 204).

Dykes vary from less than a foot to 70 feet or upward in breadth, and run in nearly straight courses sometimes for many miles. They consist most usually of diabase, andesite, basalt, or allied rock. Sometimes they have risen along lines of fault; but in hundreds of instances in Great Britain, they do not appear to be connected with any faults, but actually cross some of the largest

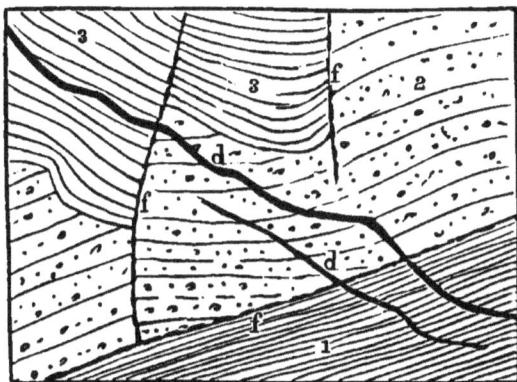

Fig. 110.—Map of Dykes near Muirkirk, Ayrshire. 1. Silurian rocks. 2. Lower Old Red Sandstone. 3. Carboniferous rocks. *f, f, f.* Faults. *d, d.* Dykes.

faults in the country without being deflected. The remarkable way in which dykes have risen through a complicated series of rocks and faults and have preserved their courses is exemplified in Fig. 110.

Like intrusive sheets, but in a less degree, dykes harden or otherwise alter the rocks on either side of them; they likewise present a similar closeness of grain along their margins where the molten rock was most rapidly chilled by coming in contact with the cold walls of the fissure. Sometimes, indeed, their sides are coated with a thin crust of black glass, as if they had been painted with tar. This glass represents the effect of rapid cooling (see Basalt-glass, p. 165). No doubt the whole rock of the dyke, at the time when it rose from below and filled up the space between the two walls of its opened fissure, was a molten glass. The portions that were at once chilled by contact with the walls adhered as a layer of glass. But inside this layer, the molten rock had more

time to cool.   In cooling, its various minerals crystallised and the
present crystalline structure was developed.   But even yet, though
most of the rock is formed of crystalline· minerals, portions of the
original glass may not infrequently be detected between them
when thin sections are placed under the microscope (p. 144).

Necks.—These are the filled-up pipes or funnels of former
volcanic vents.   Their connection with volcanic action has been
already alluded to on p. 106.   They are circular or elliptical in
ground-plan, and vary in diameter from a few yards up to a mile
or more (see Figs. 40, 41, 42).   They consist of some form of
lava (quartz-porphyry, basalt, diorite, etc.) or of the fragmentary
materials that, after being ejected from the volcanic chimney, fell
back into it and consolidated there.   They occur more particu-

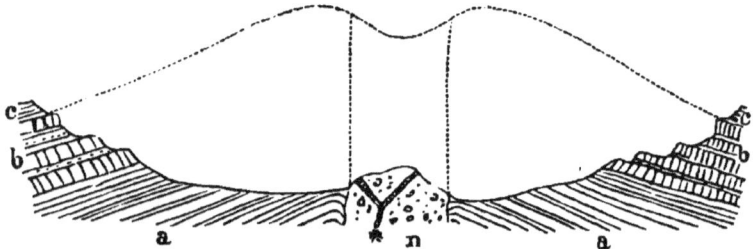

FIG. 111.—Section of a volcanic neck.   The dotted lines suggest the original form
of the volcano.

larly in districts where beds of lava and tuff are interstratified with
other rocks.   The necks represent the vents from which these
volcanic materials were ejected.   In Fig. 111, for example, the
beds of lava and tuff (*b b*) interstratified between the strata *a a*
and *c c* have been folded into an anticline.   In the centre of the
arch rises the neck (*n*) which has probably been the chimney that
supplied these volcanic sheets, and which has been filled up with
coarse tuff, and traversed with dykes and veins of basalt (*).   The
dotted lines, suggestive of the outline of the original volcano, may
serve to indicate the connection between the neck and its volcanic
sheets, and also the effects of denudation.

Necks are frequently traversed by dykes (* in Fig. 111), as we
know also to be the case with the craters of modern volcanoes.
The rocks surrounding a neck are sometimes bent down round it,
as if they had been dragged down by the subsidence of the
material filling up the vent ; they are also frequently much
hardened and baked.   When we reflect upon the great heat of

molten lava and of the escaping gases and vapours, we may well expect the walls of a volcanic vent to bear witness to the effects of this heat. Sandstones, for instance, have been indurated into quartzite, and shales have been baked into a hardened clay or porcelain-like substance.

Mineral Veins.

Into the fissures opened in the earth's crust there have been introduced various simple minerals and ores which, solidifying there, have taken the form of *Mineral Veins.* These materials are to be distinguished from the eruptive veins and dykes above

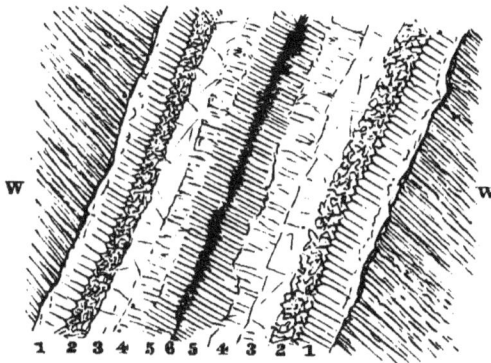

FIG. 112.—Section of a Mineral vein.

described. A true mineral vein consists of one or more minerals filling up a fissure which may be vertical, but is usually more or less inclined, and may vary in width from less than an inch up to 150 feet or more. The commonest minerals (or *veinstones*) found in these veins are quartz, calcite, barytes, and fluor-spar. The metalliferous portions (or *ores*) are sometimes native metals (gold and copper, for example), but are more usually metallic oxides, silicates, carbonates, sulphides, chlorides, or other combinations. These materials are commonly arranged in parallel layers, and it may often be noticed that they have been deposited in duplicate on each side of a vein. In Fig. 112, for instance, we see that each wall (w w) is coated with a band of quartz (1, 1), followed successively by one of blende (sulphide of zinc, 2, 2), galena (sulphide of lead, 3, 3), barytes (4, 4) and quartz (5, 5). The central portion of the vein (6) is sometimes empty or may be filled up

P

with some veinstone or ore. Remarkable variations in breadth characterise most mineral veins. Sometimes the two walls come together and thereafter retire from each other far enough to allow a thick mass of mineral matter to have been deposited between them. Great differences may also be observed in the breadth of the several bands composing a vein. One of these bands may swell out so as to occupy the whole breadth of the vein, and then rapidly dwindle down. The ores are more especially liable to such variations. A solid mass of ore may be found many feet in breadth and of great value ; but when followed along the course of the vein, may die away into mere strings or threads through the veinstones.

The duplication of the layers in mineral veins shows that the deposition proceeded from the walls inwards to the centre. In the diagram (Fig. 112) it is evident that the walls were first coated with quartz. The next substance introduced into the vein was sulphide of zinc, a layer of which was deposited on the quartz. Then came sulphide of lead, and lastly, quartz again. The way in which the quartz-crystals project from the two sides shows that the space between them was free, and, as above stated, it has sometimes remained unfilled up.

There appears to be now no reason to doubt that the substances deposited in mineral veins were mainly introduced dissolved in water. Not improbably heated waters rose in the fissures, and as they cooled in their ascent, they coated the walls with the minerals which they held in solution. These minerals may have been abstracted from the surrounding rocks by the permeating water ; or they may have been carried up from some deeper source within the crust. During the process of infilling, or after it was completed, a fissure has sometimes reopened, and a new deposition of veinstones or ores has taken place. Now and then, too, land-shells and pebbles are found far down in mineral veins, showing that during the time when the layers of mineral matter were being deposited, the fissures sometimes communicated with the surface.

Summary.—In this chapter it has been shown that, in many cases, the rents in the earth's crust have been filled up with mineral matter introduced into them, either (i) in the molten state, or (ii) in solution in water. (i) The forms assumed by the masses of eruptive rock injected into the crust of the earth have depended upon the shape of the fissures into which the melted matter has been poured, as the form of a cast-iron bar is regulated by that of the mould into which the melted metal is allowed to run. Taking this principle of arrangement, we find that eruptive rocks may be grouped into (1) Bosses, or irregularly-shaped masses, which have risen through fissures or orifices, and now, owing to the removal of the rock under which they solidified, form hills or ridges. The eruptive material sends out veins into the surrounding rocks which are sometimes considerably altered, forming a metamorphic ring round the eruptive rock. (2) Sheets or masses which have been thrust between the bedding-planes of strata. These resemble truly interstratified beds, but the difference between the two kinds of structure can be readily appreciated. Interstratified beds mark the occurrence of volcanic phenomena at the surface during the time of the formation of the strata among which they occur. Intrusive sheets, on the other hand, are always subsequent in date to the rocks between which they lie. (3) Veins and dykes, consisting of eruptive rock which has been thrust between the walls of irregular rents or straight fissures. (4) Necks, or the filled-up pipes of former volcanic vents. (ii) Mineral veins are masses of mineral matter which has been deposited, probably from aqueous solution, between the walls of fissures in the earth's crust, and consists of bands of veinstones (quartz, calcite, barytes, etc.) and ores (native metals, or oxides, sulphides, etc., of metals).

CHAPTER XV

HOW FOSSILS HAVE BEEN ENTOMBED AND PRESERVED, AND HOW THEY ARE USED IN INVESTIGATING THE STRUCTURE OF THE EARTH'S CRUST, AND IN STUDYING GEOLOGICAL HISTORY.

IN an earlier part of this volume (Chapter VIII) attention was called to the various circumstances under which the remains of plants and animals may be entombed and preserved in sedimentary accumulations. When these remains have thus been buried they are known as *Fossils*.

Nature and use of Fossils.—The word "fossil," meaning literally "dug up," was originally applied to all kinds of mineral substances taken out of the earth ; but it is now exclusively used for the remains or traces of plants and animals imbedded by natural causes in any kind of rock, whether loose and incoherent, like blown sand, or solid, like the most compact limestone. It includes not only the actual remains of the organisms. The empty mould of a shell which has decayed out of the stone that once enveloped it, or the cast of the shell which has been entirely replaced by inorganic sand, mud, calcite, silica, etc., are fossils. The very impressions left by organisms, such as the burrow or trail of a worm in hardened mud, and the footprints of birds and quadrupeds upon what is now sandstone, are undoubted fossils. In short, under this general term is included whatever bears traces of the form, structure, or presence of organisms preserved in the sedimentary accumulations of the surface, or in the rocks underneath.

In geological history fossils are of fundamental importance. They enable us to investigate conditions of geography, of climate, and of life in ancient times, when these conditions were very different from those which now prevail on the earth's surface. They likewise furnish the ground on which the several epochs of geological history can be determined, and on which the stages of that history in one country can be compared with those in another. So valuable and varied is the evidence supplied by fossils to the geologist, that he regards them as among the most precious documents accessible to him for unravelling the past history of the earth. Some knowledge of the structure and classification of plants and animals is essential for an intelligent appreciation of the use of fossils in geological inquiry. To aid the learner, a synopsis of the Vegetable and Animal Kingdoms is given in the Appendix, with especial reference to the fossil forms ; but it must be understood that for adequate information on this subject recourse should be had to text-books of Botany and Zoology.

Conditions for the preservation of Organic Remains.—It is obvious that all kinds of plants and animals have not the same chances of being preserved as fossils. In the first place, only those, as a rule, are likely to become fossils whose remains can be kept from decay and dissolution by being entombed in some kind of deposit. Hence land-animals and plants have, on the whole, less chance of preservation than those living in the sea, because deposits capable of receiving and securing their remains are exceptional on land, but are generally distributed over the floor of the sea. We should expect, therefore, that among the records of past time, traces of marine should largely preponderate over traces of terrestrial life. Now this is everywhere the case. We know relatively little of the assemblages of plants and animals which in successive epochs have lived upon the dry land, but we have a comparatively large amount of information regarding those which have tenanted the sea. For this reason, marine fossils are more valuable than terrestrial, in comparing the records of the successive epochs of geological history in different parts of the globe.

In the second place, from their own chemical composition and structure, plants and animals present extraordinary differences in their aptitude for preservation as fossils. Where they possess no hard parts, and are liable to speedy decay, we can hardly expect that they should leave behind them any enduring relic of their existence. Hence a large proportion, both of the vegetable and animal kingdoms, may at once be excluded as inherently unlikely to occur in the fossil condition. Of course, under exceptional circumstances, traces of almost any organism may be preserved, and therefore we should probably not be justified in saying that by no chance might some recognisable vestige of it be found fossil. Nothing seems more perishable than the tiny gnats and other forms of insect life that fill the air on a summer evening. Yet many of these short-lived flies have been sealed up within the resin of trees (amber), and their structure has been admirably preserved. Such exceptional instances, however, only bring out more distinctly how large a proportion of the living tribes of the land must utterly perish, and leave no recognisable record of their ever having existed.

But, where there are hard parts in an organism, and especially where, from their chemical composition, they can for some time resist decay, they may, under favourable conditions, be buried in sedimentary deposits, and may remain for indefinite ages locked up there. It is obvious, therefore, that animals possessing hard parts are much the most likely to leave permanent relics of their presence, and ought to occur most frequently as fossils. It is these animals whose remains are preserved in peat-mosses, river-gravels, lake-marls, and on the sea-floor at the present time. Yet, if we were to judge of the extent of the whole existing animal kingdom solely from the fragmentary remains so preserved, what an utterly inadequate conception of it we should form! So, too, if we estimate the variety of the living creatures of past time merely from the evidence of the fossils that have chanced to be preserved among the rocks, we shall probably arrive at quite as erroneous a conclusion. There can be no doubt that from the earliest time only an insignificant fraction of the varied life of each period has been preserved in the fossil state, as is unquestionably the case at the present day.

Durable parts of Plants.—The essential parts of the solid framework of plants consist of the substances known as cellulose and vasculose, which, when kept in dry air, or when waterlogged and buried in stiff mud, may remain undecomposed for long periods. The timber beams in the roofs and floors of old buildings are evidence that, under favourable conditions, wood may last for many centuries. Some plants eliminate carbonate of lime from solution in water, and form with it a solid substance which requires no further treatment to enable it to endure for an indefinite period, when screened from the action of water. Still more durable are the remains of those plants which abstract silica and build it up into their framework, such as the diatoms of which the frustules become remarkably permanent fossils, in the form of diatom-earth or tripoli-powder, which is made up of them (p. 83).

Durable parts of Animals.—The hard parts of animals may be preserved with little or no chemical change, and remain as durable relics. The hard horny integuments of insects, arachnids, crustacea, and some other animals, are composed essentially of the substance called *chitin*, which can long resist decomposition, and which may therefore be looked for in the sedimentary deposits of the present time, as well as of former periods. The chitin of some fossil scorpions, admirably preserved among the Carboniferous rocks of Scotland, can hardly be distinguished from that of the living scorpion. Many of the lower forms of animal life secrete silica, and their hard parts are consequently easily preserved, as in the case of radiolaria and sponges. In the great majority of instances, however, the hard parts of invertebrates consist mainly of carbonate of lime, and are readily preserved among sedimentary deposits. The skeletons of corals, the plates of echinoderms, and the shells of molluscs, are examples of the abundance of calcareous organisms, and the frequency of their remains in the fossil state shows how well fitted they are for preservation. Among vertebrates the hard parts consist chiefly of phosphate of lime. In some forms (ganoid fishes and crocodiles, for example) this substance is partly disposed outside the body (exo-skeleton) in the form of scales, scutes, or bony plates. But more usually it is confined to the internal skeleton (endo-skeleton). It is mainly by their bones and teeth that the higher vertebrates can be recognised in the fossil state. Sometimes the excrement has been preserved (*Coprolites*), and may furnish information regarding the food of the animals, portions of undigested scales, teeth, and bones being traceable in it (Fig. 65).

Fossilisation.—The process by which the remains of a plant or animal are preserved in the fossil state is termed Fossilisation. It varies greatly in details, but all these may be reduced to three leading types.

1. *Entire or partial preservation of the original substance.*— In rare instances, the entire animal or plant has been preserved, of which the most remarkable examples are those where carcases of the extinct mammoth have been sealed up in the frozen mud and peat of Siberia, and have thus been preserved in ice. Insects have been involved in the resin of trees, and may now be seen, embalmed like mummies, in amber. More usually, however, a variable proportion of the organic matter has passed away, and

its more durable parts have been left, as in the carbonisation of plants (peat, lignite, coal) and the disappearance of the organic matter from shells and bones, which then become dry and brittle and adhere to the tongue.

2. *Entire removal of the original substance and internal structure, only the external form being preserved.*—When a dead animal or plant has been entombed, the mineral matter in which it lies hardens round it and takes a mould of its form. This may be accomplished with great perfection if the material is sufficiently fine-grained and solidifies before the object within has time to decay. Carbonate of lime and silica are specially well adapted

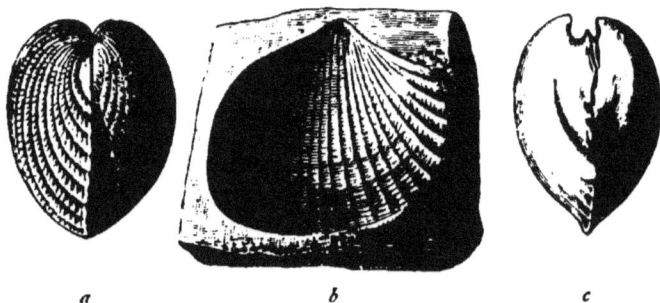

a   b   c

FIG. 113.—Common Cockle (*Cardium edule*); (*a*) side view of both valves; (*b*) mould of the external form of one valve taken in plaster of Paris; (*c*) side view of cast in plaster of Paris of interior of the united valves.

for taking the moulds of organisms, but fine mud, marl, and sand, are also effective. The organism may then entirely decay, and its substance may be gradually removed by percolating water, leaving a hollow empty space or mould of its form. Such moulds are of frequent occurrence among fossiliferous rocks, and are specially characteristic of molluscs, the shells of which are so abundant, and occur imbedded in so many different kinds of material. Sometimes it is the external form of the shell that has been taken, the shell itself having entirely disappeared ; in other cases a cast of the interior of the shell has been preserved. How different these two representations of the same shell may be is shown in Fig. 113, wherein *a* represents a side view of the common cockle, while *c* is a cast of the interior of the shell in plaster of Paris. The contrast between a mould of the outside and inside of the same shell is shown by the difference between *b* and *c*, which are both impressions taken in plaster.

After the decay and removal of the substance of the enclosed organisms, the moulds may be filled up with mineral matter, which is sometimes different from the surrounding rock. The empty cavities have formed convenient receptacles for any deposit which permeating water might introduce. Hence we find casts of organisms in sand, clay, ironstone, silica, limestone, pyrites, and other mineral substances. Of course, in such cases, though the external form of the original organism is preserved, there is no trace of internal structure. No single particle of the cast may ever have formed part of the plant or animal.

3. *Partial or entire petrifaction of organic structure by molecular replacement.*—Plants and animals which have undergone this change have had their substance gradually removed and replaced, particle by particle, with mineral matter. This transformation has been effected by percolating water containing mineral solutions, and has proceeded so tranquilly, that sometimes not a delicate tissue in the internal structure of a plant has been displaced, and yet so rapidly, that the plant had not time to rot before the conversion was completed. Accordingly, in true petrifactions, that is, plants or animals of which the structure has been more or less perfectly preserved in stone, the petrifying material is always such as may have been deposited from water. The most common substance employed by nature in the process of petrifaction is carbonate of lime, which, as we have seen, is almost always present in the water of springs and rivers. Organic structures replaced by this substance are said to be *calcified*. Frequently the carbonate of lime has assumed, more or less completely, a crystalline structure after its deposition, and in so doing has generally injured or destroyed the organic structure which it originally replaced. Where the calcareous matter of an organism has been removed by percolating water, as often happens in sands, gravels, or other porous deposits, the fossil is said to be *decalcified*. Another abundant petrifying medium in nature is silica, which, in its soluble form, is generally diffused in terrestrial waters, where humous acids or organic matter are present in solution. The replacement of organic structures by silica, called *silicification*, furnishes the most perfect form of petrifaction. The interchange of mineral matter has been so complete that even the finest microscopic structures have been faithfully preserved. Silicified wood is an excellent example of this perfect replacement. Sulphides, which are often produced by the reducing action of decaying organic matter upon sulphates, occur also as petrifying media, the

most common being the iron sulphide, usually in the less stable form of marcasite, but sometimes as pyrite. Carbonate of iron likewise frequently replaces organic structures ; the clay-ironstones of the Carboniferous system abound with the remains of plants, shells, fishes, and other organisms which have been converted into siderite (Fig. 62).

The chief value of fossils in geology is to be found in the light which they cast upon former conditions of geography and climate, and in the clue which they furnish to the relative ages of different geological formations.

1. **How Fossils indicate former changes in Geography.**— Terrestrial plants and animals obviously point to the existence of land. If their remains are found in strata wherein most of the fossils are marine, they usually show that the deposits were laid down upon the sea-floor not far from land. But where they occur in the positions in which they lived and died, they prove that their site was formerly a land-surface. The stumps of trees remaining in their positions of growth, with their roots branching out freely from them in the clay or loam underneath, undoubtedly mark the position of an ancient woodland. If, with these remains, there are associated in the same strata wing-cases of beetles, bones of birds and of land-animals, additional corroborative evidence is thereby obtained as to the existence of the ancient land. More usually, however, it is by deposits left on lake-bottoms that the land of former periods of geological history is known. As already pointed out (Chapter IV), the fine mud and marl of lakes receive and preserve abundant relics of the vegetation and animal life of the surrounding regions. As illustrations of lacustrine formations, from which most of our knowledge of the contemporary terrestrial life is obtained, reference may be made to the Molasse of Switzerland, the limestones and marls of the Limagne d'Auvergne, and the vast depth of strata from which so rich an assemblage of plant and animal remains has been obtained in the Western Territories of the United States (see Chapter XXV). Alternations of buried forests or peat-mosses, with lake deposits, show how lakes have successively increased and diminished in volume. The frequent occurrence of a bed of lacustrine marl at the bottom of a peat-bog proves how commonly shallow lakes have been filled up and displaced by the growth of marshy vegetation (pp. 4, 82).

Remains of marine plants and animals almost invariably demonstrate that the locality in which they are found was once covered by the sea. Exceptions to this rule are so few as hardly to be worthy of special notice, as, for instance, when molluscs, crustaceans, and other forms of marine life are carried up by sea-birds to considerable elevations, where, after their soft parts have been eaten, their hard shells and crusts may be preserved in truly terrestrial deposits, or when sea-shells, tossed up by breakers above the tide-line, are swept inland by wind.

Rolled fragments of shells, mingled in well-rounded gravel and sand, point to some former shore where these materials were ground down by beach-waves. Fine muddy sediment, containing unbroken shells, echinoderms, crustaceans, and other relics of the sea, indicate deeper water beyond the scour of waves, tides, and currents. Beds of limestone, full of corals and crinoids, mark the site of a clear sea, in which these organisms were allowed to flourish undisturbed for many generations. It may often be observed that the fossils, which are abundant and large in a limestone, become few in number and small in size in an overlying bed of shale or clay ; or that they wholly disappear in the argillaceous rock. The meaning of this can hardly be mistaken. The clear water in which the marine creatures were able to build up the limestone was at last invaded by some current carrying mud. Consequently, while the more delicate forms perished, others continued to live on in diminished numbers and dwarfed development, until at last the muddy sediment settled down so thickly that the animals, whose hard parts might have been preserved, were driven away from that area of the sea-bottom.

2. **How Fossils indicate former conditions of Climate.—** The remains of plants or animals characteristic of tropical countries may be taken to bear witness to a tropical climate at the time which they represent. If, for example, a deposit were found containing leaves of palms and bones of tigers, lions, and elephants, we should infer that it was formed in some tropical country, such as the warmer parts of Africa or Asia. On the other hand, were a stratum to yield leaves of a small birch and willow, with bones of reindeer, musk-ox, and lemming, we would regard it as evidence of a cold climate. Such inferences, however, must be based either upon the occurrence of the very same species as are now living, and the characteristic climate of which is known, or upon assemblages of plants or animals which may be compared with corresponding assemblages now living. We may be tolerably confident that the existing reindeer has always been restricted to a cold climate, and that the living elephants have as characteristically been confined to warm climates. But it would be rash

to assume that all deer prefer cold and all elephants choose heat. The bones of an extinct variety of elephant and one of rhinoceros, have long been known as occurring even up within the Arctic regions, and when these remains were first found the conclusion was naturally drawn that they proved the former existence of a warm climate in the far north. But the subsequent discovery of entire carcases of the animals covered with a thick mat of woolly hair, showed that they were adapted for.life in a cold climate, and their occurrence in association with the remains of animals which still live in the Arctic regions, proved beyond doubt that the original inference regarding them was erroneous. In drawing conclusions as to climate from fossil evidence, it is always desirable to base them upon the concurrent testimony of as large a variety of organisms as possible, and to remember that they become less and less reliable in proportion as the organisms on which they are founded depart from the species now living.

3. **How Fossils indicate Geological Chronology.**—As the result of careful observations all over the world, it has been ascertained that in the youngest strata the organic remains are nearly or quite the same as species now living, but that, as we proceed into older strata, the number of existing species diminishes, and the number of extinct species increases, until at last no living species is to be found. Moreover, the extinct species found in younger strata disappear as we trace them back into older rocks, and their places are taken by other extinct species. Every great series of fossiliferous rocks is thus characterised by its own peculiar assemblage of species. Not only do the species change ; the genera, too, disappear one by one as we follow them into older rocks, until among the earliest strata only a few of the living genera are represented. Whole families and orders of animals which once flourished have utterly vanished from the living world, and we only know of their existence from the remains of them preserved among the rocks.

A certain definite order of succession has been observed among the organic remains imbedded in the stratified rocks of the earth's crust, and this order has been found to be broadly alike all over the world. The fossils of the oldest fossiliferous rocks of Europe, for instance, are like those of the oldest fossiliferous rocks of Asia, Africa, America, and Australasia, and those of each succeeding series of rocks follow the same general sequence. It is obvious, therefore, that fossils supply us with an invaluable means of fixing the relative position of rocks in the series of geological formations.

Whether or not the same type of fossils was always contemporaneous over the whole planet cannot be determined ; but it generally occupied the same place in the procession of life.   Hence stratified formations, which may be quite unlike each other in regard to the nature of their component materials, if they contain similar organic remains, may be compared with each other, and classed under the same name.

Fossils characteristic of particular subdivisions of the series of geological formations are known as *type-fossils*, of which the following are examples :—

Lepidodendra and Sigillariæ, characteristic of Old Red Sandstone and Carboniferous rocks (pp. 259, 272, 273).
Cycads, characteristic of Mesozoic rocks (pp. 293, 299, 314).
Graptolites, characteristic of Silurian rocks (pp. 242, 250).
Trilobites   ,,   Cambrian to Carboniferous rocks (pp. 244, 253, 264, 277).
Cystideans, characteristic of Silurian rocks (Fig. 123).
Blastoids   ,,   Carboniferous rocks (Fig. 150).
Hippurites   ,,   Cretaceous rocks (p. 317).
Orthoceratites   ,,   Palæozoic rocks (Figs. 130, 157).
Ammonites   ,,   Mesozoic rocks (Figs. 167, 176, 190).
Cephalaspid fishes   ,,   Silurian, Old Red Sandstone (p. 261).
Ichthyosaurus and Plesiosaurus—Mesozoic rocks (Fig. 180).
Iguanodon—Cretaceous rocks (Fig. 192).
Toothed birds—Cretaceous rocks (p. 321).
Nummulites, Palæotherium, Anoplotherium, Deinocerata, characteristic of older Tertiary rocks (pp. 327-338).
Mastodon, Elephas, Equus, Cervus, Hyæna, Apes, characteristic of younger Tertiary and Recent rocks (pp. 339-365).

By attentive study and comparison, the fossiliferous rocks in different countries have been subdivided into sections, each characterised by its own facies or type of organic remains.   Consequently, beginning with the oldest and proceeding upward to the youngest, we advance through natural chronicles of the successive tribes of plants and animals which have lived on the earth's surface.   These chronicles, consisting of sandstones, shales, limestones, and the other kinds of stratified deposits, form what is called the Geological Record.   In order to establish their true sequence in time, their *Order of Superposition* must first be determined ; that is, it is requisite to know which lie at the bottom, and must have been formed first, and in what order the others succeed them.   When this fundamental question has once been settled, then the fossils characteristic of each group of strata serve as a guide for recognising that group wherever it may be found.

While fossils enable us to divide the Geological Record into chapters, they also show how strikingly imperfect this record is as a history of the plants and animals that have lived on the surface of the earth, and of the revolutions which that surface has undergone. We may be sure that the progress of life, from its earliest appearance in lowly forms of plant or animal, has been continuous up to the present condition of things. But in the Geological Record there occur numerous gaps. The fossils of one group of rocks are succeeded by a more or less completely different series in the next group. At one time it was supposed that such breaks in the continuity of the record marked terrestrial convulsions which caused the destruction of the plants and animals of the time, and were followed by the creation of new tribes of living things. But evidence has every year been augmenting to indicate that no such general destruction and fresh creation ever took place. The gaps in the record mark no real interruption of the life of the globe. They are rather to be looked upon as chapters that have been torn out of the annals, or which never were written. We have already learnt in Chapter VIII how many chances there must be against the preservation of anything like a complete record of the life of the globe at any particular time. It is also clear that even where the chronicle may have been comparatively full, it is exposed to many dangers afterwards. The rocks containing it may be hidden beneath the sea, or raised up into land and entirely worn away, or entombed beneath volcanic ejections, or so crushed and crumpled as to become no longer legible.

Taking fossils as a guide, geologists have partitioned the fossiliferous rocks into what are called *stratigraphical* subdivisions as follows :—A bed, or limited number of beds, in which one or more distinctive species of fossils occur, is called a *zone* or *horizon*, and may be named after its most typical fossil. Thus in the Lias, the zone in which the ammonite known as *Ammonites Jamesoni* occurs, is spoken of as the "zone of *Ammonites Jamesoni*," or "*Jamesoni*-zone." Two or more zones, united by the occurrence in them of a number of the same characteristic species or genera, form what are known as *Beds* or an *Assise*. Two or more of such beds or assises may be termed a *Group* or *Stage*. Where the number of assises in a stage is large they may be subdivided into *Sub-stages* or *Sub-groups*. The stage or group will then consist of several sub-stages, and each sub-stage or sub-group of several assises. A number of groups or stages is combined into a *Series*, *Section*, or *Formation*, and a number of series, sections, or formations constitute a *System*. A number of systems are connected together to form each of the great divisions of the Geological Record. This classification will be best understood if placed in tabular form, as in the subjoined subdivisions, which occur in the Cretaceous System.[1]

Stratigraphical Components.	Descriptive Names.	Examples from the Cretaceous System.
A stratum, layer, seam, or bed, or a number of such minor subdivisions, characterised by some distinctive fossil	= Zone or horizon . . .	Zone of *Pecten asper.*
Two or more zones	= Beds or an assise . .	Warminster beds.
Two or more sets of connected beds or assises =	Group or stage, which may be subdivided into sub-groups or sub-stages	Cenomanian stage, comprising the Rothomagian and Carentonian sub-stages.
Two or more groups or stages =	Series, section, or formation	Neocomian formation.
Several related formations	= System	Cretaceous System.

The names by which the larger subdivision of the Geological Record are known have been adopted at various times and on no regular system. Some of them are purely lithological; that is, they refer to the mere mineral nature of the strata, apart altogether from their fossils, such as Coal-measures, Chalk, Greensand, Oolite. These names belong to the early years of the progress of geology, before the nature and value of organic remains had been definitely realised. Other epithets have been suggested by localities where the strata occur, as London Clay, Oxford Clay, Mountain Limestone. The more recent names for the larger divisions have, in general, been chosen from districts where the strata are typically developed, or where they were first critically studied, *e.g.* Silurian, Devonian, Permian, Jurassic. In some cases, the larger subdivisions have received names from some distinguishing feature in their fossil contents, as Eocene, Miocene, Pliocene.[2] But it is mainly to the minor sections that the characters of the fossil contents have supplied names. The designation of any particular group of strata has gradually

---

[1] For an account of the Cretaceous System, see Chapter XXIV.
[2] For the meanings of these names see Chapter XXV, p. 329.

come to acquire a chronological meaning. Thus we speak of
the Oolites or Oolitic formations of England, and include under
these terms a thick series of limestones, clays, sandstones, and
other strata, replete with organic remains, and containing the
records of a long interval of geological time.   But we also speak
of the Oolitic period—a phrase which, in the strict grammatical
use of the word, is of course incorrect, but which conveniently
designates the period of geological time during which the great
series of Oolites was deposited, and when the abundant life of
which they contain the remains flourished on the surface of the
earth.   This chronological meaning has indeed come to be the
more usual sense in which the names of the major subdivisions
of the Geological Record are generally employed.   Such adjec-
tives as Devonian and Jurassic do not so much suggest to the
mind of the geologist Devonshire and the Jura Mountains, from
which they were taken, nor even the rocks to which they are
applied, as the great sections of the earth's history of which these
rocks contain the memorials.   He compares the Jurassic or
Devonian rocks of one country with those of another, studies the
organic remains contained in them, and then obtains materials
for forming some conception of what were the conditions of
geography and climate, and what was the general character of
the vegetable and animal life of the globe, during the periods
which he classes as Jurassic and Devonian.

Summary.—Fossils are the remains or traces of plants and
animals which have been imbedded in the rocks of the earth's crust.
From the exceptional nature of the circumstances in which these
remains have been entombed and preserved, only a comparatively
small proportion of the various tribes of plants and animals living at
any time upon the earth is likely to be fossilised.   Those organisms
which contain hard parts are best fitted for becoming fossils.
The original substance of the organism may, in rare cases, be
preserved ; more usually the organic matter is partially or wholly
removed.   Sometimes a mere cast of the plant or animal in
amorphous mineral matter retains the outward form without any
trace of the internal structure.   In other instances, true petrifaction
has taken place, the organic structure being reproduced in calcite,
silica, or other mineral by molecular replacement.

Fossils are of the utmost value in geology, inasmuch as they
indicate (1) former changes in geography, such as the existence
of ancient land-surfaces, lakes, and rivers, the former extension
of the sea over what is now dry land, and changes in the currents

of the ocean ; (2) former conditions of climate, such as an Arctic state of things as far south as Central France, where bones of reindeer and other Arctic animals have been found ; (3) the chronological sequence of geological formations, and, consequently, the succession of events in geological history, each great group of strata being characterised by its distinctive fossils. This is the most important use of fossils. Having ascertained the order of superposition of fossiliferous rocks, that is, the order in which they were successively deposited, and having found what are the characteristic fossils of each subdivision, we obtain a guide by which to identify the various rock-groups from district to district, and from country to country. By means of the evidence of fossils the stratified rocks of the Geological Record have been divided into sections and subsections, to which names are applied that have now come to designate not merely the rocks and their fossils, but the period of geological time during which these rocks were accumulated and these fossils actually lived.

PART IV

THE GEOLOGICAL RECORD OF THE
HISTORY OF THE EARTH

CHAPTER XVI

THE EARLIEST CONDITIONS OF THE GLOBE—THE
ARCHÆAN PERIODS

THE foregoing chapters have dealt chiefly with the materials of
which the crust of the earth consists, with the processes whereby
these materials are produced or modified, and with the methods
pursued by geologists in making their study of these materials and
processes subservient to the elucidation of the History of the
Earth. The soils, rocks, and minerals beneath our feet, like the
inscriptions and sculptures of a long-lost race of people, are in
themselves full of interest, apart from the story which they
chronicle ; but it is when they are made to reveal the history of
land and sea, and of life upon the earth, that they are put to their
noblest use. The investigation of the various processes whereby
geological changes are carried on at the present day is undoubtedly
full of fascination for the student of nature ; yet he is conscious
that it gains enormously in interest when he reflects that in watching the geological operations of the present day he is brought face
to face with the same instruments whereby the very framework of
the continents has been piled up and sculptured into the present
outlines of mountain, valley, and plain.

The highest aim of the geologist is to trace the history of the
earth. All his researches, remote though they may seem from this

aim, are linked together in the one great task of unravelling the successive mutations through which each area of the earth's surface has passed, and of discovering what successive races of plants and animals have appeared upon the globe. The investigation of facts and processes, to which the previous pages have been devoted, must accordingly be regarded as in one sense introductory to the highest branch of geological inquiry. We have now to apply the methods and principles already discussed to the elucidation of the history of our planet and its inhabitants. Within the limits of this volume only a mere outline of what has been ascertained regarding this history can be given. I shall arrange in chronological order the main phases through which the globe seems to have passed, and present such a general summary of the more important facts regarding each of them as may, I hope, convey an adequate outline of what is at present known regarding the successive periods of geological history.

As the primitive stages of mankind upon the earth and the early progress of every race fade into the obscurities of mythology and archæology, so the story of the primeval condition of our globe is lost in the dim light of remote ages, regarding which almost all that is known or can be surmised is furnished by the calculations and speculations of the astronomer. If the earth's history could only be traced out from evidence supplied by the planet itself, it could be followed no further back than the oldest portions of the earth now accessible to us. Yet there can be no doubt that the planet must have had a long history before the appearance of any of the solid portions now to be seen. That such was the case is made almost certain by the traces of a gradual evolution or development which astronomers have been led to recognise among the heavenly bodies. Our earth being only one of a number of planets revolving round the sun, the earliest stages of its separate existence must be studied in reference to .the whole planetary system of which it forms a part. Thus, in compiling the earliest chapter of the history of the earth, the geologist turns for evidence to the researches of the astronomer among stars and nebulæ.

In recent years, more precise methods of inquiry, and, in particular, the application of the spectroscope to the study of the stars, have gone far to confirm the speculation known as the Nebular Hypothesis. According to this view, the orderly related series of heavenly bodies, which we call the Solar System, existed at one time, enormously remote from the present, as a Nebula—that is, a cloudy mass of matter, like one of those nebulous, faintly luminous clouds which can be seen in the heavens.   This nebula probably extended at least as far as the outermost planetary member of the system is now removed from the sun.   It may have consisted entirely of incandescent gases or vapours, or of clouds of stones in rapid movement, like the stones that from time to time fall through our atmosphere as meteorites, and reach the surface of the earth. The collision of these stones moving with planetary velocity would dissipate them into vapour, as is perhaps the case in the faint luminous tails of comets.   At all events, the materials of the nebula began to condense, and in so doing threw off, or left behind, successive rings (like those around the planet Saturn), which, in obedience to the rotation of the parent nebula, began to rotate in one general plane around the gradually shrinking nucleus. As the process of condensation proceeded, these rings broke up, and their fragments rushed together with such force as not improbably to generate heat enough to dissipate them again into vapour. They eventually condensed into planets, sometimes with a further formation of rings, or with a disruption of these secondary rings, and the consequent formation of moons or satellites round the planets.   The outer planets would thus be the oldest, and, on the whole, the coolest and least dense.   Towards the centre of the nebula the heaviest elements might be expected to condense, and there the high temperature would longest continue.   The sun is the remaining intensely hot nucleus of the original nebula, from which heat is still radiated to the furthest part of the system.

When a planetary ring broke up, and by the heat thereby generated was probably reduced to the state of vapour, its materials, as they cooled, would tend to arrange themselves in accordance with their respective densities, the heaviest in the centre, and the lightest outside.   In process of time, as cooling and contraction advanced, the outer layers might grow quite cold, while the inner nucleus of the planet might still be intensely hot. Such, in brief, is the well-known Nebular Hypothesis.

Now the present condition of our earth is very much what, according to this hypothesis or theory, it might be expected to be.   On the outside comes the lightest layer or shell in the form of an Atmosphere, consisting of gases and vapours.   Below this gaseous envelope which entirely surrounds the globe lies an inner envelope of water, the ocean, which covers about two-thirds of the earth's surface, and is likewise composed of gases.   Underneath this watery covering, and rising above it in dry land, rests the solid part of the globe, which, so far as accessible to us, is composed of rocks twice or thrice the weight of pure water. But observations with the pendulum at various heights above the sea show that the attraction of the earth as a whole indicates that the globe probably has a density about five and a half times that of water. Hence we may infer that its inner nucleus not improbably consists of heavy materials, and may be metallic. There is thus evidence of an arrangement of the planet's materials in successive spherical shells, the lightest or least dense being on the outside, and the heaviest or most dense in the centre.

Again, the outside of the earth is now quite cool ; but abundant proof exists that at no great distance below the surface the temperature is high. Volcanoes, hot springs, and artificial borings all over the world testify to the abundant store of heat within the earth. Probably at a depth of not more than 20 miles from the surface the temperature is as high as the melting-point of any ordinary rock at the surface. By far the largest part of the planet, therefore, is hotter than molten iron. We need have no hesitation in admitting it to be highly probable that the earth was formerly in the state of incandescent vapour, and that it has ever since that time been cooling and contracting. Its present shape affords strong presumption in favour of the opinion that the globe was once in a plastic condition. The flattening at the poles and bulging at the equator, or what is called the oblately spheroidal figure of the planet, is just the shape which a plastic mass would have assumed in obedience to the influence of the movement of rotation, imparted to it when detached from the parent nebula.

At present a complete rotation is performed by the earth in twenty-four hours. But calculations have been made with the result of showing that originally the rate of rotation was much greater. Fifty-seven millions of years ago it was about four times faster, the length of the day being only six and three-quarter hours. The moon at that time was only about 35,000 miles distant from the earth, instead of 239,000 miles as at present. Since these early times the rate of rotation has gradually been diminishing, and the figure of the earth has been slowly tending to become more spherical, by sinking in the equatorial and rising in the polar regions.

Of the first hard crust that formed upon the surface of the earth no trace has yet been found. Indeed, there is reason to suppose that this original crust would break up and sink into the molten mass beneath, and that not until after many such formations and submergences did a crust establish itself of sufficient strength to form a permanent solid surface.   Even though solid, the surface may still have been at a glowing red-heat, like so much molten iron.   Over this burning nucleus lay the original atmosphere, consisting not merely of the gases in the present atmosphere, but of the hot vapours which subsequently condensed into the ocean, or were absorbed into the crust.   It was a hot, vaporous envelope, under the pressure of which the first layers of water that condensed from it may have had the temperature of molten lead.   As the steam passed into water, it would carry down with it the gaseous chlorides of sodium, magnesium, and other vapours in the original atmosphere, so that the first ocean was probably not only hot, but intensely saline.

Regarding these early ages in the earth's history we can only surmise, for no direct record of them has been preserved.   They are sometimes spoken of as pre-geological; but geology really embraces the whole history of the planet, no matter from what sources the evidence may be obtained.   Deposits from this original hot saline ocean have been supposed to be recognisable in the very oldest crystalline schists; but for this supposition there does not appear to be any good ground.   The early history of our planet, like that of man himself, is lost in the dimness of antiquity, and we can only speculate about it on more or less plausible suppositions.

When we come to the solid framework of the earth we stand on firmer footing in the investigation of geological history.   The terrestrial crust, or that portion of the globe which is accessible to human observation, has been found to consist of successive layers of rock, which, though far from constant in their occurrence, and though often broken and crumpled by subsequent disturbance, have been recognised over a large part of the globe.   They contain the earth's own chronicle of its history, which has already been referred to as the Geological Record, and the subdivision of which into larger and minor sections, according mainly to the evidence of fossils, was explained in the preceding chapter.

Had the successive layers of rock that constitute the Geological Record remained in their original positions, only the uppermost, and therefore most recent, of them would have been visible, and nothing more could have been learnt regarding the underlying layers, except in so far as it might have been possible to explore them by boring into them.   But the deepest mines do not reach greater depths than between 3000 and 4000 feet from the surface. Owing, however, to the way in which the crust of the earth has been plicated and fractured, portions of the bottom layers have

been pushed up to the surface, and those that lay above them
have been thrown into vertical or inclined positions, so that we
can walk over their upturned edges and examine them, bed by bed.
Instead of being restricted to merely the uppermost few hundred
feet of the crust, we are enabled to examine many thousand feet
of its rocks.   The total mean thickness of the accessible fossiliferous rocks of Europe has been estimated at 75,000 feet, or upwards
of 14 miles.   This vast depth of rock has been laid bare to
observation by successive disturbances of the crust.

The main divisions of the Geological Record and, we may also
say, of geological time, are five : (1) Archæan, embracing the
periods of the earliest rocks, wherein no traces of organic life
occur ; (2) Palæozoic (ancient life) or Primary, including the long
succession of ages during which the earliest types of life existed ;
(3) Mesozoic (middle life) or Secondary, comprising a series of
periods when more advanced types of life flourished ; (4) Cainozoic
(recent life) or Tertiary, embracing the ages when the existing
types of life appeared, but excluding man ; and (5) Quaternary or
Post-tertiary and Recent, including the time since man appeared
upon the earth.   It must not be supposed that each of these five
divisions was of the same duration.   The Palæozoic ages were
probably vastly more prolonged than those of any later division ;
while the Quaternary periods must comprise a very much briefer
time than any of the other four groups.

Each of these main sections is further subdivided into systems
or periods, and each system into formations as already explained.
Arranged in their order of sequence, the various divisions of the
Geological Record may be placed as in the accompanying Table.

THE GEOLOGICAL RECORD,

or, Order of Succession of the Stratified Formations of the Earth's Crust.

Post-tertiary or Quaternary.	Recent and Prehistoric. Pleistocene or Glacial.
Cainozoic or Tertiary.	Pliocene. Miocene. Oligocene. Eocene.

Cretaceous.
  Danian.
  Senonian.
  Turonian.
  Cenomanian.
  Gault (Albian).
  Neocomian.

Jurassic.
  Purbeckian.
  Portlandian.
  Kimmeridgian.
  Corallian.
  Oxfordian.
  Bathonian.
  Bajocian.
  Liassic.

Triassic.
  Rhætic.
  Keuper or Upper Trias.
  Muschelkalk.
  Bunter or Lower Trias.

Mesozoic or Secondary.

Permian.
  Upper Red Sandstones, clays, and gypsum.
  Magnesian Limestone (Zechstein).
  Marl-Slate (Kupferschiefer).
  Lower Red Sandstones, breccias, etc. (Rothliegende).

Carboniferous.
  Coal-Measures.
  Millstone-Grit.
  Carboniferous Limestone series.

Devonian and Old Red Sandstone.

Devonian Type { Upper—Cypridina and Goniatite beds.
  Middle—Stringocephalus (Eifel) Limestone.
  Lower—Spirifer Sandstone, etc.

Old Red Sandstone Type. { Upper Yellow and Red Sandstones, with *Holoptychius*, *Pterichthys major*, etc.
  Lower Sandstones, flagstones, and conglomerates, with *Cephalaspis*, *Coccosteus*, *Asterolepis*, etc.

Silurian.
  Upper { Ludlow group.
    Wenlock group.
    Upper Llandovery group.
  Lower { Lower Llandovery group.
    Caradoc and Bala group.
    Llandeilo group.
    Arenig group.

Palæozoic or Primary.

Palæozoic or Primary.	Cambrian or Primordial	{ Upper—Tremadoc Slates. Lingula Flags. Lower—Menevian group. Harlech group.
Pre-Cam-brian.		Longmyndian—Uriconian. [Dalradian.] Torridonian. Archæan—Lewisian, Hebridean.

THE PRE-CAMBRIAN PERIODS.

Owing to the revolutions which the crust of the earth has undergone, there have been pushed up to the surface, from underneath the oldest fossiliferous strata, certain very ancient crystalline rocks which form what is termed the Archæan system. As already mentioned, these rocks have by some geologists been supposed to be a part of the primeval crust of the planet, which solidified from fusion. By others they have been thought to have been formed in the boiling ocean, which first condensed upon the still hot surface of the globe. In truth, we are still profoundly ignorant as to the conditions under which they arose. We have hardly any means of ascertaining in what order they were formed. We know no method of determining whether those of one region belong to the same period as those of another. Nor can we always be sure that what have been called Archæan rocks may not belong to a much later part of the Geological Record, their peculiar crystalline structure having been superinduced upon them by some of those subterranean movements described in Chapter XIII.

Of Archæan rocks the most abundant is gneiss, passing on the one hand into granite, and on the other into micaceous and argillaceous schists, with interstratified bands of various hornblendic, pyroxenic, and garnetiferous rocks, limestone, dolomite, serpentine, quartzite, graphite, hæmatite, magnetite, etc. These various materials are more or less distinctly bedded. But the beds are for the most part inconstant, swelling out into thick zones, and then rapidly diminishing and dying out. This bedding somewhat resembles that of sedimentary rocks, and the manner in which the limestone and graphite occur, recalls the way in which limestone and coal are found in the fossiliferous formations. The inference has accordingly been drawn that the Archæan crystalline bands were really deposited as chemical precipitates or mechanical sediments on the floor of the primeval ocean, and have since been more or less crystallised and disturbed. But from what has been brought forward in Chapter XIII, regarding the totally new structures which have been developed in rocks by subterranean movement, it is evident that a bedded arrangement and a crystalline texture, like those of the Archæan gneisses and schists, have sometimes been induced in rocks by excessive crumpling, fracture, and shearing. How far, therefore, the apparent bedding of Archæan rocks is their original condition, or is the result of subsequent disturbance, is a question that cannot yet be answered.

The alternations of gneiss and other crystalline masses form bands which are usually placed on end or at high angles, and are

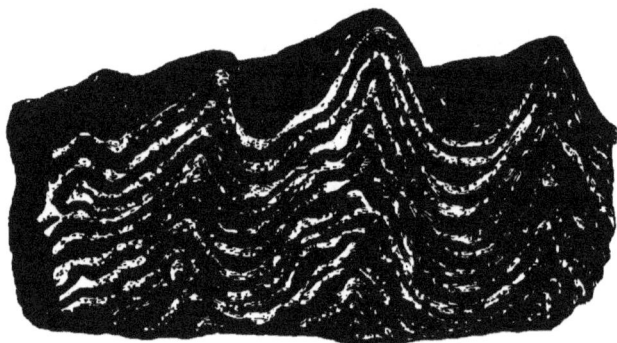

FIG. 114.—Fragment of crumpled Schist.

often intensely crumpled and puckered, having evidently undergone enormous crushing (Fig. 114). Attempts have been made to subdivide them into groups or series, according to their apparent order of succession and lithological characters. But such subdivisions, even where practicable, are probably only of local value. As a rule, those members of the system which, if the succession of beds may be trusted, are the lowest and oldest, present coarser crystalline characters than those which seem to be higher and later. They often consist of massive granitic gneiss, with abundant veins and bands of the coarsely crystalline variety of granite, known as pegmatite. The apparently higher rocks are less coarsely crystalline gneiss, and often mica-schists and other schistose masses.

No unquestionable relic of organic existence has been met with among Archæan rocks. Some of the Archæan limestones of Canada have yielded a peculiar mixture of serpentine and calcite, with a structure which is regarded by some able naturalists as that of a reef-building foraminifer. It occurs in masses, and is supposed by these writers to have grown in large, thick sheets or reefs over the sea-bottom. By most observers, however, this supposed organism (to which the name of *Eozoon* has been given) is now regarded as merely a mineral segregation, and various undoubted mineral structures are pointed to in illustration and confirmation of this view.

Archæan rocks cover a large area in Europe. Among the Hebrides and along the north-west coast of the Scottish Highlands, where they are most largely developed, they consist of a very ancient group of rocks, of which the most conspicuous are various forms of gneiss (Lewisian, Hebridean, Fundamental), overlain by a much younger series of dull red sandstones, conglomerates, and dark grey shales (Torridon), which lie on the gneiss unconformably, and are covered unconformably by the oldest Cambrian strata, containing annelid tracks and the trilobite *Olenellus*. The gneiss gives rise to a singular type of scenery. Over much of that region it forms hummocky bosses of naked rock, with tarns and peat-bogs lying in the hollows, seldom rising into mountains, but forming the platform which supports a singular group of red (Torridon) sandstone mountains. Here and there, it mounts up into solitary hills or groups of hills. The highest point it reaches on the mainland is among the mountains on the east side of Loch Maree, in Ross-shire, where it attains an elevation of 3000 feet. Some of its masses in that region were mountains at the time of the deposition of the overlying Torridon sandstone, which when removed by denudation reveals a very ancient system of hills and valleys. In the Island of Harris the gneiss sweeps upwards into rugged mountainous ground, of which the highest summits rise more than 2600 feet out of the Atlantic, and are visible far and wide as a notable landmark. The different varieties of gneiss are associated with dykes of various basic igneous rocks, and bands of pegmatite. Rocks of similar character appear likewise in the west of Ireland ; while in Anglesey, and possibly in the south-west of England, other scattered bosses of them rise to the surface.

Later than the Archæan gneiss, but older than the lowest Cambrian strata, are certain groups of sedimentary (partly also igneous) rocks which at present can only be grouped under the common term Pre-Cambrian. The Torridon sandstone reaches a thickness of 8000 or 10,000 feet, and is almost entirely confined to the west of the counties of Sutherland and Ross. It there forms a remarkable group of pyramidal mountains, to which their nearly horizontal stratification gives a characteristic architectural aspect. Traces of organic remains (annelid tracks, etc.) have been found in these strata.

The term " Dalradian " has recently been applied to a thick series of metamorphosed sedimentary and igneous rocks forming the Central and Southern Highlands of Scotland. They must be of great thickness, but their true geological position is not yet ascertained. They may possibly contain altered representatives of the old gneiss, Torridon sandstone and Cambrian quartzites and limestones of the north-west.

On the borders of Wales and Shropshire a thick series of sedimentary rocks (Longmyndian) forms the Longmynd country. It appears to be Pre-Cambrian, and may be partly the equivalent of the Torridon sandstone of the north-west. It is underlain by a group of felsitic lavas and tuffs named Uriconian.

On the continent of Europe, Archæan rocks have their greatest extension in Scandinavia, where they evidently belong to the same ancient land as that of which the Hebrides and Scottish Highlands are fragments. They range through Finland far into Russia, appearing in the centre of the chain of the Ural Mountains. They form likewise the nucleus of the Carpathians and the Alps, and appear in detached areas in Bavaria, Bohemia, France, and the Pyrenees. They are estimated to occupy an area of more than 2,000,000 of square miles in the more northerly part of North America, stretching from the Arctic regions southwards to the great lakes. Both in the Old and New World, the Archæan rocks are chiefly exposed in the northern tracts of the continents. The areas which they there overspread were probably land at an early geological period, and it was the waste of this land that mainly supplied the original materials out of which the enormous masses of stratified rocks were formed. Various thick accumulations of sedimentary and igneous rocks have been ascertained to lie in North America, as in Europe, between the Archæan gneisses and the base of the Cambrian system.

In the southern hemisphere also ancient gneisses and other schists rise from under the oldest fossiliferous formations. In Australia and in New Zealand they cover large tracts of country, and appear in the heart of the mountain ranges. It thus appears that all over the world the oldest known rocks are gneisses and similar or allied crystalline masses, having a remarkable uniformity of character.

CHAPTER XVII

THE PALÆOZOIC PERIODS—CAMBRIAN

THE portion of geological history which treats of those ages in which the earliest known types of plants and animals lived is termed Palæozoic. Of the first appearance of organic life upon our planet we know nothing. Whether plants or animals came first, and in what forms they came, are questions to which as yet no satisfactory answer can be given. The oldest discovered fossils are assuredly not vestiges of the first living things that peopled the globe. There is every reason, indeed, to hope that as researches in all parts of the world are pushed into older and yet older rocks, still more ancient organisms may be discovered. But it is in the highest degree improbable that any trace of the earliest beginnings of life will ever be found. The first plants and the first animals were probably of a lowly kind, with no hard parts capable of preservation in the fossil state. Moreover, the sedimentary rocks which may have chronicled the first advent of organised existence are hardly likely to have escaped the varied revolutions to which all parts of the crust of the earth have been exposed. They have more probably been buried out of sight, or have been so crushed, broken, and metamorphosed, that their original condition, together with any fossils they may have enclosed, is no longer to be recognised. The first chapters have been, as it were, torn out from the chronicle of the earth's history.

The Palæozoic rocks, which contain the earliest record of plant and animal life, consist mainly of the hardened mud, sand, and gravel of the sea-bottom. Here and there they include beds, or thick groups of beds, of limestone composed of marine shells, crinoids, corals, and other denizens of salt water. They are thus essentially the chronicles of the sea. But they also contain occasional vestiges of shores, and even of the jungles and swamps of the land, with a few rare glimpses into the terrestrial life of the time. Everywhere they abound in evidence of shallow water; for though chiefly marine, they appear to have been accumulated not far from land. We may believe that in the earliest periods, as at the present day, the sediment washed away from the land has been deposited on the sea-floor, for the most part at no great distance from the coast.

The land from the waste of which the Palæozoic rocks were formed lay in Europe and North America chiefly towards the north. It no doubt consisted of Archæan rocks, such as still rise out from under the oldest Palæozoic formations. As already mentioned, the north-west Highlands of Scotland, part of the table-land of Scandinavia, and most of North America to the north of the great lakes, are probably portions of that earliest land, which, after being deeply buried under later geological accumulations, have once more been laid bare to the winds and waves. We can form some conception of the bulk of the primeval northern land by noting the thickness of sedimentary rocks that were formed out of its detritus during the Palæozoic periods. The older half of the Palæozoic rocks in the British Islands, for example, is at least 16,000 feet or 3 miles thick, and covers an area of not less than 60,000 square miles. This material, derived from the waste of the Archæan rocks, would make a table-land larger than Spain, with an average height of 5000 feet, or a mountain chain 1800 miles long, with an average elevation of 16,000 feet. Of the general form and height of the northern land that supplied this vast mass of sedimentary matter nothing is known. Perhaps it was lofty; but it may have been slowly uplifted, so that its rise compensated for the ceaseless degradation of its surface.

Abundant evidence of volcanic action has been preserved among the Palæozoic rocks in the form of piles of lavas and tuffs. We find also many indications of upward and downward movements of the crust of the earth. The mere fact of the superposition of many thousands of feet of shallow-water strata, one above another, is a proof of gradual subsidence. For it is evident that the accumulation of such a thickness of sediment, and the continuance of a shallow sea over the area of deposition, could only take place during a progressive subsidence (see p. 181).

The life of the Palæozoic periods, so far as known from the fossils which have been obtained from the rocks, appears to have been far more uniform over the whole globe than at any subse-

quent epoch in geological history. For instance, the same species
of fossils are found in corresponding rocks in Britain, Russia,
United States, China, and Australia. The climate of the globe
at that ancient date was doubtless more uniform than it afterwards
became, and was probably also generally warmer. Palæozoic
fossils, obtained from high northern latitudes, are precisely similar
to those that abound in England, whence it may be inferred that not
only was there a greater uniformity of climate, but that the great
cold which now characterises the Arctic regions did not then exist.

In the earlier Palæozoic periods, the animal life of the globe
appears to have been entirely invertebrate, the highest known
types being chambered shells, of which our living nautilus is a
representative. In the middle periods vertebrate life appeared.
The earliest known vertebrate forms are fishes akin to some
modern sharks and to the sturgeon, the polypterus of the
Nile, and the gar-pike of American lakes. The most highly
organised forms of existence upon the earth's surface in the
later Palæozoic periods were amphibians—a class of animals
represented at the present day by frogs, toads, newts, and
salamanders. It is evident, however, that the number and
kinds of animal remains preserved in Palæozoic rocks afford
only an imperfect record of the animal life of these early ages.
Whole tribes of creatures no doubt existed of which no trace
whatever has yet been recovered. An accidental discovery may
at any moment reveal the former presence of some of these vanished forms. For example, the examination of a fossil tree-trunk
imbedded among the coal-strata of Nova Scotia, led to the finding
of the first and as yet almost the only traces of Palæozoic land-
shells, though thousands of species of marine shells, belonging to
the same period, had long been known. Every year is enlarging
our knowledge in these respects, but from the very nature of the
circumstances in which the records of the rocks were formed, we
cannot expect this knowledge ever to be more than fragmentary.

The Palæozoic rocks are divided into five systems which in
the order of their age have been named; (1) Cambrian; (2)
Silurian; (3) Devonian; (4) Carboniferous; (5) Permian.

CAMBRIAN.

The strata containing the earliest organic remains were formerly
known as Greywacke, from the rock which is specially abundant
among them. They were also termed Transition, from the sup-

position that they were deposited during a transitional period, between the time when no organic life was possible on the earth's surface, and the time when plant and animal life abounded. But the late Sir Roderick Murchison, who first explored them, showed that they contain a series of formations, each characterised by its own assemblage of organic remains. He called them the Silurian system, after the name of the old British tribe—the Silures, who lived on the borders of England and Wales, where these rocks are especially well developed. This name has now been adopted all over the world as the designation of those stratified formations which contain the same or similar organic remains to those found in the typical region described by Murchison.

While the succession of the rocks and fossils was established by that geologist in South Wales, and in the border counties of Wales and England, Professor Sedgwick was at work among similar rocks in North Wales. These were at first believed to be all older than those called Silurian, and were accordingly named CAMBRIAN, after the old name for Wales, Cambria. In the end, however, it was found that throughout a large part of them the same fossils occurred, as in the Silurian series, and they were accordingly claimed as Silurian. Much controversy has since been carried on regarding the limits and names to be assigned to these rocks, and geologists are not yet agreed upon the nomenclature that should be followed. Murchison and his followers claimed the Cambrian as the lowest portion of the Silurian system, while Sedgwick and his disciples maintained that the lower half of the Silurian system should be included in his Cambrian series. There can be no doubt that the first succession of organic remains established among these ancient members of the great Palæozoic series of formations was that worked out by Murchison and named by him Silurian. But it has been found convenient to retain the name Cambrian for the oldest group of fossiliferous formations. It may be well to repeat that these words, like all those adopted by geologists to distinguish the successive rock-groups of the earth's crust, have acquired a chronological meaning. We speak, not only of Cambrian and Silurian strata and Cambrian and Silurian fossils, but of Cambrian and Silurian time. The terms are used to denote those particular periods in the history of the earth when Cambrian and Silurian strata were respectively deposited, and when Cambrian and Silurian fossils were the living denizens of sea and land.

The rocks of which the Cambrian system is composed, like those

of the whole of the Lower Palæozoic formations, present considerable uniformity over the whole globe.   They consist of grey and
reddish grits, sandstones, greywackes, quartzites, and conglomerates,
with thick groups of shale, slate, or phyllite.   These sedimentary
accumulations attain a great thickness in some countries.   In Wales
they have been estimated by some observers to be at least 20,000
feet in depth.   Their ripple-marks, pebble-beds, and frequent
alternations of coarse and fine sediment, point to their having
probably been laid down in comparatively shallow water, during
a period of prolonged subsidence of the sea-bottom.   They include
tuffs and basic lavas which indicate contemporaneous submarine
eruptions.

With regard to the occurrence of fossils among the older
Palæozoic formations, and indeed among stratified rocks in general,
it is worthy of notice that they are far from being equally distributed ; that, on the contrary, they occur by preference in certain
kinds of material rather than in others.   Grits and sandstones, for
instance, are comparatively unfossiliferous, while fine shales, slates,
and limestones are often crowded with fossils.   It is not that life was
probably on the whole more abundant at the time of the deposition
of some kinds of strata, but that the local conditions for its growth
and for the subsequent entombment and preservation of its remains
were then more favourable.   At the present time, for example,
dredging operations show the most remarkable variations between
different and even adjacent parts of the sea-bottom as regards the
abundance of marine life.   Some tracts are almost lifeless, while
others are crowded with a varied and prolific fauna.   We can
easily understand that if, from the nature of the bottom, plants and
smaller animals cannot flourish on a particular tract, the larger
kinds that feed on them will also desert it.   Even if organisms
live and die in some·numbers over a part of the sea-bed, the conditions may not be suitable there for the preservation of their
remains.   The rate of deposit of sediment, for instance, may be
so slow that the remains may decay before there is time for them
to be covered up ; or the sediment may be unfitted for effectually
preserving them, even when they are buried in it.   We must not
lose sight of these facts in our explorations of the Geological Record.
A relation has always existed between the abundance or absence
of fossils in a sedimentary rock and the circumstances under which
the rock was originally formed.

The oldest fossiliferous strata (Cambrian or Primordial) contain
a remarkable assemblage of animal remains, which, being the

R

earliest traces of the animal life of the globe, might have been anticipated to belong to the very lowest tribes of the animal kingdom. But they are by no means of such humble organisation. On the contrary, they include no representatives of many of the groups of simpler invertebrates, which we may be sure were

nevertheless living at the same time. Not only so, but some of the fossils belong to comparatively high grades in the scale of invertebrate life, such as chambered molluscs. From this incompleteness, and from the wide differences in the organic grade of the forms actually preserved in the rocks, we may reasonably infer that only a most meagre representation of the life of the time has come down to us in the fossil state. Some of the fossils, moreover, have been so indistinctly preserved that considerable difficulty is experienced in deciding to what sections of the animal or vegetable kingdoms they should be assigned.

Among the markings which have given rise to much discussion allusion may be

FIG. 115.—Fucoid-like impression (*Eophyton Linneanum*) from Cambrian rocks (⅓).[1]

made to plant-like impressions, some of which, like *Eophyton* (Fig. 115), have been claimed as sea-weeds. Others, however, may only be irregular wrinklings of the surfaces of deposit, and of no organic origin at all (see p. 249). Another puzzling impression is that called *Oldhamia* (Fig. 116), which has been variously referred to the Hydrozoa, the Sertularia, the Polyzoa, and the calcareous Algae.

Some of the most characteristic older Palæozoic organisms belong to the Hydrozoa, and are embraced under the general title of Graptolites—a name given to them from their fancied resemblance to quill-pens. They were composed of a horny or chitinous substance, and hence they commonly present themselves merely as black streaks upon the stone. Each graptolite was a colony comprising many individuals which occupied each its own cell. The cells are in some kinds placed in a row on one side of

[1] The fractional numbers inserted within parentheses in the titles of the figures of fossils indicate how much the figures have been reduced or magnified. Thus ⅓ = reduced one-third ; ⁴⁄₁ = magnified four times.

a supporting rod or axis ; in other kinds there is a row of cells on both sides (see Fig. 121). Some varieties are straight, others curved

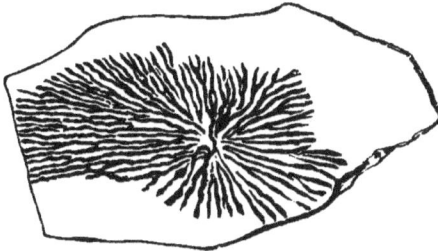

FIG. 116.—*Oldhamia radiata* (natural size), Ireland.

or spiral. Some are simple branches, others are composed of two or more branches, while in certain types a large number of separate branches is united in one common centre. One of the most

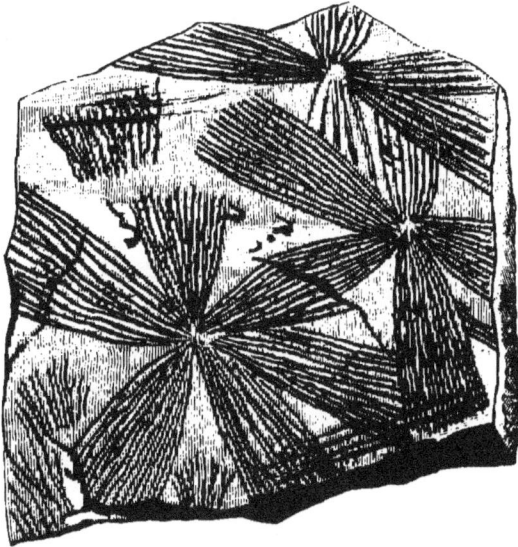

FIG. 117.—Hydrozoon from the Cambrian rocks (*Dictyograptus* (*Dictyonema*) *sociale*), natural size.

ancient hydrozoa is *Dictyograptus* (*Dictyonema*, Fig. 117)—a characteristic fossil of the Cambrian rocks of Scandinavia. The graptolites are more especially characteristic of the Silurian system (p. 250).

The Echinodermata had their representatives in Cambrian

time, though the remains of these are few and infrequent. The great tribe of the Crinoids or Sea-lilies (p. 251) had already established itself on the floor of the Cambrian sea, where also there were representatives of Cystideans and star-fishes (see p. 252).

Numerous kinds of Sea-worms (Annelids) crawled over the sandy and muddy bottom and shores of the Cambrian ocean. These creatures have left no trace of their bodies, which, like those of their representatives in the present ocean, were soft and unfitted for preservation. But the burrows they made in wet sand or mud, and the trails they left upon the soft surfaces over which they moved, have been abundantly preserved (see Fig. 124). These

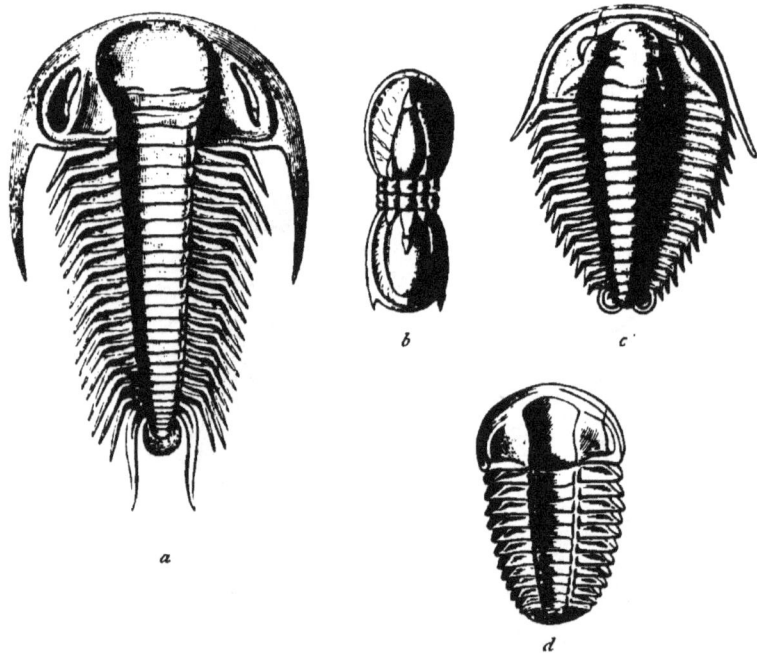

FIG. 118.—Cambrian Trilobites. (*a*) *Paradoxides bohemicus* (natural size); (*b*) *Agnostus princeps* (?); (*c*) *Olenus micrurus* (natural size); (*d*) *Ellipsocephalus Hoffi* (natural size).

markings afford unquestionable proof of the presence of creatures which have otherwise utterly disappeared.

Among the most abundant and characteristic fossils of the older stratified rocks of the earth's crust are those to which the general name of Trilobites has been given. These long-extinct animals were crustaceans, having a more or less distinctly three-lobed body, at one end of which was the head or cephalic shield, usually with a pair of fixed compound eyes ; at the other end the caudal shield or tail ; while between the two shields was the ringed or jointed body, the rings of which were movable, so that the animal could bring the two shields together or coil itself up. It will be seen from the different genera represented in Figs. 118 and 125 how varied were the forms which they assumed. In the shapes and relative sizes of the shield and segmented body, in the number of the body-rings, in the development of spines, and in other features, the most wonderful variety is traceable among the trilobites even of the oldest fossiliferous strata. Some of the earliest genera were also the largest ; *Paradoxides* sometimes reaching a length of nearly two feet. Yet contemporaneous with this large creature were some diminutive forms. A few genera (among them *Agnostus*) were blind ; but most possessed eyes furnished with facets, which in some forms are fourteen in number, while in others they are said to amount to 15,000. The peculiar crescent-shaped eye on each side of the head is well shown in some of the forms represented in Figs. 118 and 125. The trilobites appear to have particularly swarmed on sandy and muddy bottoms, for their remains are abundant in many sand-stones and shales.

Another form of crustacean life represented in the early Palæozoic ocean was that of the Phyllopods—animals furnished with bivalve shell-like carapaces, which protected the head and upper part of the body, while the jointed tail projected beyond it. Most of them were of small size (see Fig. 126). The character-istic Cambrian genus is *Hymenocaris*.

Of all the divisions of the animal kingdom none is so important to the geologist as that of the Mollusca. When one walks along the shores of the sea at the present time, by far the most abundant remains of the marine organisms to be there observed are shells. They occur in all stages of fresh-ness and decay, and we may trace even their comminuted fragments forming much of the white sand of the beach. So in the geological formations, which represent the shores and shallow sea-bottoms of former periods, it is mainly remains of the marine shells that have been preserved. From their abundance and wide diffusion, they supply us with a basis for the comparison of the strata of different ages and countries, such as no other kind of organic remains can afford.

It is interesting and important to find that among the fossils of the oldest fossiliferous rocks the remains of molluscan shells occur, and that they are of kinds which can be satisfactorily referred to their place in the great series of the Mollusca. The most abundant of them are representatives of the Brachiopods or Lamp-shells. Among these are species of the genera *Lingula* (*Lingulella*, Fig.

FIG. 119.—Cambrian bra-chiopod (*Lingulella Davisii*, natural size).

119) and *Discina* which have a peculiar interest, inasmuch as they are the oldest known molluscs, and are still represented by living species in the ocean. They have per-sisted with but little change during the whole of geological time, from the early Palæozoic periods downwards, for the living shells do not appear to indicate any marked divergence from the earliest forms. They possess horny shells which are not hinged together by teeth. A more highly organised order of brachiopods possesses two hard calcareous shells articulated by teeth on the hinge-line. These forms, apparently later in their advent, soon vastly outnumbered the horny lingulids and discinids. So abundant are they both in individuals and in genera and species among the older Palæozoic rocks, that the period to which these rocks belong is sometimes spoken of as the " Age of Brachiopods."

The ordinary bivalve shells or Lamellibranchs had their re-presentatives even in Cambrian times. From that early period they have gradually increased in numbers, till they have attained their maximum at the present time. Among the known Cambrian genera are *Ctenodonta* allied to the living " ark-shells," and *Modio-lopsis*, probably representing some of the modern mussels.

The Gasteropods or common univalve shells, now so abundant in the ocean, made their advent not later than Cambrian time, for the remains of the genus *Bellerophon* (Fig. 129) are found in the group of strata known as the Lingula-flags in Wales.

The highest division of the molluscs, the Cephalopods, to which the living nautilus and cuttle-fish belong, is but poorly re-presented at the present time. But during the Palæozoic and Secondary periods it flourished exuberantly, both as regards number of individuals and variety of forms. It is divisible into two great families. In one of these the shell is usually internal and is never chambered ; in the other the shell is chambered and external, the chambers being connected by a tube or siphuncle. The former family includes all the living cuttle-fishes, squids, and

the paper-nautilus ; the latter comprises only one living representative—the pearly nautilus. It is to the family of chambered
cephalopods that the Palæozoic forms are all referable. In some
the shell was straight, in others it was variously curved. Only
scanty traces of cephalopodan life have yet been found among
the Cambrian rocks. But occasional examples of the important
genus *Orthoceras* (see Fig. 130) show that this great division of the
molluscs had even in the earliest Palæozoic ages appeared upon
the earth.

As the term Cambrian denotes, the rocks to which this
name is applied are well developed in Wales. There, and in
the border English counties, they attain a depth of perhaps more
than 20,000 feet. They are found also in the east of Ireland,
while in the north-west of Scotland they appear to be represented
by massive red sandstones and conglomerates. The peculiar
Primordial fauna has been widely recognised in both hemispheres.
It occurs at intervals from France to Russia, and from Sweden to
Bohemia and Sardinia. It is well represented in the United States
and Canada, and it is met with even as far east as China.

The following Table gives the commonly accepted subdivisions
of the Cambrian rocks in Britain.

UPPER, distinguished by the prevalence of trilobites like *Olenus*.	Tremadoc group—dark grey slates. Lingula Flags—bluish and black slates, flags, and sandstones.
LOWER, marked by abundance of trilobites of the genus *Paradoxides*.	Menevian group—sandstones, shales, slates, and grits. Harlech group—purple, red, and grey flags, sandstones, slates, and conglomerates. (Horizon of *Olenellus*.)

Recently the existence of *Olenellus*—a genus of trilobite characteristic of the
lowest platform of the Cambrian system has been found in the midland counties
of England and in the north-west Highlands of Scotland. In the latter region
the occurrence of this fossil fixes the stratigraphical position of the well-known
quartzites as Lower Cambrian.

CHAPTER XVIII

SILURIAN

THE origin and use of the term SILURIAN have already been
given (p. 240). The rocks embraced under this term form a mass
of strata which in some countries (Wales and Scotland) must be
many thousand feet thick. Like the Cambrian system below, into
which they graduate downward, they consist mainly of greywackes,
sandstones, shales, or slates ; but they are marked by the occasional occurrence of bands of limestone—a rock which from this
part of the geological record appears in increasing quantity onwards to recent times. Some highly characteristic bands of dark
carbonaceous shale are in some countries persistent for long distances, and contain abundant graptolites. Not infrequently these
dark shales are full of pyritous impregnations, which, when the
rock weathers, give rise to the efflorescence of alum or the formation of chalybeate springs ; such bands are sometimes called *alum-
schists*. In Wales, the Lake District of the north of England, and
to a less degree in the south of Scotland, there are remains of
submarine volcanic eruptions of Silurian time in the form of inter-
. calated sheets of tuff and beds of different lavas.

In certain regions (Russia, New York) Silurian rocks have
undergone little change since the time of their deposition ; but, as
a rule, they have been more or less indurated, plicated, and dislocated (Wales, Lake District, etc.), while in some countries
(Norway, Scotland) they have been so crushed and metamorphosed
as to have assumed the character of schistose rocks (phyllites,
mica-schists, etc.)

Murchison subdivided his Silurian system into two great sections,
Lower and Upper. This classification still holds, though the
limits and nomenclature of the several component groups have
not been exactly maintained. The arrangement of the various

subdivisions, as followed in Britain, is shown in the table on
p. 257.

Taking the fossils of the Silurian system as a whole, we find
that they prolong and amplify the peculiar type of life found to
characterise the Cambrian system.   They include both flora and
fauna.   The flora, however, is exceedingly meagre.   It consists
almost entirely of sea-weeds, which occur usually in the form of
fucoid-like impressions.   But, as already remarked in reference to
the so-called plants of the Cambrian rocks, many of the supposed
vegetable remains are almost certainly not such (see p. 242).   Some
of them may be tracks left by
worms, crustaceans, or other
marine creeping or crawling
creatures, upon soft mud or sand;
others may be casts of hollows
made by trickling water or
yielding sediment; while others
seem to be the result of some
peculiar crumpling or puckering
of the strata.   But undoubted
remains of sea-weeds do occur.
Some of these are delicate
branching forms, like some still
living, as shown in the organism figured in Fig. 120 from the

FIG. 120.—An Upper Silurian sea-weed
(*Chondrites verisimilis*), natural size.

Upper Silurian series.   Among the Upper Silurian strata, also,
traces of land-vegetation have been detected in the form of spores
and stems of cryptogamous plants.   Lycopods or club-mosses and
ferns appear to have been the chief types in the earliest terrestrial
floras; at least, it is remains referable to them that chiefly occur
in the older Palæozoic rocks.   They reached a great development
in the Carboniferous period, in the account of which a fuller
description of them will be given.   We can dimly picture the
Silurian land with its waving thickets of fern, above which lycopod
trees raised their fluted and scarred stems, threw out their scaly
moss-like branches, and shed their spiky cones.

The fauna of the Silurian period has been more abundantly
preserved than that of the Cambrian, and appears to have been
more varied and advanced.   Among its simpler forms were Foraminifera and sponges.   A foraminifer (of which there were no
doubt representatives in Cambrian times, and there are still many
living types in the present ocean, see Fig. 33) is generally a

minute animal, composed of a jelly-like substance which, possessing no definite organs, has in some kinds the power of secreting a hard calcareous or horny shell, through openings or pores (*foramina*) in which filaments from the jelly-like mass are protruded. By other kinds, grains of sand are cemented together to form a protecting shell. It is these calcareous and sandy coverings which occur in the fossil state and prove the presence of foraminifera in the older oceans of the globe. Sponges also are known to have existed in the Cambrian and Silurian seas, and their remains have been met with in all parts of the Geological Record down to the present day. It is, of course, only where these animals secrete hard durable parts that they can be detected as fossils. A sponge is a mass of soft, transparent, jelly-like substance, perforated by tubes or canals, and supported on an internal network of minute calcareous or siliceous spicules, or of interlacing horny fibres. Most fossil sponges are calcareous or siliceous, and their hard parts, being durable, have been preserved sometimes in prodigious numbers and in wonderful perfection. The common sponge of domestic use is an example of the horny type.

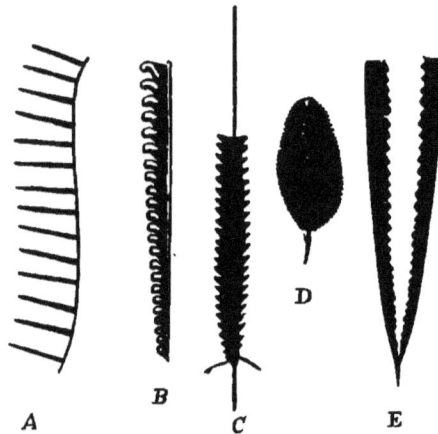

FIG. 121.—Graptolites from Silurian rocks. A, *Rastrites Linnæi*. B, *Monograptus priodon*. C, *Diplograptus pristis*. D, *Phyllograptus typus*. E. *Didymograptus Murchisonii* (all natural size).

The Hydrozoa were abundantly represented in the Silurian seas by graptolites (see p. 242), of which there were many kinds. Some of the more characteristic of these are shown in Fig. 121. They abound in certain bands of shale, both in the Lower and

Upper Silurian series, the double forms (such as C, Fig. 121) being more characteristic of the Lower division, while the single forms run throughout the system.

Corals abound in some parts of the Silurian seas. Their remains chiefly occur in the limestones, doubtless because these rocks were formed in comparatively clear water, in which the corals could flourish. But they differed in structure from the familiar reef-building corals of the present day. The great majority of them belonged to the family of the Rugose corals, now only sparingly represented in the waters of the present ocean. As their name denotes, they were particularly marked by their thick rugged

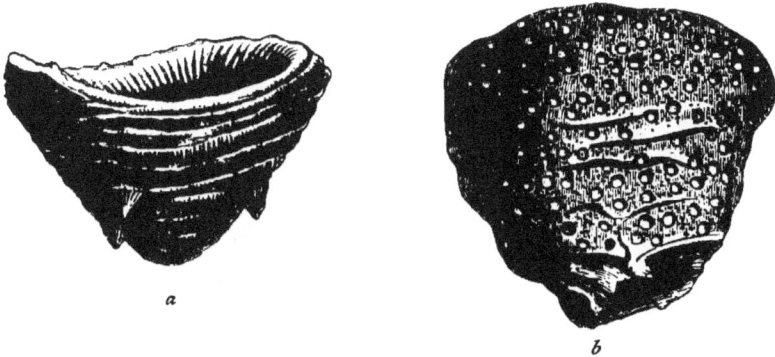

FIG. 122.—Silurian Corals. (a) Rugose Coral (*Omphyma turbinatum*, ½). (b) Alcyonarian Coral (*Heliolites interstinctus*, natural size).

walls. Many of them were single independent individuals ; some lived together in colonies ; while others were sometimes solitary, sometimes gregarious. A typical example of these rugose forms is *Omphyma*, shown in Fig. 122 (a). Other genera were *Cyathaxonia, Cyathophyllum*, and *Zaphrentis*. There were likewise less numerous and more delicate compound forms belonging to what are known as the Tabulate corals (*Favosites, Halysites*), while another type (*Heliolites*, Fig. 122, b) represented in ancient times the Alcyonarian corals (*Heliopora*) of the present time.

Crinoids or stone-lilies played an important part in the earlier seas of the globe. In some regions they lived in such abundance on the sea-floor that their aggregated remains formed solid beds of limestone hundreds of feet thick, and covering thousands of square miles. As their name denotes, crinoids are lily-shaped animals, having a calcareous, jointed flexible stalk fixed to the bottom and supporting at its upper end the body, which is composed of calcareous plates furnished with branched calcareous arms (see Figs. 149, 165, 173). It is these hard calcareous parts which have been so abundantly preserved in the fossil state.   Remains of crinoids are found in various parts of the Silurian system (*Dendrocrinus, Glyptocrinus*) chiefly in the limestones, but not in such abundance and variety as in later portions of the Palæozoic formations (compare pp. 265, 277, and Figs. 149, 165, and 173). Allied to the crinoids were the Cystideans, a curious order of echinoderms, with rounded or oval bodies en-

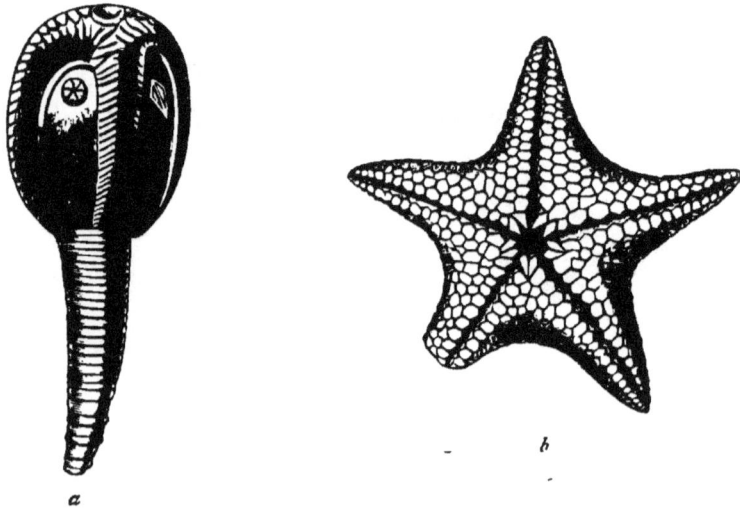

FIG. 123.—Silurian Echinoderms. (*a*) Cystidean (*Pseudocrinites quadrifasciatus,* natural size). (*b*) Star-fish (*Palæasterina stellata*, 3½).

closed in calcareous plates, possessing only rudimentary arms, and a comparatively small and short jointed stalk.   They first appeared in the Cambrian period (*Protocystites*), but attained their chief development during Silurian time, thereafter diminishing in numbers.   They are thus characteristically Silurian types of life.   One of them is represented in Fig. 123 (*a*).   Star-fishes and brittle-stars likewise occur as fossils among the Silurian rocks.   These marine creatures, still represented in our present seas, possess hard calcareous plates and spines, which, being imbedded in a tough leathery integument, have not infrequently been preserved in their natural position as fossils.   Some of the genera of star-fishes found in the Silurian system are *Palæaster, Palæasterina* (Fig. 123, *b*), *Palæochoma*.   Brittle-stars were represented by *Protaster*.

In the Silurian system are found many tracks and burrows like those of the Cambrian rocks, indicative of the presence of different kinds of sea-worms.   Throughout great
thicknesses of strata, indeed,
these markings are sometimes
the only or chief fossils to be
found.   Names   have   been
given to the different kinds of
burrows   (*Arenicolites,   Scoli-
thus, Lumbricaria,* Fig. 124),
and of trails (*Palæochorda,
Palæophycus*).   There   were
likewise representatives of the
familiar *Serpula,* which is found
so abundantly on the present
sea-bottom, encrusting shells

Fig. 124.—Filled-up Burrows or Trails left by
a sea-worm on the bed of the Silurian sea
(*Lumbricaria antiqua,* ⅓).

and stones with a calcareous protecting tube, inside of which the
annelide lives.   This tube has been preserved in the fossil state
in rocks of all ages.

The Trilobites, which had already appeared in Cambrian time,
attained their maximum development during the Silurian period.
A few of the primordial or Cambrian types continued to live into
this period, but many new genera appeared.   In the Lower
Silurian series some of the more abundant genera are *Asaphus,
Ampyx, Ogygia,* and *Trinucleus*; in the Upper Silurian division
characteristic genera are *Calymene, Phacops, Encrinurus, Illænus,*
and *Homalonotus* (Fig. 125).'   Trilobites continued to flourish,
but in gradually diminishing variety, during the Devonian and
Carboniferous periods, after which they seem to have died out.
They are thus a distinctively Palæozoic type of life, each great
division of the Palæozoic rocks being characterised by its own
varieties of the type.

Phyllopod crustaceans likewise attained to greater variety
during the Silurian period; some of the more frequent genera are
Ceratiocaris (Fig. 126), *Discinocaris,* and *Caryocaris.*   The
Phyllopods attained their maximum development during Palæozoic
time, but they have continued in existence ever since, and are
at present represented by a number of genera, some of which
live in the sea, others in fresh water.   ·

The Mollusca are far more abundant and varied in the Silurian
than in the Cambrian rocks.   Among the more characteristic
Silurian genera of Brachiopods are *Atrypa, Leptæna, Orthis,*

Pentamerus, Rhynchonella, and *Strophomena* (Fig. 127). Among the lamellibranchs we find the Cambrian genera *Ctenodonta* and *Modiolopsis,* with new forms such as *Orthonota* (Fig. 128), *Cleido-phorus* and *Ambonychia.*

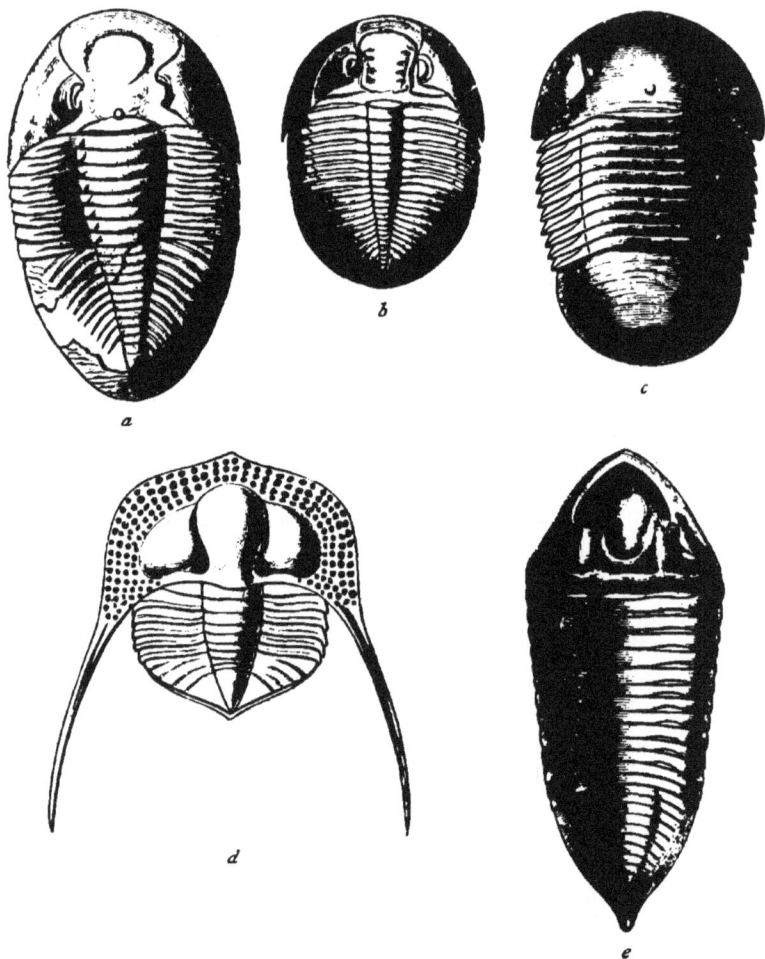

FIG. 125.—Lower and Upper Silurian Trilobites. (*a*) *Asaphus tyrannus* (½); (*b*) *Ogygia Buchii* (½); (*c*) *Illænus barriensis* (½); (*d*) *Trinucleus concentricus* (natural size); (*e*) *Homalonotus delphinocephalus* (½).

The Gasteropods played an important part in the fauna of the Silurian sea, for upwards of 1300 species of them have been found in Silurian rocks. Among the more frequent genera are

Bellerophon (Fig. 129), *Ophileta*, *Holopca*, *Murchisonia*, *Platy-schisma.*

Numerous representatives of the chambered cephalopods have been found in the Silurian rocks, especially in the upper division. Among the more frequent genera are *Orthoceras* (straight, Fig. 130 *a*), *Cyrtoceras* (curved), *Ascoceras* (globular or pear-shaped), *Lituites* (coiled, Fig. 130 *b*), and also *Nautilus*, a genus which has persisted through the greater part of geological time to the present day, and now remains the only representative of the chambered cepha-lopods formerly so abundant.

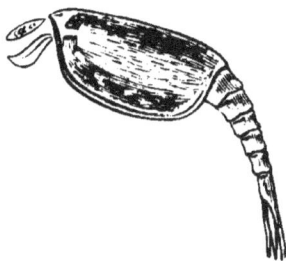

FIG. 126.—Silurian Phyllopod Crustacean (*Ceratiocaris papilio*).

Remains of Fishes detected in the Upper Silurian rocks are the earliest traces of vertebrate life yet known. They consist partly of plates which·are regarded

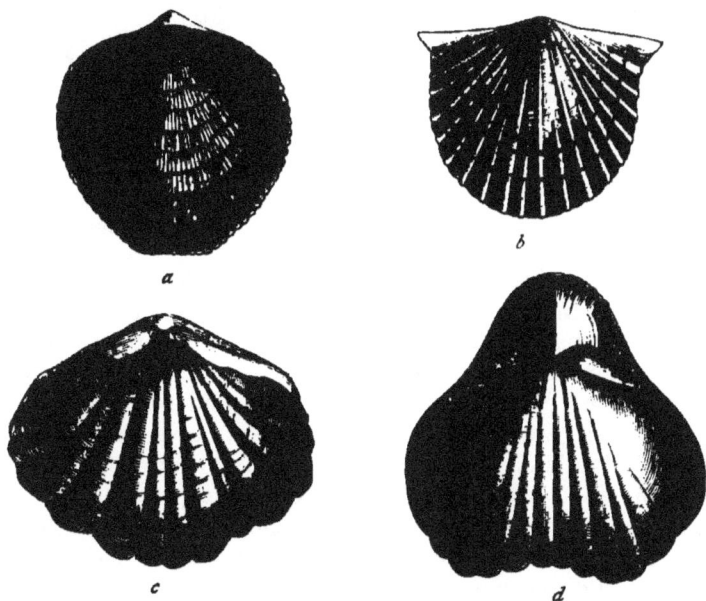

FIG. 127.—Silurian Brachiopods. (*a*) *Atrypa reticularis* (natural size), Caradoc beds to Lower Devonian; (*b*) *Orthis actoniæ* (natural size); (*c*) *Rhynchonella borealis* (natural size); (*d*) *Pentamerus galeatus* (natural size).

as portions of the bony covering of certain placoderms or bone-plated forms (*Pteraspis, Cephalaspis, Auchenaspis*); partly

of curved spines and shagreen-like fragments. The creatures
of which these are relics appeared as forerunners of the remarkable
assemblage of fishes in the next geological period (see p. 260).
All the animal remains hitherto.enumerated are relics of the
inhabitants of the sea. Of the land-animals of the time nothing
was known until the year 1884, when, by a curious coincidence,
the discovery was made of the remains of scorpions in the
Silurian rocks of Sweden, Scotland, and .the United States, and
of an insect allied to the living cockroach (*Palæoblattina*) in
those of France. If scorpions and insects existed during this
ancient period we may be sure that other forms of terrestrial
life were also present. A new interest is thus given to the
prosecution of the search for fossils among the older formations.

FIG. 128.—Silurian Lamellibranch (*Orthonota semisulcata* (natural size).

FIG. 129.—Silurian Gasteropod
(*Bellerophon dilatatus*, ⅓).

Putting together the evidence furnished by the rocks and
fossils of the Silurian system, we get a glimpse of the aspect of
the globe during the early geological period which they represent.
The rocks bring before us the sand, mud, and gravel of the
bottom of the sea, and tell of some old land from which these
materials were worn away. The detritus carried out from the
shores of that land was laid down upon the sea-bottom just as
similar materials are being disposed of at the present day. The
area occupied by Silurian rocks marks out the tracts then covered
by the sea. Following these upon a map we perceive that vast
regions of the existing continents were then parts of the ocean-
floor. In Europe, for example, Silurian rocks underlie the greater
part of the British Islands, whence they stretch northwards across
a large part of Scandinavia and the basin of the Baltic. They
rise to the surface in many places on the continent from Spain
to the Ural Mountains. They are found forming parts of some

of the great mountain-chains of the globe, as, for instance, in the Cordilleras of South America, in the Alps, and in the Himalayas. Even at the antipodes they are met with as thick masses in Australia and New Zealand. It is evident that the geography of the globe in Silurian times was utterly unlike what it is now. A large part of the present land was then covered with shallow seas, in which the Silurian sedimentary rocks were laid down. There would seem to have been extensive masses of land in the boreal part of the northern hemisphere connecting the European, Asiatic, and American continents. Along the coast-line of the

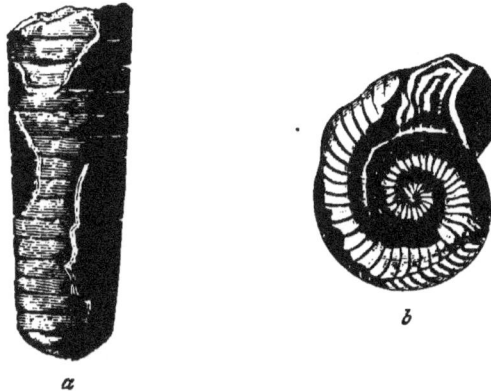

Fig. 130.—Silurian Cephalopods. (*a*) *Orthoceras emeritum* (⅓); (*b*) *Trochoceras* (*Lituites*) *cornu-arietis* (⅓).

northern land and across the shallow seas lying to the south of it, the same species of marine organisms migrated freely between the Old and the New Worlds.

The following Table shows the subdivisions which have been made in the Silurian system of Britain.

Upper Silurian {
Ludlow group (mudstone and Aymestry limestone)—Kirkby Moor and Bannisdale flags and slates.
Wenlock group (shales and limestones)—Denbighshire and Coniston grits and flags.
Upper Llandovery group—May Hill sandstones.
}

Lower Silurian {
Lower Llandovery group—grits and sandstones.
Bala and Caradoc group—sandstones, slates, and grits, with Bala (Coniston) limestone.
Llandeilo group—dark argillaceous and sometimes calcareous flagstones and shales.
Arenig group—dark slates, flags, and sandstones.
}

S

CHAPTER XIX

DEVONIAN AND OLD RED SANDSTONE

THE DEVONIAN system, which comes next in order, was named by Sedgwick and Murchison after the county of Devon where they studied its details. In Europe, and likewise in the eastern part of North America, it occurs in two distinct types, which bring before us the records of two very different conditions in the geography of these regions during the time when the rocks composing the system were being deposited. The ordinary type, which occurs all over the world, represents the tracts that were covered by the sea, and has preserved the remains of many forms of the marine life of the period. It is that to which the name Devonian is more particularly applicable. The less frequent type is characterised by thick accumulations of sandstones, flagstones, and conglomerates that were laid down in lakes and inland seas, and contain a distinct assemblage of land and fresh-water fossils. This lacustrine type is known by the name of OLD RED SANDSTONE.

In their general character the Devonian rocks resemble those of the Silurian system underneath. In Central Europe, where they attain a thickness of many thousand feet, their lower division consists mainly of sandstones, grits, greywackes, slates, and phyllites. The central zone contains thick masses of limestone, often full of corals and shells, while the upper portions comprise thin-bedded sandstones, shales, and limestones. These various strata represent the sediments intermittently laid down upon the bottom of the sea which then covered the greater part of Europe. Here and there, they include bands of diabase and tuff, which show that submarine volcanic eruptions took place during their deposition.

In the north-west of Europe, however, the floor of the Silurian sea was irregularly ridged up into land, and large lakes were formed, into which rivers from the ancient northern continent poured enormous quantities of gravel, sand, and silt. The sites of these lakes can be traced in Scotland, the north of England, and Ireland. Similar evidence of land and lake-waters is found in New Brunswick and Nova Scotia. That some of the larger lakes were marked by lines of active volcanoes is well shown in Central Scotland, where the piles of lava and ashes left by the eruptions are more than 6000 feet thick.

The occurrence of both marine and lacustrine deposits is of the highest interest, for, on the one hand, we learn what kinds of

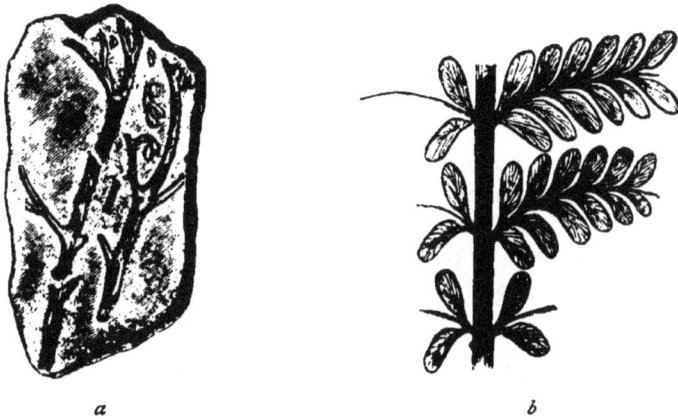

FIG. 131.—Plants of the Devonian period.   (a) *Psilophyton* (⅓); (b) *Palæopteris* (¼).

animals lived in the sea in succession to those that peopled the Silurian waters, and, on the other hand, we meet with the first abundant remains of the vegetation that covered the land, and of the fishes that inhabited the fresh waters. The terrestrial flora of the Devonian period has been only sparingly preserved in the marine strata; but occasional drifted specimens occur to show that land was not very distant from the tracts on which these strata were laid down. In the lacustrine series or Old Red Sandstone of Britain more abundant remains have been met with, but the chief sources of information regarding this flora are to be sought in New Brunswick and Gaspé, where upwards of 100 species of plants have been discovered. Both in Europe and in North America, the Devonian vegetation was characterised by the predominance of ferns, lycopods (*Lepidodendron*, etc.), and calamites. It was essentially acrogenous — that is, it consisted mainly of flowerless plants like our modern ferns, club-mosses, and horse-tail reeds. Traces of coniferous plants, however, show that on the upland of the time pine-trees grew, the stems of which were now and then swept down by floods into the lakes or the sea.

While the general aspect of the flora was uniformly green and somewhat monotonous, the fauna had now become increasingly varied. We know that these early woodlands were not without insect life, for neuropterous and orthopterous wings have been preserved in the strata. Some of these remains indicate the existence of ancient forms of ephemera or May-fly, one of which was so large as to have a spread of wing measuring 5 inches across. There were likewise millipedes, which fed on the decayed wood of the forests. Traces of land-snails too have been detected among the fossil vegetation. It is evident, however, that the plant and animal life of the land has only been sparingly preserved ; and though our knowledge of it has in recent years been largely increased, we shall probably never discover more than a mere fragmentary representation of what the original terrestrial flora and fauna really were.

The lake-basins of the Old Red Sandstone have yielded large numbers of remains of the fishes of the time. These are members of the remarkable order of Ganoids — the earliest known type of fishes—which, though so abundant in early geological time, is represented at the present day by only a few widely scattered species, such as the sturgeon, the polypterus of the Nile, and the bony pike or garpike of the American lakes. These modern forms are denizens of fresh water, and there is reason to believe that their early ancestors were also inhabitants of lakes and rivers, though many of them may also have been able to pass out to the sea. The ganoids are so named from the enamelled scales and plates of bone in which they are encased. In some of the fossil forms, this defensive armour consisted of accurately fitting and overlapping scales (Figs. 132, 133) ; in others, the head with more or less of the body was protected by large and thick plates of

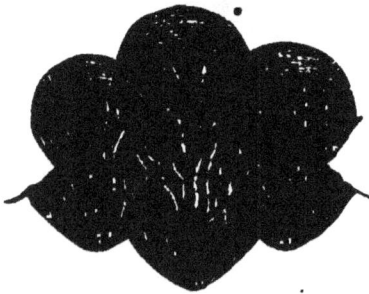

FIG. 132.—Overlapping scales of an Old Red Sandstone fish (*Holoptychius Andersoni*, natural size).

bone (Fig. 134). Examples of both these kinds of armature are to be observed among the fishes of the Old Red Sandstone. Some of the most characteristic scale-covered genera are *Osteolepis*, *Diplopterus*, *Glyptolæmus*, *Holoptychius*, *Acanthodes*

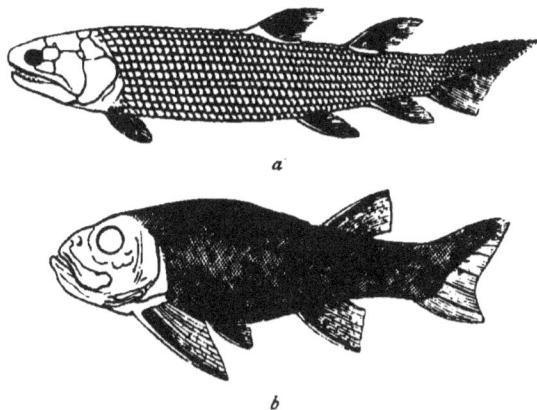

FIG. 133.—Scale-covered Old Red Sandstone fishes.   (*a*) *Osteolepis* ; (*b*) *Acanthodes* (both reduced).

(Figs. 132, 133). The acanthodians (Fig. 133, *b*), distinguished by the thorn-like spines supporting their fins, reached their

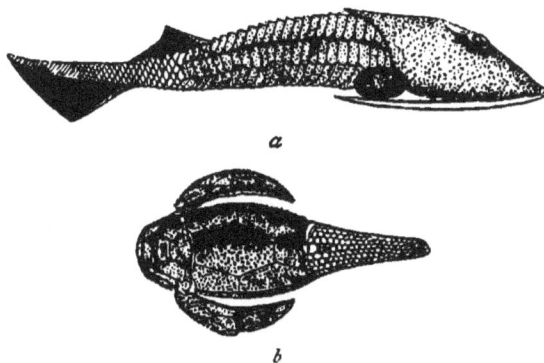

FIG. 134.—Plate-covered Old Red Sandstone fishes.   (*a*) *Cephalaspis*; (*b*) *Pterichthys* (both reduced).

greatest development during the Devonian period. Of the plate-covered ganoids or placoderms some of the most characteristic were the curious *Cephalaspis* (Fig, 134, *a*), with its head-buckler shaped like a saddler's awl, the *Pteraspis*, which, with *Cephalaspis*, had already appeared in the Silurian period, the *Coccosteus* and *Pterichthys* (Fig. 134, *b*). Some of the contemporaries of these creatures attained a great size. Thus the *Asterolepis* had its head and shoulders encased in a buckler, which in some examples is 20 inches long by 16 broad. Still larger were some of its American allies, one of which, the *Dinichthys*, had a head-buckler 3 feet long armed with formidable teeth.

One of the fishes of the Old Red Sandstone, named *Dipterus*, has recently been found to have a singular modern representative in the barramunda or mud-fish (*Ceratodus*) of the Queensland rivers in Australia. *Dipterus* resembled the ganoids in its external enamel and strong bony helmet, but its jaws present the characteristic teeth, and its scales have the rounded or "cycloid" form of *Ceratodus*. That some of these fishes swarmed in the waters of the Old Red Sandstone is shown by the prodigious numbers of their remains occasionally preserved in the sandstones and flagstones. Their bodies lie piled on each other in such numbers, and often so well preserved, as to show that probably the animals were suddenly killed, and were covered up with sediment before their remains had time to decay and to be dispersed by the currents of water. Perhaps earthquake shocks, or the copious discharge of mephitic gases, or other sudden baneful influence, may have been the cause of the extensive destruction of life in these ancient waters.

FIG. 135.—Devonian Eurypterid Crustacean (*Pterygotus*, reduced).

That some of the fishes found their way to the sea, as our modern salmon does, is indicated by the occasional occurrence of their remains among those of the truly marine fauna of the Devonian rocks. But the rarity of their presence there, compared with their prodigious abundance in some parts of the Old Red Sandstone, probably serves to show that they were essentially inhabitants of the lakes and rivers of the land.

Among the animals that appear to have been migratory between the sea and the terrestrial waters, were the curious forms known as Eurypterids, which, though generally classed with the crustaceans, had many affinities with the arachnids or scorpions. One of the most remarkable of these creatures was the *Pterygotus*, of which the general form is shown in Fig. 135.   Most of the species are small, though one of them found in Scotland must have attained a length of 5 or 6 feet.

But it is the marine or Devonian fauna which is most widely

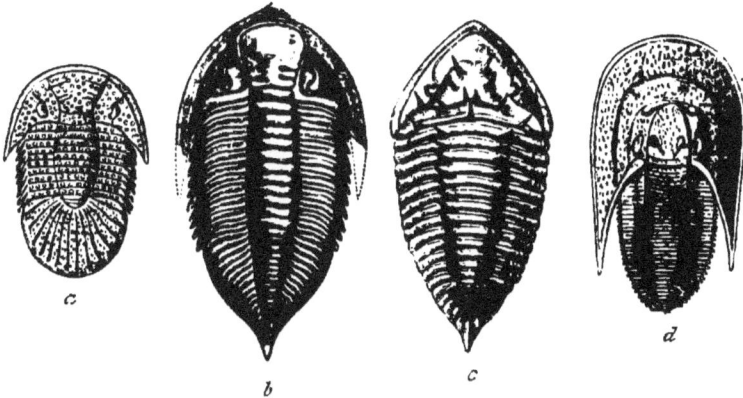

FIG. 136. – Devonian Trilobites.   (*a*) *Bronteus flabellifer* (½); (*b*) *Dalmanites rugosa* (½); (*c*) *Homalonotus armatus* (¼): (*d*) *Harpes macrocephalus* (½).

spread over the globe, and from its extensive distribution is of most importance to the geologist.   Taken as a whole, it presents a general resemblance to that of the Silurian period which it succeeded.   Some of the Silurian species survived in it, and new

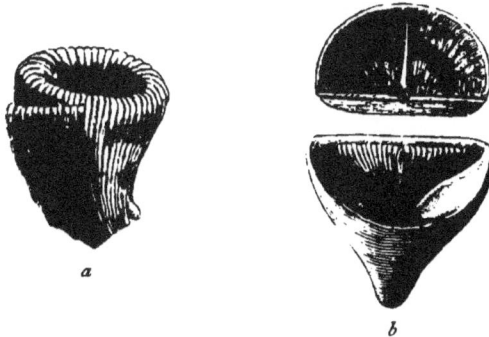

FIG. 137.—Devonian Corals.   (*a*) *Cyathophyllum ceratites* (½); (*b*) *Calceola sandalina* (¾).

species of the old genera made their appearance.   But important differences are to be observed between the faunas of the two systems, showing the long lapse of time, and the changes which it brought about in the life of the globe.

It is specially interesting to mark how some of the characteristic Silurian types dwindle and finally die out in the Devonian system. One of the best examples of this survival and disappearance is supplied by the Graptolites.   It will be remembered how pro-digiously abundant these creatures were in the Silurian seas.   They are met with also in scattered specimens in the lower and middle divisions of the Devonian system, but their rarity there affords a

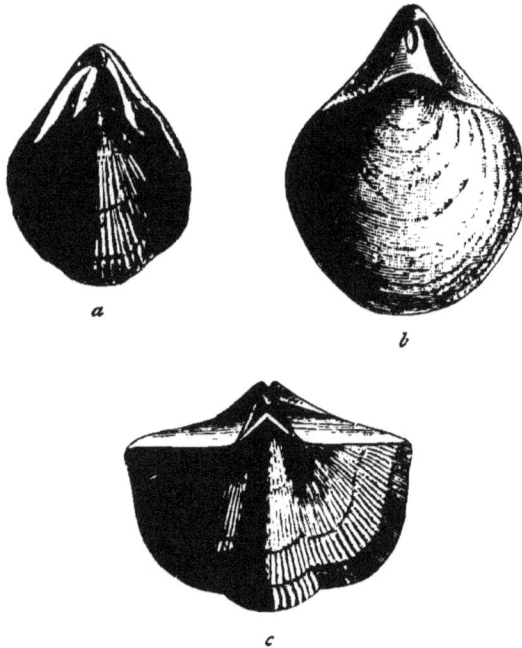

FIG. 138.—Devonian Brachiopods.   (*a*) *Uncites gryphus* (⅔); (*b*) *Stringocephalus Burtini* (⅔); (*c*) *Spirifera disjuncta* (*Verneuillii*) (¼).

striking contrast to their profusion among the Silurian strata, and they seem to have entirely died out before the end of the Devonian period, for no traces of them occur in the later parts of the system, and they have never been met with in any later geological forma-tion.

Again, Trilobites, which form such a predominant and striking feature of the Silurian fauna, occur in greatly diminished number and variety among the Devonian rocks.   Most of the Silurian genera are absent, some of the most frequent Devonian types being *Phacops, Cryphæus, Homalonotus, Dalmanites,* and *Bronteus* (Fig.

136).   We shall find that this peculiarly Palæozoic type of
crustacea finally died out in the next or Carboniferous period.
But while the trilobites were waning, the eurypterids, already referred to, appeared and attained a great development.

In the clearer parts of the sea vast numbers of rugose corals
flourished, and, with other calcareous organisms, built up solid
masses of limestone.   Some of the characteristic genera were
Cyathophyllum (Fig. 137), *Acervularia, Cystiphyllum,* and the
curious *Calceola* which, after being successively placed among the
lamellibranchs and the brachiopods, is now regarded as a rugose
coral with an opercular lid.   With these were likewise associated

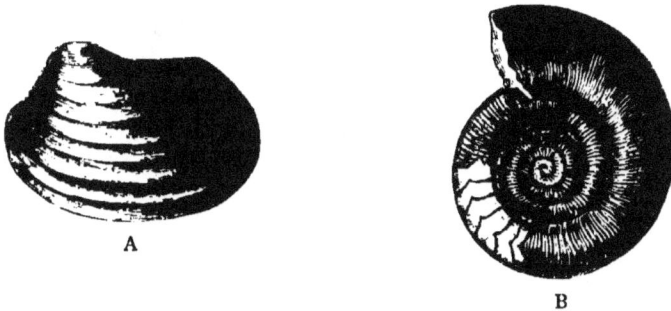

Fig. 139.—A, Devonian Lamellibranch (*Cucullæa Hardingii,* ⅓).   B, Devonian
Cephalopod (*Clymenia Sedgwickii,* ⅔).

vast numbers of crinoids, of which the genera *Cyathocrinus* and
Cupressocrinus were especially characteristic.

The Brachiopods reached their maximum of development in
the Devonian seas, upwards of 60 genera and 1100 species having
been described from Devonian rocks.   Comparing them with
those of the Silurian system, we notice that some of the most
characteristic Silurian types, such as forms of *Orthis* and *Strophomena,* became fewer in number, while forms of *Productus* and
Chonetes increased.   The most abundant families were those of
the Spirifers (*Uncites, Cyrtia, Athyris, Atrypa*) and Rhynchonel-
lids (Fig. 138).   Two distinctively Devonian brachiopods were
Stringocephalus and *Rensseleria,* allied to the still living *Terebratula.*   The former is especially characteristic of one of the
Middle limestones (see Table on next page).

The other mollusca appear to have been well represented in
the Devonian seas.   Of the Lamellibranchs, *Pterinea* is particularly abundant in the lower part of the system, *Cucullæa* (Fig.

139, A) in the upper part.   The Devonian cephalopods included many species of the genera *Orthoceras*, *Cyrtoceras*, *Clymenia*, *Goniatites*, and *Bactrites* (Fig. 139, B).

The Devonian system in Europe is subdivided as in the subjoined Table :—

Upper	Pilton and Pickwell-Down group of England — Upper Old Red Sandstone of Scotland. Famennian and Frasnian sandstones, shales, and limestones of the North of France and Belgium—Psammites de Condroz. Cypridina-shales, Spirifer sandstone, *Rhynchonella cuboides* beds of Germany.
Middle	Ilfracombe and Plymouth limestones, grits, and conglomerates of Devonshire. [No middle Old Red Sandstone.] Limestone of Givet, and Calceola shales of North of France. Stringocephalus - limestone of the Eifel — Calceola - group of Germany.
Lower	Linton slates and sandstones of Devon and Cornwall — Lower Old Red Sandstone of Scotland and Wales. Coblenzian, Taunusian, and Gedinnian rocks of the Ardennes and Taunus.

CHAPTER XX

THE next great division of the Geological Record has received the name of CARBONIFEROUS, from the beds of coal (Latin *Carbo*) which form one of its most conspicuous features. The rocks of which it consists reach sometimes a thickness of fully 20,000 feet, and contain the chronicle of a remarkable series of geographical changes which succeeded the Devonian period. They include limestones made up in great part of corals, crinoids, polyzoa, brachiopods, and other calcareous organisms which swarmed in the clearer parts of the sea ; sandstones often full of coaly streaks and remains of terrestrial plants ; dark shales not infrequently charged with vegetation, and containing nodules and seams of clay-ironstone ; and seams of coal varying from less than an inch to several feet or yards in thickness, and generally resting on beds of fire-clay.

These various strata are disposed in such a way as to afford clear evidence of the physical geography of large areas of the earth's surface during the Carboniferous period. The limestones attain a thickness of sometimes several thousand feet, with hardly any intermixture of sedimentary material. They consist partly of aggregated masses of corals and coralloid animals, which grew on the sea-floor somewhat after the manner of modern coral-reefs ; partly of aggregated stems and joints of crinoids, which must have flourished in prodigious numbers on the bottom, mixed with fragments of other organisms, the whole being aggregated into sheets of solid stone. The Carboniferous or Mountain Limestone, which forms the lower part of the Carboniferous system, stretches from the west of Ireland eastwards for a distance of 750 miles, across England, Wales, Belgium, and Rhineland into Westphalia. In the basin of the Meuse it is not less than 2500 feet thick, and in Lancashire, where it attains its maximum development, it exceeds

6000 feet. Such an enormous accumulation of organic remains shows that, during the time of its deposition, a wide and clear sea extended over the centre of Europe. But as the limestone is traced northwards, it is found to diminish in thickness. Beds of sandstone, shale, and coal begin to make their appearance in it, and rapidly increase in importance, as they are followed away from the chief limestone area ; while the limestone itself is at last reduced in Scotland to a few beds, each only a yard or two in thickness. From this change in the character of the rocks, the inference may be drawn that, while the sea extended from the west of Ireland eastwards into Westphalia, land lying to the north supplied sand, mud, and drifted plants, which, being scattered over the sea-floor, prevented the thick limestone from extending northwards. These detrital materials now form the masses of sandstone and shale that take the place of the limestone in the north of England and in Scotland. The northward extension of a few limestone beds full of marine organisms serves to mark a time when, for a longer or shorter interval, the water cleared, sand and mud ceased to be carried so far southward, and the corals, crinoids, and other limestone-building creatures were able to spread themselves farther over the sea-floor. But the thinness of such intercalated limestones also indicates that the intervals favourable for their formation were comparatively short, the sandy and muddy silt being once again borne southward from the land, killing off or driving away the limestone-builders and spreading new sheets of sand and mud over the site.

There can be no doubt that, while these changes were in progress, the whole wide area of deposition in Western and Central Europe was undergoing a gradual depression. The sea-bottom was sinking, but so slowly that the growth of limestone and the deposit of sediment probably on the whole kept pace with it. The actual depth of the water may not have varied greatly even during a subsidence of several thousand feet. That this was the case may be inferred from the structure of the limestone itself. We have seen that this rock sometimes exceeds 6000 feet in thickness. Had there been no subsidence of the sea-floor during the accumulation of so thick a mass of organic debris, it is evident that the first beds of limestone must have been begun at a depth of at least 6000 feet below the surface of the sea, and that, by the gradual increase of calcareous matter, the sea was eventually filled up to that amount, if it was not filled up entirely. But we can hardly suppose that the same kinds of organisms could live at a depth of 6000 feet and also at or near the surface. We should expect to find the organic contents of the lower parts of the limestone entirely different from those in the upper parts. But though there are differences sufficient to admit of the limestone being separated into stages, each marked by its own distinctive assemblage of fossils, the general character or facies of the organisms remains so uniform and persistent throughout, as to make it quite certain that the conditions under which the creatures lived on the bottom and built up the limestone continued with but little change during the whole time when the 6000 feet of rock were being deposited. As this could not have been the case had there been a gulf of 6000 feet to fill up, we are led to conclude that the bottom slowly subsided until its original level, on which the limestone began to form, had sunk at least 6000 feet.

This conclusion is borne out by many other considerations. Thus the sedimentary strata that replace the limestone on its northern margin are also several thousand feet thick. But from bottom to top they abound with evidence of shallow-water conditions of deposition. Their repeated alternations of sandstone, grit (even conglomerate), and shale ; the presence in them of constant current-bedding ; the frequent occurrence of ripple-marked and sun-cracked surfaces ; the preservation of abundant remains of terrestrial vegetation — some of it evidently in its position of growth — prove that the mass of sediment was not laid down in a deep hollow of the sea-bottom, but in shallow waters not far from the margin of the land.

But probably the most interesting evidence of long-continued subsidence during the Carboniferous period is furnished by the history of the coal-seams. Coal is composed of compressed and mineralised vegetation. In Britain each layer of coal is usually underlain by a bed of fire-clay, or at least of shale, through which roots and rootlets, descending from the under surface of the coal-seam, branch freely. There can be little doubt that each bed of fire-clay is an old soil, while the coal lying upon it represents the matted growth of vegetation which that soil supported. Hence the association of a fire-clay and a coal-seam furnishes distinct evidence of a terrestrial surface.[1]

In many regions the Carboniferous system comprises a series

[1] In some Continental coal-fields there is evidence that coal has likewise been formed out of matted vegetation which has been swept down by floods and been buried under sand, gravel, and other sediment.

of sandstones, shales, and other strata, many thousands of feet in thickness, throughout which, on successive platforms, there lie hundreds of seams of coal. If each of these seams marks a former surface of terrestrial vegetation, how is this succession of buried land-surfaces to be accounted for? There is obviously but one solution of the problem. The area over which the coal-seams extend must have been slowly sinking. During this subsidence, sand, mud, and silt were transported from the neighbouring land, and in such quantity as to fill up the shallow waters. On the muddy flats thus formed, the vegetation of the flat marshy swamps spread seaward. There may not improbably have been pauses in the downward movement, during which the maritime jungles and forests continued to flourish and to form a thick matted mass of vegetable matter. When the subsidence recommenced, this mass of living and dead vegetation was carried down beneath the water and buried under fresh deposits of sand and mud. As the weight of sediment increased, the vegetable matter would be gradually compressed and would slowly pass into coal. But eventually another interval of rest or of slower subsidence would allow the shallow sea once more to be silted up. Again the marsh-loving plants from the neighbouring swampy shores would creep outward and cover the tract with a new mantle of vegetation, which, on the renewal of the downward movement, would be submerged and buried.

FIG. 140.—Section of part of the Cape Breton coal-field, showing a succession of buried trees and land-surfaces. (*a*) sandstones; (*b*) shales; (*c*) coal-seams; (*d*) under-clays or soils.

In the successive strata of a coalfield, therefore, we are presented with the records of a prolonged period of subsidence, probably marked by longer or shorter intervals of rest. These more stationary periods are indicated by the coal-seams, and perhaps their relative duration may be inferred from the thickness of the coal. A thick coal-bed not

improbably marks a time of rest, when the vegetation was allowed
to flourish unchecked, or when at least the sinking was so imperceptible that the successive generations of plants, springing up on
the remains of their predecessors, contrived to keep themselves
above the level of the water.

In the present world there is no vegetable growth now in
progress quite like that of the coal-seams of the Carboniferous
period. Perhaps the nearest analogy is supplied by the mangrove-
swamps of tropical coasts (p. 83). In these tracts, the mangrove
trees grow seaward, dropping their roots and radicles into the
shallow waters, and gradually forming a belt of swampy jungle
several miles broad. That the coal-jungles extended into the sea
is shown by the occurrence of marine shells and other organisms
in the coal itself. But there were probably also wide swamps
wherein the water was fresh. A single coal-seam may sometimes
be traced over an area of more than 1000 square miles, showing
how widespread and uniform were the conditions in which it was
formed.

During the subterranean movements that marked the Carboniferous period, the Devonian physical geography was entirely
remodelled. The lake-basins of the Old Red Sandstone were
effaced, and the sea of the Carboniferous limestone spread over
their site. Much of the Devonian marine area was upridged into
land, and the rocks eventually underwent that intense compression and plication which have given them their cleaved, crumpled,
and metamorphic aspect, and in connection with which they were
invaded by granite and intersected with mineral veins. It is
deserving of remark that volcanic action, which played so notable
a part in Devonian time, was continued, but with diminished
vigour, in the Carboniferous period. During the earlier half of
the period, volcanic outbursts were frequent in different parts of
Britain, particularly in Derbyshire, the Isle of Man, central and
southern Scotland, and the south-west of Ireland. The lava and
ashes ejected in some of these areas during the time of the
Carboniferous Limestone form conspicuous groups of hills.

Of the plant and animal life of the Carboniferous period much
is now known from the abundant remains which have been
preserved of the terrestrial surfaces and sea-floors of the time.
Beginning with the flora, we have first to notice its general
resemblance to that of the Devonian period. Many of the genera
of the older time survived in the Carboniferous jungles ; but other
forms appear in vast profusion, which have not been met with in

any Devonian or Old Red Sandstone strata. The Carboniferous flora, like that which preceded it, must have been singularly monotonous, consisting as it did almost entirely of flowerless plants. Not only so, but the very same species and genera appear to have then ranged over the whole world, for their remains are found in Carboniferous strata from the Equator to the Arctic Circle. Ferns, lycopods, and equisetaceæ, constituted the main mass of the vegetation. The ferns recall not a few of their modern allies, some of the more abundant kinds being *Sphenopteris*, *Neuropteris*, and *Pecopteris* (Fig. 141). Among the

FIG. 141.—Carboniferous Ferns. (*a*) *Neuropteris macrophylla* (⅓); (*b*) *Sphenopteris artemisiæfolia* (⅓); (*c*) *Alethopteris* (*Pecopteris*) *lonchitica* (⅓).

lycopods the most common genus is *Lepidodendron*, so named from the scale-like leaf-scars that wind round its stem (Fig. 142). Its smaller branches, closely covered with small pointed leaves, and bearing at their ends little cones or spikes (*Lepidostrobus*), remind one of the club-mosses of our moors and mountains ; but instead of being low-growing or creeping plants, like their modern representatives, they shot up into trees, sometimes 50 feet or more in height. . Equisetaceæ abounded in the Carboniferous swamps, the most frequent genus being *Calamites*, the jointed and finely-ribbed stems of which are frequent fossils in the sandstones and shales (Fig. 143, *a*). This plant probably grew in dense thickets in the sandy and muddy lagoons, and bore as its foliage slim branches, with whorls of pointed leaves set round the joints (*Asterophyllites*, Fig. 143, *b*). The Sigillarioids were among the most abundant, and, at the same time, most puzzling members of the Carboniferous flora. They do not appear to have any close modern allies, and their place in the botanical scale has been a subject of much controversy. The stem of these trees, sometimes reaching a height of 50 feet or more, was fluted, each of the parallel ribs being marked by a row of leaf-scars, hence the name *Sigillaria*, from the seal-like impressions of the scars (Fig. 144). These surface-markings disappeared as the tree grew, and in the lower part of the trunk they passed down into the pitted and tubercular surface characteristic of the roots (*Stigmaria*), still so abundant in their position of growth in fire-clay, and also as drifted broken specimens in sandstones and shales. Another plant that took a prominent part in the Carboniferous flora was that named *Cordaites* (Fig. 145). Its true botanical place is still matter of dispute ; some writers placing it with the lycopods, others with the cycads, or even among the conifers. It bore parallel - veined leaves somewhat like those of a yucca, which, when they fell off, left prominent scars on the stem, and it also carried spikes or buds (*Carpolithes*, Fig. 145). All the plants now enumerated probably grew on the lower grounds and swamps. But on the higher and drier tracts of the interior there grew araucarian pines (*Dadoxylon, Araucarioxy-lon*), the trunks of which, swept down by floods, were imbedded in some of the sands of the time and now appear petrified in the sandstones.

Fig. 142. — Carboniferous Lycopod (*Lepidodendron Sternbergii*, ¼).

While the terrestrial vegetation of the Carboniferous period has been so abundantly entombed, the fauna of the land has been but scantily preserved. That air-breathers existed, however, has been made known by the finding of specimens of scorpions, myriapods, true insects, and amphibians. Within the last few years vast numbers of the remains of scorpions have been discovered in the Carboniferous rocks of Scotland. These ancient forms (*Eoscorpius*) presented a remarkably close resemblance to the living scorpion, and so well have they been preserved among the shales that even the minutest parts of their structure can be recognised. They possessed stings like their modern descendants, whence we may infer the presence of other forms of life which they killed. The

Carboniferous woodlands had plant-eating millipedes, and their silence was broken by the hum of insect-life ; for ancestral forms

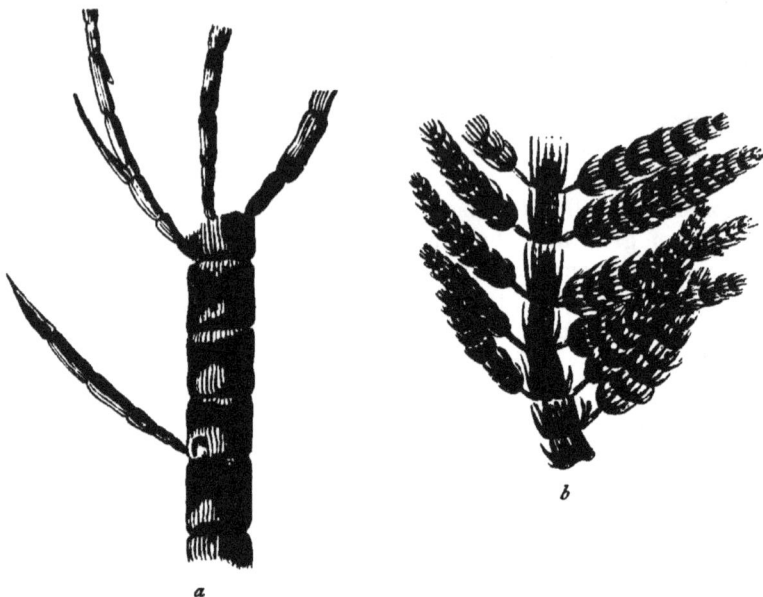

FIG. 143.—Carboniferous Equisetaceous Plants. (a) *Calamites Lindleyi* (= *C. Mougeoti*, *Lindl.*, ⅓) ; (b) *Asterophyllites densifolius* (⅓).

of dragon-flies (*Libellulæ*), May-flies (*Ephemeridæ*), stone-flies (*Perlidæ*), white-ants (*Termidæ*), cockroaches (*Blattidæ*), spectre-

FIG. 144.—Sigillaria with Stigmaria roots (much reduced).

insects (*Phasmidæ*), crickets (*Gryllidæ*), locusts (*Acrydiidæ*), and curious transitional forms between modern types that are quite distinct have been detected, chiefly among the shales and coals of the Coal-measures. Some of these insects attained a great size ; a single wing of one of them (*Megaptilus*) must have measured between 7 and 8 inches in length. While detached wings and more or less complete bodies have been found as rare and precious discoveries in many coal-fields in Europe and America, it is at Commentry in France that remains of insects have been met with in largest numbers—no fewer than 1300 specimens having there been disinterred, most of them admirably preserved. In the interior of decaying trees early forms of land-snails lived, having a

FIG. 145.—*Cordaites alloidius* (⅓), with *Carpolithes* attached.

striking resemblance to some kinds that are still to be found in our present woodlands (*Pupa*).

The lagoons in which the coal-growths flourished were tenanted by numerous forms of animal life. Among these were various mussel-like molluscs (*Anthracomya*, Fig. 154, *Anthracosia*), which were possibly restricted to fresh water. But wherever the sea-water penetrated, it carried some of its characteristic life with it, particularly *Lingula, Discina, Aviculopecten, Goniatites*, and other marine shells. The fishes of the lagoons were chiefly ganoids (*Megalichthys, Rhizodus*, Fig. 158, *Cheirodus, Strepsodus*, etc.) But some of the rays and sharks from the sea made their way into these waters, for their spines are occasionally found among the coal-seams and shales (*Gyracanthus, Pleuracanthus*, Fig. 158). That the larger fishes lived upon the smaller ones is shown by a curious and interesting piece of evidence. Many of the shales are full of small oblong bodies which contain in their interior the broken and undigested scales and bones of small fishes. From their contents, their peculiar external form and markings, and their phosphatic composition, these bodies (*coprolites*) are recognised as the excrement of some of the larger fishes, and the teeth and scales within them serve to show what were the smaller forms on which these fishes fed (see Fig. 65).

During the Carboniferous period, and indeed throughout the later parts of the Palæozoic ages, the most highly organised creatures living on the globe, so far as we at present know, belonged to the Amphibia—the great class which includes our modern frogs, toads, and salamanders. They belonged, however, to an order that has long been entirely extinct—the Labyrinthodonts, so named from the labyrinthine folds of the internal substance of their teeth. They were somewhat like the existing salamander in form, with weak limbs and a long tail. Their skulls were encased in strong plates of bone, and they likewise carried protective bony scutes on the under sides of their bodies. Those found in Carboniferous rocks are mostly small in size, but some of them, measuring perhaps 7 or 8 feet in length, must have been the monsters of the lagoons, in which they lived. Some of the leading genera are *Archegosaurus, Anthracosaurus, Loxomma, Dendrerpeton, Baphetes.*

The marine life of the Carboniferous period has been extensively preserved in the Carboniferous Limestone, which, as already stated, consists of little else than aggregated remains of organisms. In walking over the surface of the beds of this limestone, one treads upon the floor of the sea in Carboniferous times, with its corals, crinoids, and shells crowded and crushed upon each other. Beginning with the most lowly of

FIG. 146.—Carboniferous Foraminifer (*Fusulina cylindrica,* ?).

these organisms, we may observe abundant remains of foraminifera, which in some portions of the limestone constitute the greater part of the rock. One of their most characteristic forms, named *Fusulina* (Fig. 146), enters largely into the structure of the limestone across the Old World from Russia to China and Japan, and likewise in North America. Another, called *Sacammina*, abounds as aggregates of little globular bodies in some parts of the limestone of Britain. Corals have been preserved in prodigious numbers ; indeed, some parts of the limestone are almost entirely made up of them. Most of them are rugose kinds, characteristic genera being *Zaphrentis, Lithostrotion,* (Fig. 147), *Clisiophyllum, Lonsdaleia.* With these there occur also tabulate forms, including *Chœtetes*, *Alveolites*, *Favosites*, etc. Of the sea-urchins, the plates and spines of the genus *Archœocidaris* (Fig. 148) are specially frequent. But the most common echinoderms are members of the great order of crinoids, which must have grown in thick groves over many square miles of the sea-bottom. So prodigiously numerous were they that their remains have been aggregated into beds of limestone hundreds of feet in thickness, hence known as crinoidal or encrinite limestone (Fig. 76). The general plant-like form of these animals is shown in Fig.

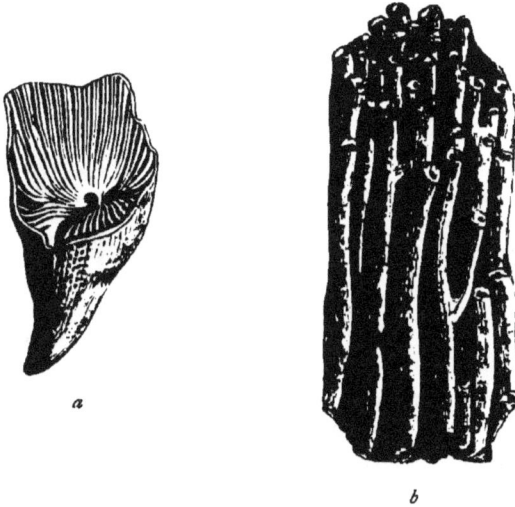

FIG. 147.—Carboniferous Rugose Corals. (*a*) *Zaphrentis Enniskilleni* (½); (*b*) *Lithostrotion junceum* (natural size).

149. But usually the calcareous joints and plates fell asunder. Frequent genera are named *Platycrinus*, *Poteriocrinus*, *Cyathocrinus*. The Carboniferous seas were tenanted by a peculiar extinct order of echinoderms known as Blastoids or Pentremites (Fig. 150), distinguished from true crinoids by the want of free arms, and by the arrangement of the plates forming the cup. These creatures are characteristically Carboniferous, though they are found also in the higher part of the Silurian system and in Devonian rocks.

The Crustacea of the Carboniferous period presented a strong contrast to those of earlier geological time. In particular, the great family of the Trilobites, so characteristic of the older Palæozoic systems, now died out altogether. Instead of its numerous types in the Silurian and Devonian rocks, it is represented in the Carboniferous system by only four genera, all the species of which are small (*Phillipsia*, Fig. 151, *Griffithides*,

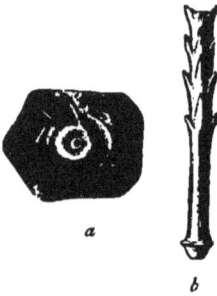

FIG. 148. — Carboniferous Sea - Urchin (*Archæocidaris Urei*, natural size). (*a*) Single plate ; (*b*) Portion of spine.

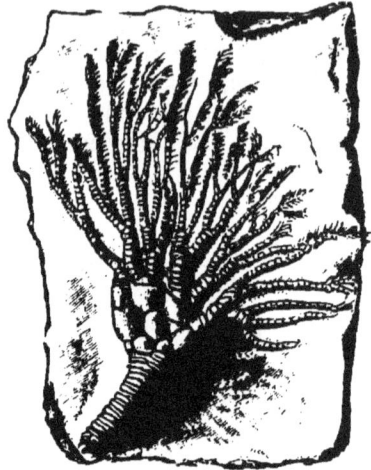

FIG. 149.—Carboniferous Crinoid (*Woodocrinus expansus*, ⅓).

Brachymetopus), and none of which rises into the next succeeding system. The most abundant crustaceans were ostracods—an order still abundantly represented at the present day. They are

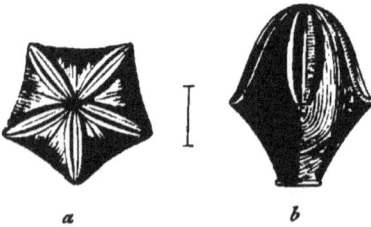

FIG. 150.—Carboniferous Blastoid (Cup of Pentremite, magnified). (*a*) View from above ; (*b*) Side view.

FIG. 151.—Carboniferous Trilobite (*Phillipsia derbiensis*, natural size).

minute forms enclosed within a bivalve shell or carapace which entirely invests the body. Many of these live in fresh water ; the *Cypris*, for example, being abundant in ponds and ditches. Others are marine, while some are brackish-water forms. In the

Carboniferous lagoons, as at the present time, they lived in enormous numbers; their little seed-like valves are crowded together in some parts of the shale which represents the mud of these lagoons; sometimes they even form beds of limestone. Doubtless, they served as food to the smaller fishes whose remains are usually to be found where the ostracod valves are plentiful. One of the principal genera is *Leperditia.* There were likewise long-tailed shrimp-like crustaceans (*Anthrapalæmon, Palæocrangon*), and king-crabs (*Prestwichia*); while in the earlier part of the period Eurypterids still survived in the waters.

Some of the most delicately beautiful fossils of the Carboniferous limestone belong to the Polyzoa. These animals, of which familiar living examples are the common sea-mats of our shores, are characterised by their compound calcareous or horny framework studded with minute cells, each of which is occupied by a separate individual, though the whole forms one united colony. One of the most abundant Carboniferous genera is *Fenestella* (Fig. 152). So numerous are the polyzoa in some bands of limestone as to constitute the main part of the stone. Their delicate lace-like fronds are best seen where

FIG. 152.—Carboniferous Polyzoon (*Fenestella Morrisii*, natural size).

the rock has been exposed for a time to the action of the weather; they then stand out in relief and often retain perfectly their rows of cells.

The Brachiopods, so preponderant among the molluscs of the earlier division of Palæozoic time, now decidedly wane before the great advance of the more highly organised lamellibranchs and gasteropods. Some of the most characteristic genera (Fig. 153) are *Productus, Spirifera, Streptorhynchus, Rhynchonella, Athyris, Chonetes, Terebratula, Lingula, Discina.* Some of the species appear to range over the whole world, for they have been met with across Europe, in China, Australia, and North America. Among these cosmopolitan forms are *Productus semireticulatus, Productus longispinus, Streptorhynchus crenistria, Spirifera glabra, Terebratula hastata* (Fig. 153).

Some of the more common Lamellibranch molluscs (Fig. 154) belong to the genera *Aviculopecten, Leda, Nucula, Edmondia, Modiola, Anthracomya.* Among the Gasteropods *Euomphalus,*

Pleurotomaria, *Loxonema*, and *Bellerophon* (Fig. 155) are not

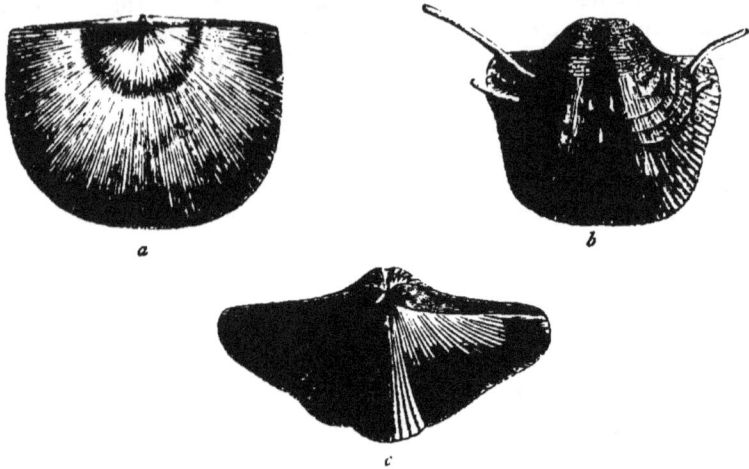

FIG. 153.—Carboniferous Brachiopods. (*a*) *Productus semireticulatus* (½);
(*b*) *Streptorhynchus crenistria* (½); (*c*) *Spirifera striata* (½).

infrequent. A pteropod (*Conularia*, Fig. 156) may be gathered
in great numbers in some parts of the Carboniferous Limestone.

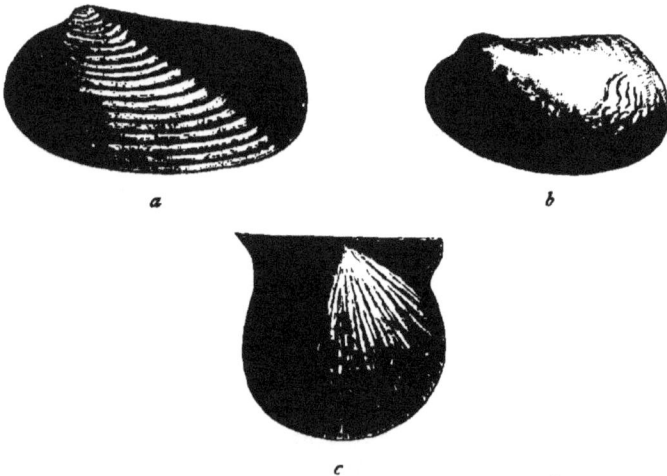

FIG. 154.—Carboniferous Lamellibranchs. (*a*) *Edmondia sulcata* (½); (*b*) *Anthracomya Adamsii* (⅔); (*c*) *Aviculopecten fallax* (¾).

The Cephalopods were represented by numerous species of *Ortho-ceras*, *Nautilus*, and *Goniatites* (Fig. 157).

Remains of fishes are not infrequent in the Carboniferous
Limestone. But they present a striking contrast to those of the
black shales and ironstones of the Coal-measures. They consist for the most part of teeth or of spines belonging to large
predatory sharks. These teeth were placed as a kind of pavement

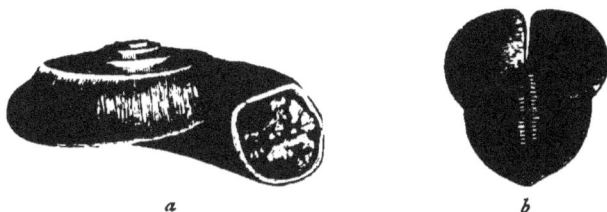

a   *b*
FIG. 155.—Carboniferous Gasteropods. (*a*) *Euomphalus pentangulatus* (½);
(*b*) *Bellerophon tenuifascia* (⅔).

and roof in the mouth, and were used as effective instruments for
crushing the hard parts of the animals, on which these larger
creatures preyed. If, as is probable, the sharks fed upon the
ganoid fishes of the time, they must have required a powerful
apparatus of teeth for crushing the hard, bony armour in which

FIG. 156.—Carboniferous
Pteropod (*Conularia
quadrisulcata* (⅔).

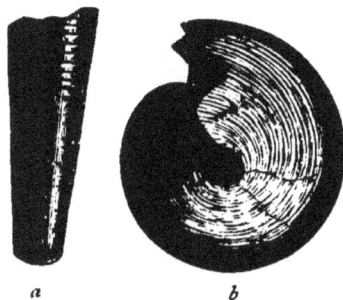

a   *b*
FIG. 157.—Carboniferous Cephalopods.
(*a*) *Orthoceras goldfussianum* (½) ; (*b*)
Goniatites sphæricus (natural size).

these fishes were encased. Of the commoner genera of sharks,
which have been named from the forms of their teeth—the only
hard parts of their structure that have survived—the following may
be mentioned : *Cochliodus, Orodus, Psammodus, Petalodus*. The
small ganoids that so abound in the black shales, ironstones, and
coal-seams which represent the deposits of the sheltered lagoons
of the coal-jungles, are hardly to be found in the thick limestone,

whence we may infer that they were inhabitants of the quiet shore-
waters, and did not venture out into the open sea, where the sharks
found their congenial element. But the occasional occurrence of
the teeth and spines of sharks in the Coal-measure shales and
coal-seams shows that these monsters now and then made their
way into the inland waters, where they would find abundant food.

The Carboniferous system in Europe presents at least two
well-marked subdivisions. In the lower section the strata are in

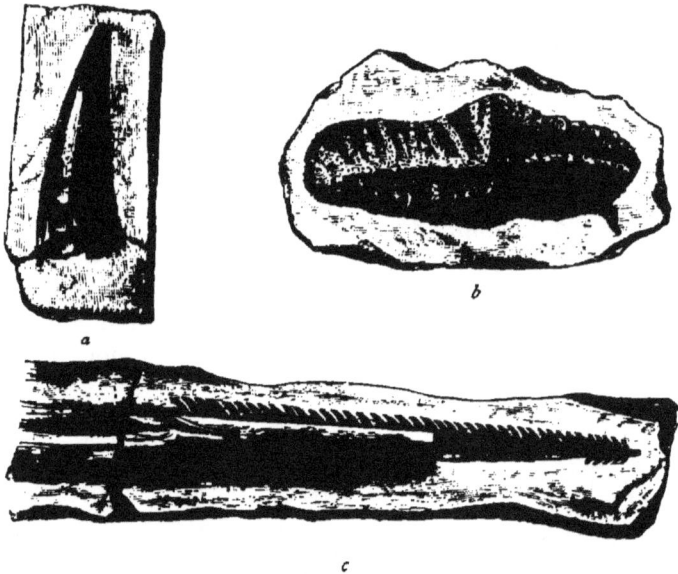

FIG. 158.—Carboniferous Fishes. (a) Tooth of *Rhizodus Hibberti* (½); (b) Tooth of
Orodus ramosus (½); (c) *Ichthyodorulite* or Fin-spine of *Pleuracanthus lævissimus* (½).

large measure marine, for they include the Carboniferous Limestone; in the upper part they consist mainly of sandstones, shales,
fire-clays, and coal-seams, constituting what are called the Coal-
measures, or coal-bearing division of the system. The subjoined
Table shows the order of succession of the rocks in Britain :—

Lagoon type.

Coal-Measures.—At the top, red and grey sandstones,
clays, and thin limestone, resting upon a great thickness of white, grey, and yellow sandstones, clays,
shales, and fire-clays, with numerous workable coal-
seams, and with a lower subdivision of coal-bearing
beds, among which there occur marine fossils
(*Orthoceras, Posidonomya*, etc.) Thickness in South
Wales, 12,000 feet ; South Lancashire, 8000 feet ;
Central Scotland, 3000 feet.

Millstone Grit.—Grits, flagstones, sandstones, and shales, with thin seams of coal and occasional bands containing marine fossils. Thickness 400-1000 feet, increasing in Lancashire to 5500 feet.

Marine type, but passing northwards into that of the lagoons.

Carboniferous Limestone.—Consisting typically of massive marine limestones and shales, but passing laterally into sandstones and shales, with thin coal-seams, which indicate alternations of marine and brackish water conditions. Thickness in South Wales, 500 feet, increasing northwards to more than 4000 feet in Derbyshire, and to upwards of 6000 feet in Lancashire, but diminishing northwards into Scotland.

The base of the Carboniferous Limestone series passes down conformably into the Upper Old Red Sandstone.

The Carboniferous system occupies a number of detached areas on the European continent. Its largest tract extends from the north of France, through Belgium, into Westphalia. The most important coal-fields of Europe belonging to this system are those of Belgium, Westphalia, the north of France, Saarbrücken, St. Etienne in Central France, Bohemia, and the Donetz in Southern Russia. In North America, the Coal-measures of the eastern United States reach a thickness of 4000 feet in Pennsylvania, and contain many valuable seams of coal. They increase in thickness northwards, reaching a maximum of 8000 feet in Nova Scotia. They are underlain by bands of conglomeratic strata, answering to the English Millstone grit, below which comes a group of beds with marine fossils (sub-carboniferous), probably representing the Carboniferous Limestone of Europe. In Australia and New Zealand also thick masses of sedimentary strata contain recognisable Carboniferous organic remains. In New South Wales they include a valuable succession of coal-seams.

CHAPTER XXI

THE prolonged subsidence during which the Coal-measures were accumulated was at last brought to an end by a series of great terrestrial disturbances, whereby the lagoons and coal-growing swamps were in great measure effaced from the geography of Europe. So abrupt in some regions is the discordance between the Coal-measures and the next series of strata, that geologists have naturally been led to regard this break as one of great chronological importance, serving as the boundary between two distinct systems. Nevertheless, so far as the evidence of fossils goes, there is no such interruption of the Geological Record as might be supposed from this stratigraphical unconformability, many of the Carboniferous types of life having survived the terrestrial disturbances. Again, though the discordance among the strata is, in many parts of Europe, particularly in England, most striking, yet it is by no means universal. On the contrary, some localities (Autun in France, and the Bohemian coal-field, for example) escaped the upheaval and prolonged denudation which elsewhere have produced so marked a hiatus in the chronicle. And in these places a gradual passage can be traced from the strata and fossils of the Coal-measures into those of the next succeeding division of the series, no sharp line being there discoverable, nor any evidence to warrant the separation of the overlying strata as an independent system distinct from the Carboniferous. Hence, by many geologists, the rocks now to be described are regarded as the upper part of the Carboniferous system.

To these overlying rocks the name of PERMIAN was given, from the Russian province of Perm, where they are well developed. They consist of red sandstones, marls, conglomerates, and breccias, with limestones and dolomites. In Germany they are often called

Dyas, because they are there easily grouped in two great divisions. The coarsest strata—breccias and conglomerates—are composed of rounded and angular fragments of granite, diorite, gneiss, grey-wacke, sandstone, and other crystalline and older Palæozoic rocks, which must have been upheaved and exposed to denudation before Permian time. The sandstones are usually bright brick-red in colour, owing to the presence of earthy peroxide of iron which serves to cement the particles of sand together. The shales or marls are coloured by the same pigment. So characteristic indeed is the red colour of the rocks that they form part of a great series of strata, originally known as the New Red Sandstone. Generally, greenish or whitish spots and streaks occur in the red beds, marking where the iron-oxide has been reduced and removed by decaying organic matter. Red strata are, as a rule, singularly barren of organic remains, probably because the water from which the iron-peroxide was precipitated must have been unfitted for the support of life. The red Permian rocks are therefore generally unfossiliferous. Among them, however, occur dark shales or "marl-slate," which have yielded numerous remains of fishes. The limestones too are fossiliferous, but they are associated with unfossiliferous dolomite, gypsum, anhydrite, and rock-salt. In some places seams of coal also occur.

These various rocks tell distinctly the story of their origin. They could not have been deposited in the open sea, but rather in basins more or less shut off from it, wherein the water was charged with iron and was liable to concentration, with the consequent precipitation of its solutions. The beds of anhydrite, gypsum, and rock-salt are memorials of these processes. The dolomite may at first have been laid down as limestone which afterwards was converted into dolomite by the action of the magnesian salts in the concentrated water. In such intensely saline and bitter solutions, animal life would not be likely to flourish, and hence, no doubt, the poverty of fossils in the Permian series of rocks. But it is observable that where evidence occurs of the cessation of ferruginous, saliferous, and gypseous deposition, fossils not infrequently appear. The brown Marl-Slate, for example, and the thick beds of limestone are sometimes abundantly fossiliferous, and indeed are almost the only bands of rock in the whole series where organic remains occur. They were probably deposited during intervals when the barriers of the inland seas or salt-lakes were broken down, or, at least, when from some cause the waters came to be connected with the open sea, and when a portion of the ordinary marine fauna swarmed into them. Volcanic action showed itself during Permian time in many parts of Western, Central, and Southern Europe. There was a group of small volcanoes in the south of Scotland. Great eruptions took place in Germany, notably in the area of the present Vosges mountains, and the region of volcanic activity extended across the region where now the Alps stand, as far south as Cannes on the shores of the Mediterranean.

Hence, from the very circumstances in which their remains have been entombed and preserved, the flora and fauna of Permian

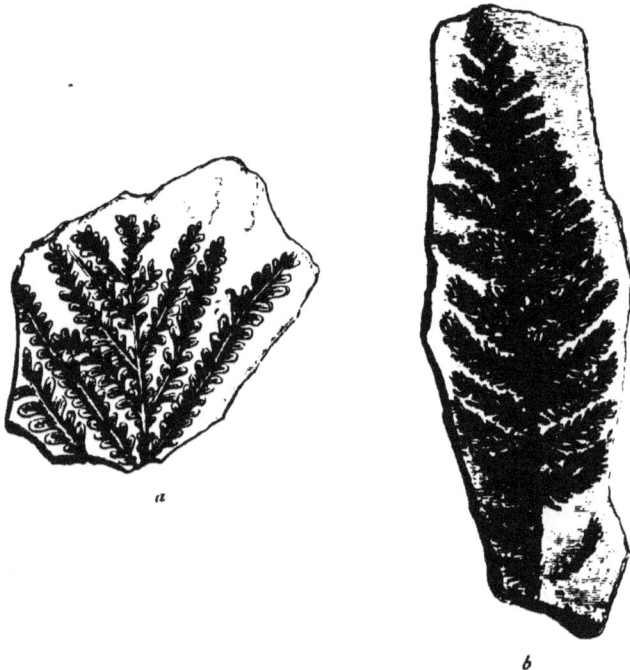

FIG. 159.—Permian Plants. (*a*) *Callipteris conferta* (½) ; (*b*) *Walchia piniformis* (½).

times are comparatively little known. The total number of species and genera obtained from Permian rocks, hardly more than 300 in all, forms a singular contrast to the ample assemblages which have been recovered from the older systems. But that the land of these times was still richly clothed with vegetation and the sea abundantly stocked with animal life, there can be no doubt. The flora appears to have closely resembled that of the Carboniferous period, a considerable proportion of the species of plants being survivals from the Carboniferous jungles and forests. The Lepidodendra, Sigillariæ, and Calamites, which had been such conspicuous members of all the Palæozoic floras, now appear in diminishing number and variety, and finally die out. With their cessation, new features arise in the vegetation. Among these may be mentioned the abundance of tree-ferns, which, though they

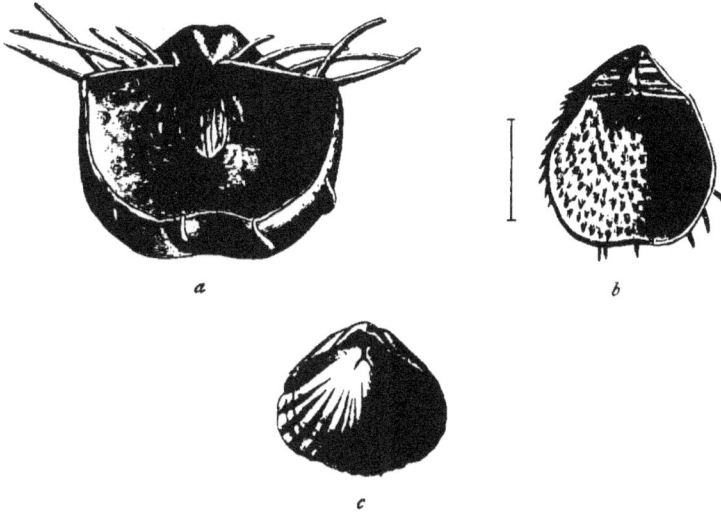

FIG. 160.—Permian Brachiopods. (a) *Productus horridus* (reduced); (b) *Strophalosia Goldfussi*; (c) *Camarophoria humbletonensis* (⅔).

sparingly existed even as far back as Devonian times, now attained a conspicuous development (*Psaronius, Caulopteris*). The genus of ferns called *Callipteris* likewise played a prominent part in the Permian woodlands (Fig. 159, *a*). But perhaps the most remarkable feature in the flora was the abundance of its conifers, and the appearance of the earliest forms of cycads. The yew-like conifer *Walchia* (Fig. 159, *b*) if we may judge from the abundance of its remains, flourished in great profusion on the drier grounds, mingled with others that bore cones (*Ullmannia*). The cycads, which now made their advent, continued during Mesozoic time to give the leading character to the vegetation of the globe.

The scanty relics of the Permian fauna, as above stated, have been almost wholly preserved in those strata which were deposited during temporary irruptions of the open sea into the inland salt-basins of the time.   Some of the Carboniferous genera of brachio-pods still survived—*Productus*, *Spirifera*, and *Strophalosia* being conspicuous (Fig. 160).   Among the lamellibranchs *Axinus*, *Bakevellia*, and *Schizodus* are frequent forms (Fig. 161).   Among

FIG. 161.—Permian Lamellibranchs.   (*a*) *Bakevellia tumida* (natural size);
(*b*) *Schizodus Schlotheimi* (natural size).

the higher molluscs, which have been but sparingly preserved in the rocks, the old types of *Orthoceras*, *Cyrtoceras*, and *Nautilus* are still to be noticed.   In Europe, the fishes of the time have been chiefly sealed up in the marl-slate or copper-shale (Kupfer-

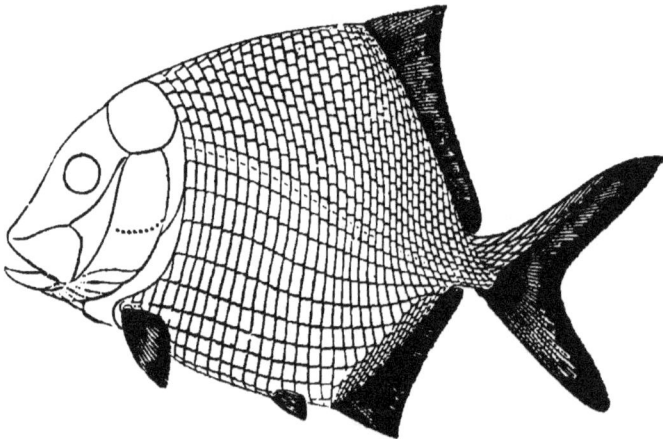

FIG. 162.—Permian Ganoid Fish (*Platysomus striatus*, ½).

schiefer); two of the most frequent genera being *Palæoniscus* and *Platysomus* (Fig. 162).

Labyrinthodonts continued to abound in the waters.   Some of the Carboniferous genera still survived, but with these were associated many new forms, most of which have been discovered in the strata overlying the true Coal-measures of Bohemia (Fig. 163).   But a great onward step in the advance of animal organ-

isation was made in Permian time by the appearance of the
earliest known lizard—*Proterosaurus*, which, like the living
crocodile, had its teeth implanted in distinct sockets.

FIG. 163.—Permian Labyrinthodont (*Branchiosaurus salamandroides*, natural size).

In Britain the Permian strata rest unconformably on the
Carboniferous system, which must have been greatly disturbed
and enormously denuded before they were deposited. They consist of the following subdivisions :—

Upper red sandstones, clays, and gypsum (50 to 100 feet thick in the east of
England, but swelling out west of the Pennine Chain to 600 feet).
Magnesian limestone—a mass of dolomite ranging up to 600 feet in thickness,
and the chief repository of the Permian fossils ; remarkable for the curious
concretionary forms assumed by many of its beds on the coast of Durham
(Fig. 75). [Zechstein of Germany.]
Marl-slate—a hard brown shale with occasional limestone bands. [Kupfer-
schiefer].
Lower red and variegated sandstones with conglomerates and breccias. This
division attains a thickness of 3000 feet in Cumberland, but is hardly
represented in the east of England. [Rothliegende of Germany.]

In Germany, where the Dyas or twofold development of the
Permian rocks is so well displayed, the lower subdivision, called
Rothliegende, consists of great masses of conglomerate with sandstones, shales, thin limestones, and important intercalations of
contemporaneous volcanic rocks, both lavas and tuffs. The upper
section is composed chiefly of limestone called Zechstein, and
answering to the Magnesian limestone of England. With it are
associated beds of anhydrite, gypsum, rock-salt, and bituminous
limestone, and underneath it lies the celebrated Kupferschiefer or
copper-shale—a black bituminous shale, about two feet thick, which
has long been extensively worked on the flanks of the Harz
Mountains for the ores of copper with which it is impregnated,
and which is the great repository for the fossil fishes of the
Permian period. This remarkable band of rock was probably

U

deposited in one of the inland basins, which at first may have maintained a free communication with the open sea. But eventually mineral springs, not improbably connected with the volcanic action of the time, brought up such an abundant supply of dissolved metallic salts as to kill the fish and render the water unsuitable for their existence. The metallic salts were reduced and precipitated as sulphides round the organisms, and impregnated the surrounding mud. In the overlying succession of strata, we see how the area was once more overspread by the clearer and opener water, which brought in the fauna of the Zechstein, and then how the basin gradually came to be shut off once more, until from its concentrated waters the various beds of anhydrite, gypsum, and rock-salt were thrown down.

In the heart of France, at Autun, the Coal-measures pass upward into Permian strata, as already stated. That area appears to have escaped the disturbance which in Western Europe placed the Permian unconformably upon the Carboniferous rocks. It presents a mass of sandstones, shales, coal-seams, and some bands of magnesian limestone, the whole having a thickness of more than 3000 feet referred to the Permian system. The plants in the lower part of this group of strata are unmistakably Carboniferous, but Permian forms appear in increasing numbers as the beds are followed upwards until the highest stage presents a predominant Permian flora. Besides the characteristic Permian fishes, these strata have yielded remains of several salamander-like animals (*Protriton* or *Branchiosaurus, Melanerpeton*), and of some labyrinthodonts (*Actinodon, Euchirosaurus,* etc.)

CHAPTER XXII

THE MESOZOIC PERIODS—TRIASSIC

THE great series of red strata referred to in the foregoing chapter as overlying the Carboniferous system in England was called "New Red Sandstone," to distinguish it from the "Old Red Sandstone" which underlies that system. But the progress of geology on the European continent eventually proved that, notwithstanding their general similarity of lithological character, two series of rocks had been comprised under the general title of New Red Sandstone. The older of these, separated from the rest under the name of Permian, was placed at the top of the great succession of Palæozoic formations. The younger division (still sometimes spoken of as New Red Sandstone) was called Trias, and was regarded as the first system in the great Mesozoic or Secondary succession.

Essentially the Permian strata form merely the upper part of the Carboniferous system. Their types of life are fundamentally Palæozoic, but, as we have seen, both the flora and fauna are marked by a decrease in the number and variety of old forms, and by the advent of the precursors of a new order of things. Conifers and cycads now began to replace the early types of lepidodendron and sigillaria ; amphibians became more abundant, and saurians now took their place at the head of the animal world.

But when we ascend into the Trias, though in Europe the physical conditions of deposition remained much the same as in Permian time, we meet with a decided contrast in the organic remains. A new and more advanced phase of development presents itself in that richer and more varied assemblage of plant and animal life which characterised Mesozoic or Secondary time.

The word TRIAS has reference to the marked threefold division of the rocks of this system in Germany. In that country, and generally in Western Europe, the rocks consist of bright red sandstones and marls or clays, with beds of gypsum, anhydrite, rock-salt, dolomite, and limestone. These rocks, so closely resembling the Permian series below, had evidently a similar origin. They were in large part deposited in inland seas or salt-lakes, wherein, by evaporation and concentration of the water, the dissolved salts were precipitated upon the bottom, and where, consequently, the conditions must have been extremely unfavourable for the presence of living things. The sites of these inland basins can still be partially traced. They extended at least as far west as the north of Ireland. One or more of them lay across the site of the plains of Central England. Others were dotted over the lowlands of middle Europe. The largest of them occupied an extensive area now traversed by the Rhine. It stretched, on the one hand, from Basel to the plains of Hanover, and, on the other, from the highlands of Saxony and Bohemia across the site of the Vosges Mountains westward to the flank of the Ardennes. The continent must then have been somewhat like the steppes of Southern Russia—a region of sandy wastes and salt-lakes, with a warm and dry climate. Probably higher land rose to the north, as in earlier geological times, for traces of its vegetation have been found in Sweden. But southwards lay the more open sea, spreading over part at least of the site of the modern Alps, and thence probably across much of Asia to the Indian and Pacific Oceans.

So long as only the deposits of the salt-basins had been explored, it was but natural that comparatively little should be known of the flora and fauna of the Triassic period. The climate around these lakes was perhaps not a very salubrious one, and hence there may have been only a scanty terrestrial fauna in their immediate vicinity, while the waters of the lakes themselves were unsuited for the support of life. It is not surprising, therefore, that the strata deposited in these tracts are on the whole unfossiliferous ; that, indeed, fossils only abound where there are indications that, owing to some temporary depression or breaking down of the barriers, the open sea spread into these basins, and carried with it the organisms whose remains gathered into beds of limestone. But over the tracts that lay under the open sea, a more abundant marine fauna lived and died. It is in the records of that sea-bottom, rather than in those of the salt-basins, that we must seek for the evidence of the general character of the life over the globe, and for the fossil data with which to compare together the Triassic rocks of distant regions.

There are traces of contemporaneous volcanic action among the
Triassic strata.   A little group of volcanoes appears to have existed during Triassic time in South Devonshire ; but in the region
of the Eastern Alps, especially around Predazzo in the Tyrol,
evidence of far more extensive eruptions exists.

The flora of the Triassic period has been preserved chiefly in
the dark shales and thin coal-seams formed in some of the inland

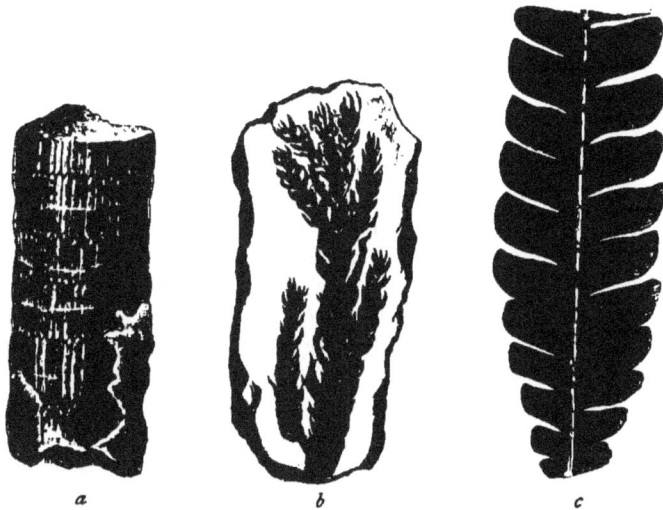

FIG. 164.—Triassic Plants.   (a) Horse-tail Reed (*Equisetum columnare*, ⅓); (b) Conifer
(*Voltzia heterophylla*, ⅓); (c) Cycad (*Pterophyllum Blasii*, ⅓).

basins.   So far as known to us it consisted chiefly of ferns,
equisetums or horse-tails, conifers, and cycads.   Among the ferns
a few Carboniferous genera still survived, but some of the most
characteristic forms were tree-ferns.   The oldest known true horse-
tails are met with in the Trias (Fig. 164, a).   The most abundant
conifer is the cypress-like *Voltzia* (Fig. 164, b).   Cycads, already
a feature in the vegetation of the Permian system, now increase
in number and variety.   During the Mesozoic ages they continued
to be the most characteristic members of the terrestrial flora,
insomuch that this division of geological time is sometimes spoken
of as the "Age of Cycads."   Some of the more common cycads
in the Triassic rocks are *Pterophyllum*, *Zamites*, and *Podozamites*
(Fig. 164, c).

The red gypseous and saliferous strata, for the reason already

given, are on the whole unfossiliferous.   Here and there, footprints of amphibians, preserved on the sandstones, give us a
glimpse of the higher forms of life that moved about on the margin
of the salt-lakes.   The beds of limestone, which represent intervals
when, for a time, the sea overspread the lakes, contain sometimes
abundant fossils.   But they are numerous in individuals rather
than in species or genera, as if the conditions for life in those
waters were still somewhat unfavourable.   On
the other hand, the limestones laid down in the
opener sea are crowded with a varied fauna.
One of the most typical fossils of the Trias is the
crinoid *Encrinus liliiformis* (Fig. 165), one of
the most familiar fossils of the limestones (Muschelkalk) which in Germany form the central
division of the system.   Among the lamellibranchs, *Myophoria, Avicula, Pecten, Cardium,
Pullastra, Daonella,* and *Monotis* are characteristic genera (Fig. 166), some species such as
Avicula contorta, Pecten valoniensis, and *Cardium
rhæticum* being eminently useful in tracing the
upper parts of the Trias (Rhætic) all over
Europe from Italy to Scandinavia.   One ' of
the most distinctive features of the Triassic

FIG. 165. — Triassic
Crinoid (*Encrinus
liliiformis*, ⅓).

fauna is its development of cephalopod life.   In the limestones of the middle subdivision in Germany, a few species of
cephalopods occur, the two prevalent forms being species of
Nautilus and the ammonite *Ceratites* (Fig. 167).   But when we
turn to the Trias of the Eastern Alps, which represents the deposits
of the more open sea, we meet with a remarkable abundance and
variety of cephalopods, and with a striking admixture of ancient
and more modern types.   For example, the venerable genus
Orthoceras, which occurs even down in the Cambrian rocks, is
found also here as a survival from Palæozoic time.   But new types
now appeared.   In particular, the tribe of Ammonites, so pre-
eminently typical of the molluscan life of the Mesozoic seas, is
represented by numerous genera and species (*Arcestes, Trachyceras,
Pinacoceras, Phylloceras,* besides *Ceratites* above referred to).
Among the fishes of the Trias, the genera *Acrodus, Ceratodus,
Gyrolepis, Hybodus,* and *Pholidophorus* may be mentioned.   Labyrinthodonts still haunted the lagoons and sandy shores (*Mastodonsaurus, Trematosaurus*) ; but they no longer remained the most
important members of the animal world.   Various early types of

lizards now took their places in the ranks of creation (*Hyperoda-pedon, Telerpeton*, Fig. 168). A strange order of Triassic reptiles was characterised by the jaws having the form of a beak, somewhat

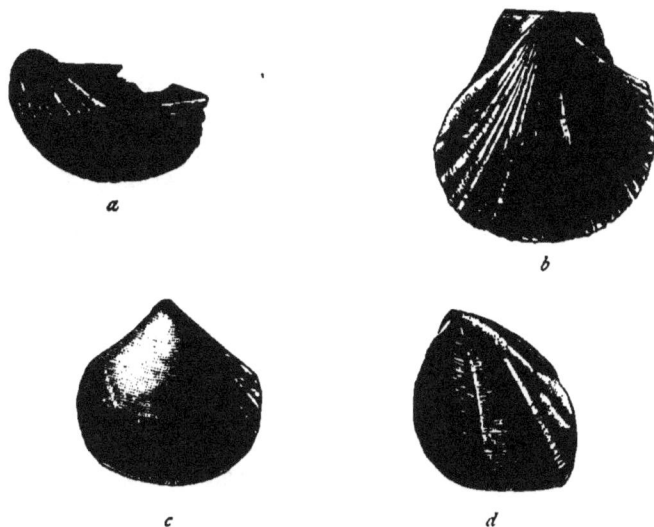

Fig. 166.—Triassic Lamellibranchs. (*a*) *Avicula contorta* (natural size); (*b*) *Pecten veloniensis* (½); (*c*) *Cardium rhæticum* (natural size); (*d*) *Myophoria vulgaris* (½).

like that of a turtle; *Dicynodon*, one of these forms, carried two huge tusks in the upper jaw. A remarkable and long-extinct order of reptiles, that of the Deinosaurs, made its first appearance in

Fig. 167.—Triassic Cephalopods. (*a*) *Nautilus bidorsatus* (½); (*b*) *Ceratites nodosus* (reduced).

Triassic time. These creatures were marked by peculiarities of structure that linked them both with true reptiles and with birds, while in size they sometimes resembled elephants and rhinoceroses. They seem to have walked mainly on their hind feet, the three-toed or five-toed bird-like imprints of which are numerous on some beds of sandstone. They are characteristically Mesozoic types of life. Another not less typically Mesozoic form, that of the Plesiosaurs,

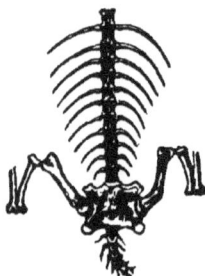

FIG. 168.—Triassic Lizard
(*Telerpeton elginense*, ½).

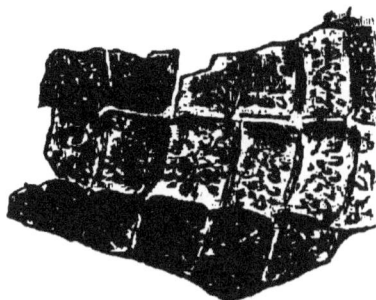

FIG. 169.—Triassic Crocodile (Scutes of *Stagonolepis elginensis*, ⅓).

likewise began in Triassic time; but it will be more particularly alluded to in the following chapter. The earliest known crocodiles have been found in Triassic rocks; some of the scutes or scales of one of these animals are shown in Fig. 169. But the most important advance in the fauna of the globe during the Triassic period was the first appearance of mammalian life. Detached teeth and lower jaws have been met with in the uppermost parts of the Triassic system, which have been identified as possessing structures like those of some of the marsupial animals of Australia (*Microlestes*, Fig. 170). It is interesting to know that the earliest representatives of the great class of the mammalia belonged to one of its lowest divisions. They were small creatures probably resembling the *Ornithorhynchus* and *Echidna* of Australia.

FIG. 170. — Triassic Marsupial (*Microlestes Moorei*). (a) Lower molar tooth, outer side (⅓); (b) Ditto (nat. size); (c) Ditto, front side (⅓).

The Triassic strata of the inland basins (England, Germany, France, etc.) have been subdivided into the following groups :—

Rhætic.	Red, green, and grey marls, black shales, sandstones, bone-beds, and in Germany sometimes thin seams of coal. Characteristic fossils are *Cardium rhæticum, Avicula contorta, Pecten valoniensis, Pullastra arenicola, Acrodus, Ceratodus, Hybodus,* Saurians, *Microlestes.*

Keuper or Upper Trias.	⎧ Red, grey, and green marls, with beds of rock-salt and ⎪ gypsum. ⎨ Red sandstones and marls (England); grey sandstones ⎪ and dark marls and clays, with thin seams of earthy ⎩ coal (Germany).
Muschelkalk or Middle Trias.	⎧ Limestones and dolomites, with bands of anhydrite, ⎨ gypsum, and rock-salt. The limestones are the ⎪ great repository of the fossils. This subdivision is ⎩ absent or only feebly represented in England.
Bunter or Lower Trias.	⎰ Mottled red and green sandstones, marls, and some- ⎱ times pebble-beds.

The salt-beds of Cheshire have long been worked for commercial purposes. The lower bed is sometimes more than 100
feet thick ; but the salt deposits of Germany are much more
important. Thus at Sperenberg, 20 miles south of Berlin, a
boring was put down through about 290 feet of gypsum, and then
through upwards of 5000 feet of rock-salt, without reaching the
bottom of the deposit.

The alternation of bands of rock-salt with thin layers of
anhydrite or of gypsum no doubt marks successive periods of
desiccation and inflow ; in other words, each seam of sulphate
of lime (which is the least soluble salt, and is therefore thrown
down first) seems to indicate a renewed supply of salt water from
outside, probably from the open sea, while the overlying rock-salt
shows continued evaporation, during which the water became
a concentrated solution and deposited a thicker layer of sodium
chloride. Sometimes the concentration continued until still more
soluble salts, such as chlorides of potassium and magnesium, were
also eliminated. These phenomena are well displayed at the
great salt-mines of Stassfurt, on the north flank of the Harz
Mountains. The lowest rock there found is a mass of pure, solid,
crystalline rock-salt of still unknown thickness, but which has
been pierced for about 1000 feet. This rock is separated into
layers, averaging about $3\frac{1}{2}$ inches in thickness, by partings of
anhydrite $\frac{1}{4}$ inch thick or less. If each of these " year rings," as
the German miners call them, represented the deposit formed
during the dry season of a single year, then the mass of 1000 feet
would have taken more than 3000 years for its formation. But
there do not appear to be any good grounds for believing that
each band marks one year's accumulation. Above the rock-salt lie
valuable deposits of the more soluble salts, particularly chlorides
of potassium and magnesium, with sulphates of lime and magnesia.
The compound known as Carnallite (a double chloride of potassium

and magnesium) is now the chief source of the potash salts of commerce.

In the Rhætic beds of England, one of the most interesting bands is the so-called "bone-bed"—a thin layer of dark sandstone, charged with bones, teeth, and scales of fishes and saurians. A thin seam of limestone in the same group contains wings and wing-cases of insects.

The Trias of the Eastern Alps reaches a thickness of many thousand feet, and forms great ranges of mountains. The lower division runs throughout the Alps with considerable uniformity of character, so that it forms a useful platform from which to investigate the complicated geological structure of these mountains. The Upper Trias consists of several thousand feet of shales, marls, limestones, and dolomites, while the Rhætic group swells out into a great succession of limestones and dolomites. During the time when the Triassic sea stretched over the site of the Alps there were evidently considerable oscillations of level, and there likewise occurred extensive volcanic eruptions, whereby large masses of lavas and tuffs were ejected. These rocks now form conspicuous hills in the Tyrol.

Triassic rocks have been traced in Beloochistan, the salt range of the Punjab, Northern Kashmir, and Western Thibet. They cover a large area of North America, and have been recognised in Australia and New Zealand. Rocks which have been assigned to the same geological period occur in South Africa, and have there yielded a remarkable series of reptilian remains.

CHAPTER XXIII

THE system which follows the Trias has received its name, JURASSIC, from the Jura Mountains, where it is well developed. It contains the record of a great series of geographical changes, which in Europe entirely effaced the inland basins and sandy wastes of the previous period, and during which sedimentary rocks were accumulated that now extend in a broad belt across England, from the coasts of Dorset to those of Yorkshire, cover an enormous area of France and Germany, and sweep along both sides of the Alps and the Apennines. These strata vary greatly in composition and thickness as they are traced from country to country. In one district they present a series of limestones which, as they are followed into another area, pass into shales or sandstones. The widespread uniformity of lithological character, so marked among the Palæozoic systems, gives place in the Mesozoic series to greater variety. Sandstones, shales, and limestones alternate more rapidly with each other, and are more local in their extent. They indicate greater vicissitudes in the process of deposition, more frequent alternations of sea and land, and not improbably greater differences of climate than in Palæozoic time.

The flora of the Jurassic period is marked by the same general characters as that of the Trias—ferns, equisetums, conifers, and cycads, being its distinguishing elements. Cycads now abound (*Pterophyllum, Zamites, Cycadites*, and many others, Fig. 171). Among the conifers are the remote ancestors of our "Puzzlemonkeys," introduced from Chili and now so common as ornamental garden shrubs (*Araucaria imbricata*), and of our pines and firs. This vegetation flourished luxuriantly over the area of Britain; on the site of Yorkshire it grew so densely as to give rise to thick peaty accumulations, which now form beds of coal. It went far northward, for its remains have been abundantly preserved in Spitzbergen, where numerous cycads have been found among them. These plants unquestionably grew and flourished within the Arctic Circle, so that, though the climates of the globe were already beginning to emerge from the greater uniformity of

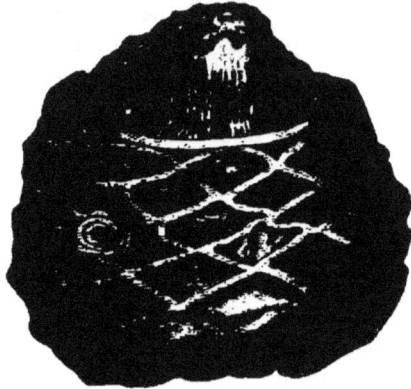

FIG. 171.—Jurassic Cycad (*Cycadeoidea microphylla*, ⅓).

Palæozoic time, the Arctic regions still enjoyed a temperature like that of sub-tropical countries at the present time.

The animal world during the Jurassic period, if we may judge of it from its fossil remains, must have been much more varied alike on land and in the sea than during the previous ages of the earth's history. From the circumstances in which the strata were deposited, relics of the life of the land are frequently met with, besides abundant records of that of the sea. A characteristic feature of the period was the profusion of corals, which at different times spread over much of the site of modern Europe. They were no longer the rugose forms so distinctive of the Palæozoic seas, but true reef-building astræids, belonging to the genera *Isastræa, Thamnastræa, Montlivaltia, Thecosmilia*, etc. (Fig. 172). Crinoids were still abundant, though less so than in the Carboniferous limestone sea ; the old forms were now replaced by

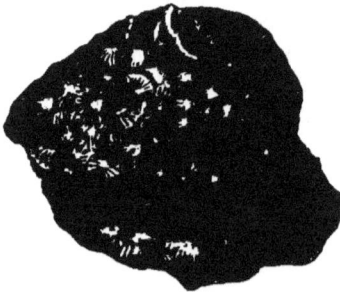

FIG. 172.—Jurassic reef-building Coral (*Isastræa explanata*, ⅓). From the Corallian.

others, among which the most conspicuous was the Pentacrinite
(Fig. 173)—a genus still living in our present seas.   Sea-urchins

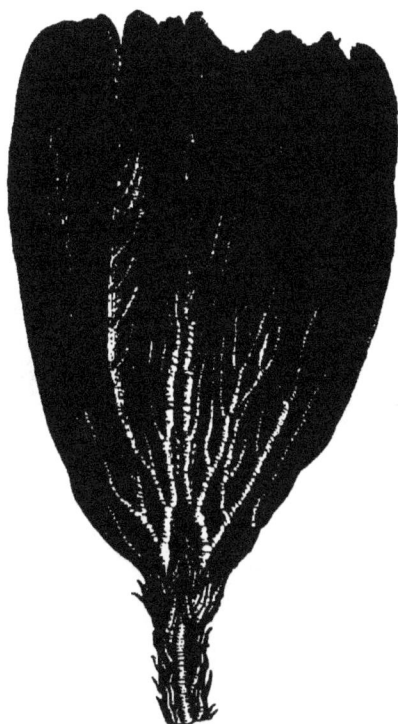

FIG. 173.—Jurassic Crinoid (*Pentacrinus fasciculosus*, ⅓).

swarmed on some parts of the sea-floor; among their more
frequent genera are *Cidaris* (Fig. 174), *Diadema*, *Hemicidaris*,
Acrosalenia, *Glyptichus*, *Pygaster*.   Of
the contrasts between the Mesozoic
and Palæozoic faunas, one of the most
marked is to be found among the
brachiopods.   Except the persistent
inarticulate types which have lived on
from Cambrian time to the present
day (*Crania*, *Lingula*, *Discina*), the
numerous and varied forms which
played so important a part in the life

FIG. 174. — Jurassic Sea - urchin
(*Cidaris florigemma*, ⅓), Corallian.

of the Palæozoic seas died out almost entirely at the close
of the Palæozoic period.   The ancient Spirifers and Leptænids

lingered on until the Jurassic period, and then disappeared. On the other hand, the genera *Rhynchonella* and *Terebratula*, which occupied a subordinate place in earlier ages, now became the chief representatives of the brachiopods. They abounded throughout Mesozoic time, but they have gradually diminished in number since then, and at the present day each genus survives only in a small number of species. With the decay of the brachiopods, the other divisions of the mollusca

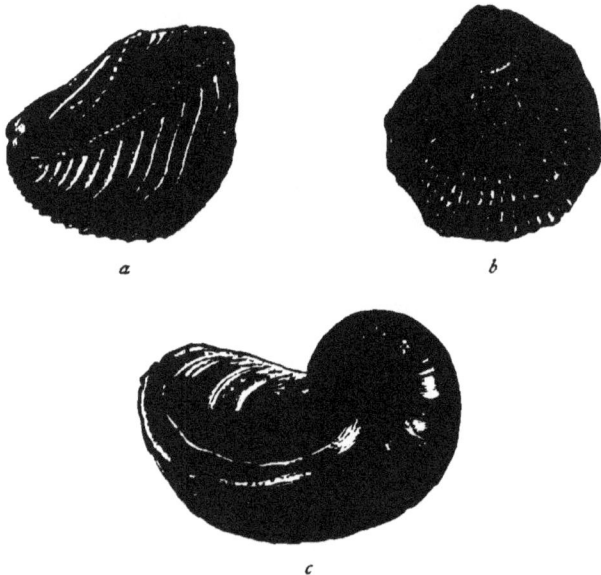

FIG. 175.—Jurassic Lamellibranchs. (*a*) *Trigonia monilifera* (¾), Kimmeridge Clay ; (*b*) *Plicatula spinosa* (¾), Middle Lias ; (*c*) *Gryphæa arcuata* (*incurva*) (¾), Lower Lias.

proportionately advanced. The lamellibranchs attained a great development in Mesozoic time, some characteristic genera being *Gervillia*, *Exogyra*, *Lima*, *Ostrea*, *Pecten*, *Pinna*, *Astarte*, *Hippopodium*, *Trigonia* (Fig. 175). Some of the oysters were particularly abundant, *Gryphæa*, for instance, being so plentiful in some bands of limestone as to give the name of " Gryphite lime-stone " to them. But undoubtedly the distinctive feature of the molluscan fauna of Mesozoic time was the great development of the cephalopods. The chambered division was represented by an extraordinary variety of Ammonites (Fig. 176), and the cuttle-fishes by many species of Belemnite (Fig. 177). The ammonites have been made use of to mark off the formations into distinct zones; for, as a rule, the vertical range of each species is comparatively small. The band of strata characterised by a particular species

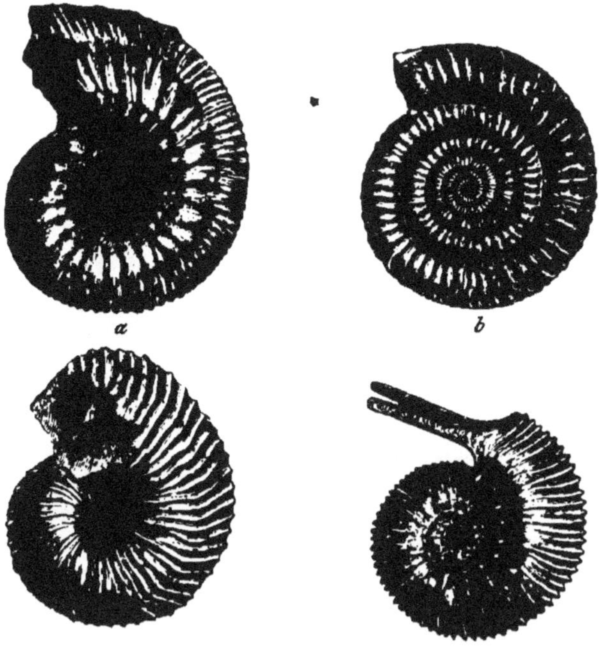

FIG. 176.—Jurassic Ammonites. (a) *Ammonites striatus* (½), Middle Lias; (b) *A. communis* (⅔), Upper Lias; (c) *A. cordatus* (⅗), Lower Calcareous Grit; (d) *A. Jason* (½), Oxford Clay.

of ammonite is called the zone of that species, *e.g.* Zone of *Amm. planorbis*, which is the lowest zone of the Lower Lias.

Another striking contrast is presented by the Jurassic crustacea

FIG. 177.—Jurassic Belemnite (*B. hastatus*, natural size), Middle Oolite.

when compared with those of the Palæozoic ages. The ancient order of trilobites, so abundant in the seas of the older time, had now entirely disappeared; the eurypterids, which took their place

upon the scene as the trilobites were on the wane, had likewise
vanished.   In their stead there now came abundant ten-footed
crustacea, including both long-tailed forms—the ancestors of our
lobsters, prawns, shrimps, and cray-fish—and short-tailed forms
that heralded the coming of our living crabs (Fig. 178).   Among

FIG. 178.—Jurassic Crustacean (*Scapheus ancylochelis*).

the Jurassic strata there occasionally occur thin bands, which
have received the name of "insect-beds" from the numerous
insect-remains which they contain.   The neuroptera are most
frequent, but orthoptera and coleoptera also occur.   Among these
remains are forms of dragon-fly, May-fly, grasshopper, and cockroach.   The wing-cases of beetles also are not uncommon ; and
there has been found the wing of a butterfly—the oldest example
of a lepidopterous insect yet known.

FIG. 179.—Jurassic Fish (*Pholidophorus Bechei*, ⅓), Lower Lias.

Fishes abounded in the waters of the Jurassic time.   Those of
which the remains have been preserved are chiefly small ganoids
(*Pholidophorus, Dapedius, Lepidotus, Pycnodus*, Fig. 179), with
no representatives of the huge bone-cased placoderms of earlier

time.   There were likewise various tribes of sharks and rays (*Hybodus, Acrodus, Squaloraia*).

But, taking the Jurassic fauna as a whole, undoubtedly its most striking character was given by the extraordinary development of its reptiles.   So remarkably varied was the reptilian life throughout the Mesozoic period that this part of the earth's history has been called the "Age of Reptiles."   There were forms which haunted the sea, others that frequented the rivers, some that lived on the land, some that flew through the air.   Never before or since has there been such a profusion of reptilian types.   Some of these are still represented at the present time.   The Jurassic *Teleosaurus* and *Steneosaurus*, for example, have their counterparts now in the living crocodile and alligator.   The modern turtles, too, are descendants of those which lived in Jurassic times.   But it is the long-extinct types that fill us with astonishment.   One of the

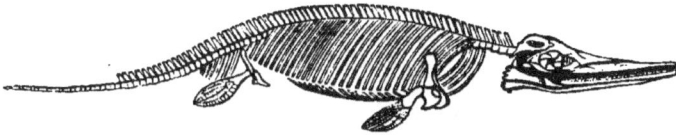

FIG. 180.—Jurassic Sea-lizard (*Ichthyosaurus communis*, $\frac{1}{8}$), Lias.

most abundant of them is that of the enaliosaurs or sea-lizards, of which the two leading forms were the *Ichthyosaurus* and *Plesiosaurus*.   The former creature (Fig. 180), occasionally more than 24 feet long, somewhat resembled a whale in shape and bulk, its head being joined by no distinct neck to the body, which tapered into a long tail.   It swam by means of two pairs of strong paddles, and probably steered itself by a fin on the tail.   Its eyes were large, and had a ring of bony plates round the eyeball, which remain distinct in the fossil state.   Its jaws were armed with numerous strong pointed teeth, not set in distinct sockets.   This reptile probably lived chiefly in the sea, feeding there upon the abundant ganoid fishes which its huge protected eyes enabled it to track even into the deeper water.   But it no doubt also sought the land, and was able to waddle along the shore or to lie there basking in the sunshine.   The Plesiosaurus, in many respects like the Ichthyosaurus, was distinguished by its proportionately shorter tail, longer neck, smaller head, larger paddles, and the insertion of the teeth in distinct sockets.   It probably haunted the lagoons, rivers, and shallow seas of the time.   Its

X

long swan-like neck enabled it to lie at the bottom and raise its head to the surface to breathe, or, when at the surface, to send down its powerful jaws and catch its prey at the bottom.

Still more extraordinary were the Pterosaurs or flying reptiles— strange bat-like creatures with disproportionately large heads, and large eyes like those of Ichthyosaurus. The outermost finger of each forefoot was prolonged to a great length, and supported a membrane with which the animals could fly. The bones were hollow and filled with air like those of birds. Various forms of these winged lizards are found in the Jurassic rocks, the most typical being

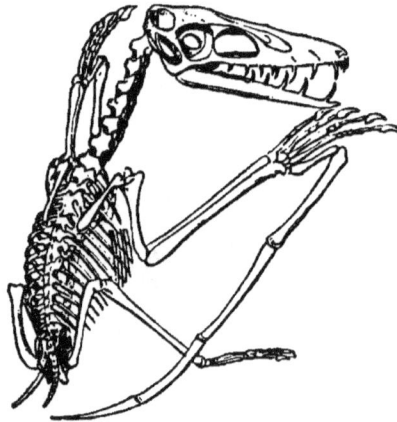

FIG. 181.—Jurassic Pterosaur, or flying reptile (*Scaphognathus crassirostris*), Middle Oolite.

Pterodactylus and *Scaphognathus* (Fig. 181); others are *Dimorphodon*, *Rhamphorhynchus*, *Rhamphocephalus*, and *Dorygnathus*.

The Deinosaurs, which have already been noticed as having appeared in Triassic time, attained a far greater development in the Jurassic period. These hugest land-reptiles now reached their maximum in size and variety. One of their genera, *Megalosaurus*, is believed to have been 25 feet long, and to have walked on its massive hind legs along the margins of the shallow waters in search of the smaller animals on which it preyed. Another form, *Ceteosaurus*, which may have been as much as 50 feet from the snout to the tip of the tail, and stood some 10 feet high, fed on the vegetation that shaded the rivers and lagoons where it lived. Still more gigantic were some deinosaurs, of which the remains have been found in the Jurassic rocks of North America. *Brontosaurus*, about 50 feet or more in length with a short body, long neck and tail, and small head, had enormous feet, each of which made an imprint measuring about a square yard in area. *Stegosaurus*, another sluggish deinosaur, was protected by numerous huge plates and spines of bone on its back, some of the latter more than 3 feet long.   The largest of all these monsters, and, so far as yet known, the most colossal animal that ever walked on the earth, was the *Atlantosaurus*, which is believed to have been not much less than 100 feet in length, and 30 feet or more in height.

In another respect, the fauna of the Jurassic period stands out

FIG. 182.—Jurassic Bird (*Archæopteryx macroura*, about ⅓), Solenhofen Limestone. Middle Jurassic.

from those that preceded it ; it contained the earliest known birds. These interesting prototypes differed much from modern birds, more particularly in the possession of certain peculiarities of structure that linked them with reptiles.   They had teeth in their jaws, and some of them carried long lizard-like tails, each vertebra of which bore a pair of quill-feathers.   The best known genus is *Archæopteryx* (Fig. 182), found in the lithographic limestone of Solenhofen.

Marsupials, which, so far as yet known, made their appearance in Triassic time, continued to be the only representatives of the mammalia during the Jurassic period, at least no other types have yet been discovered among the fossils.   Lower jaws and detached teeth (Fig. 183) have been obtained from two distinct

Fig. 183.—Jurassic Marsupial (*Phascolotherium Bucklandi*).   (*a*) Teeth magnified ;
(*b*) Jaw, natural size.

platforms in England—the Stonesfield Slate and Purbeck beds— and have been referred to a number of genera which find their nearest modern representatives in the Australian bandicoots and in the American opossums (*Phascolotherium, Stereognathus, Spalacotherium, Plagiaulax*).

The Jurassic system has been arranged in the following subdivisions :—

UPPER OR PORTLAND OOLITES.	8. Purbeckian . .	Upper fresh-water beds (Purbeck). Middle marine beds ,, Lower fresh-water beds ,,
	7. Portlandian .	Limestones and calcareous freestones (Portland Stone) ; *Cerithium portlandicum, Ammonites giganteus, Trigonia gibbosa*. Sandstones and marls (Portland Sand) ; *Ammonites biplex, Exogyra bruntrutana*.
	6. Kimmeridgian	Dark shales and clays (Kimmeridge Clay); *Ammonites decipiens, Exogyra virgula*.
MIDDLE OR OXFORD OOLITES.	5. Corallian .	Coral rag (limestone with corals), clays, and calcareous grits ; *Thamnastræa, Isastræa, Cidaris florigemma, Ammonites cordatus* (Fig. 176).
	4. Oxfordian .	Blue and brown clay (Oxford Clay) ; *Ammonites Jason* (Fig. 176). Calcareous sandstone (Kellaways Rock — Callovian); *Ammonites calloviensis*.

LOWER OR BATH OOLITES.	3. Bathonian . .	Shelly limestones, clays, and sands (Cornbrash, Bradford Clay, and Forest Marble). *Ammonites discus.* Shelly limestones (Great or Bath Oolite), Stonesfield Slate ; *Ammonites gracilis.* Fuller's Earth.
	2. Bajocian (Inferior Oolite)	Marine calcareous freestones and grits (Cheltenham), containing zones of *Ammonites Parkinsoni, A. Humphriesianus, A. Sowerbyi, A. Murchisonæ* ; represented in Yorkshire by 800 feet or more of estuarine sandstones, shales, and limestones, with beds of coal.
LIAS.	1. Liassic . . .	Sandy beds and clays (Upper Lias, Toarcian) ; *Ammonites communis, A. jurensis, A. serpentinus.* Limestones, sands, clays, and ironstones (Middle Lias, Marlstone) ; *Ammonites margaritatus, A. spinatus, A. Jamesoni.* Thin blue and brown limestones, and dark shales (Lower Lias, Sinemurian and Hettangian) ; *Ammonites planorbis, A. raricostatus, A. Bucklandi.*

1. The **Lias**, so called originally by the Somerset quarrymen from its marked arrangement into "layers," extends completely across England from Lyme Regis to Whitby. It can be divided into three distinct sections : (*a*) A lower group of thin blue limestones and dark shales with limestone nodules, the limestones being largely used for making cement. This is one of the chief platforms for the reptilian remains, entire skeletons of ichthyosaurus, plesiosaurus, etc., having been exhumed at Lyme Regis ; (*b*) Marlstone or Middle Lias—hard argillaceous or ferruginous limestones which form a low ridge or escarpment rising from the plain of the Lower Lias ; in Yorkshire contains a thick series of beds of earthy carbonate of iron, which are extensively mined as a source for the manufacture of iron ; (*c*) Clays and shales surmounted by sandy beds (Upper Lias Sands). The organic remains of the Lias are abundant and well preserved. They are chiefly marine ; but that the rocks containing them were deposited near land is indicated by the numerous leaves, branches, and fruits imbedded in them, and by the various insect-remains that have been obtained from them. In Germany, where the Lias is well developed and presents a general resemblance to the English type, it is known as the Lower or Black Jura. It is still better shown in France, where its three stages attain in Lorraine a united thickness of more than 600 feet. To the south,

however, in Provence, it reaches the great thickness of 2300 feet.

2. The **Bajocian** stage, so named from Bayeux in Normandy, where it is well displayed, has long been known in England under the name of Inferior Oolite. ·It presents two distinct types in this country, being a thoroughly marine formation in the south-western counties and passing northward into a series of strata which were accumulated in an estuary, and which contain the chief repositories of the Jurassic flora. Among the estuarine beds of Yorkshire a few thin coal-seams occur, which have been worked to some extent. On the continent, this division is characteristically marine ; it reaches its greatest development in Provence, where it is 950 feet thick. It runs through the Jura Mountains, where it is made up of more than 300 feet of strata, chiefly limestone. In Germany the strata from the top of the Lias to the base of the Callovian group—that is, the two stages of Bajocian and Bathonian—are classed together as the Middle or Brown Jura, or Dogger.

3. The **Bathonian** stage is named from Bath, where its subdivisions are admirably exposed. At its base is a local argillaceous band known as Fuller's Earth, because long used for fulling cloth. The chief member of the stage in the south-west of England is the Great or Bath Oolite, a succession of limestones, often oolitic, with clays and sands. The Stonesfield Slate is the name locally given to some thin-bedded limestones and sands forming the lower part of the Great Oolite, and of high geological interest from having supplied among their fossils remains of land-plants, numerous insects, bones of enaliosaurs and deinosaurs, and of small marsupials. The Great Oolite abounds in corals, and contains numerous genera of mollusca, fishes, and reptiles. The Cornbrash (so named from its friable (*brashy*) character, and from its forming good soil for corn) is one of the most persistent bands in the English Jurassic system, retaining its characters all the way from the south-western counties to near the Humber.

4. The **Oxfordian** stage, sometimes called the Middle or Oxford Oolite, consists of a lower zone of calcareous sandstone, known as the Kellaways rock or Callovian, from the name of a place in Wiltshire, and of a thick upper stiff blue and brown clay, called, from the locality where it is well developed, the Oxford Clay, and containing numerous ammonites, belemnites, and oysters, but no corals. In Germany, the strata from the base of the Callovian to the top of the Purbeckian group are known as the Malm or White Jura.

5. The **Corallian** stage, so named from the corals with which it abounds, is one of the most distinctive in the Jurassic system. It is traceable across the greater part of England, over the continent of Europe from Normandy to the Mediterranean, through the east of France, and along the whole length of the Jura Mountains and the flank of the Swabian Alps. While it was being formed, the greater part of Europe lay beneath a shallow sea, the floor of which was clustered over with reefs of coral.

6. The **Kimmeridgian** group or stage is typically displayed at Kimmeridge on the coast of Dorsetshire, whence its name. It there consists of dark shales, some of which are so highly bituminous as to burn readily, and which will probably be eventually of commercial value as a source for the distillation of mineral oil. This group of strata has yielded a larger number of reptilian genera and species than any other in the Mesozoic systems of Britain—plesiosaurs, ichthyosaurs, pterosaurs, deinosaurs, turtles, and crocodiles. It is well developed in France and Germany.

7. The **Portlandian** stage, so called from the Isle of Portland where it is well seen, consists of a lower set of sandy beds (Portland Sand), and a higher and thicker series of limestones and calcareous freestones, some of the beds containing abundant nodules and layers of flint. These rocks are prolonged into France near Boulogne-sur-Mer.

8. The **Purbeckian** group or stage is best seen in the Isle of Purbeck, hence its name. It lies on an upraised surface of Portlandian beds, showing that after the deposition of these beds there was some disturbance of the sea-bed, portions of which were uplifted partly into land and partly into shallow brackish and fresh waters. The Purbeck beds are subdivided into three sub-stages : the lowest consisting of fresh-water limestones, with layers of ancient soil ("dirt-beds"), in which the stumps of cycadaceous trees still stand in the positions in which they grew (Fig. 171); the middle sub-stage contains oysters and other marine shells which prove that owing to subsidence the area sank under the sea ; while in the higher subdivision fresh-water fossils reappear. Among the more interesting organisms yielded by the Purbeck beds are the remains of numerous insects and of the marsupials already referred to, which chiefly occur as lower jaws in a stratum about 5 inches thick. When the bodies of dead animals float out to sea the first bones likely to drop out of the decomposing carcases are the lower jaws ; hence the greater frequency of these bones in the fossil state. Strata belonging to the Purbeckian stage and

including red and green marls, with dolomite and gypsum, are found in north-western Germany, showing in that region also the elevation of the floor of the Jurassic sea into detached basins.

In India, a mass of strata 6300 feet thick is found in Cutch, and from its fossils is believed to represent the European Jurassic system from the Bajocian up to the top of the Portlandian stage. In Australia and New Zealand, recognisable Jurassic fossils have also been found, showing the extension of the Jurassic system even to the Antipodes. In North America, Jurassic rocks are not largely developed ; but in Colorado they have yielded an abundant series of organic remains, including fishes, tortoises, pterosaurs, deinosaurs, crocodiles, and marsupials.

CHAPTER XXIV

CRETACEOUS

THE CRETACEOUS system received its name in Western Europe, because in England and in Northern France its most conspicuous member is a thick mass of white chalk (Latin, *Creta*). It covers a far more extensive area of the surface of this continent than any of the preceding systems. Its western extremity reaches to the north of Ireland and the Western Islands of Scotland. It covers a large part of the east and south of England, stretching thence into France, where it forms a broad band, encircling the Tertiary basin of Paris. It sweeps across Belgium into Westphalia, underlies the vast plain of Northern Germany and Denmark, whence it is prolonged into Southern Russia, where it overspreads many thousands of square miles. It flanks most of the principal mountain-chains of Europe—the Pyrenees, Alps, Apennines, and Carpathians. It spreads far and wide over the basin of the Mediterranean Sea, extending across vast tracts of Northern Africa, and from the Adriatic athwart Greece and Turkey into Asia Minor, whence it is prolonged through the Asiatic continent.

As most of the rocks of the system are of marine origin, we at once perceive how entirely different the Cretaceous geography must have been from that of the present day, and to what a great extent the existing land of the Old World lay then below the sea. But in tracing out the distribution of the rocks, geologists have found that the Cretaceous sea did not sweep continuously across Europe. On the contrary, as they have ascertained, the old northern land still rose over the site of Northern Britain and Scandinavia, while to the south of it a wide depression extended across the area of Southern Britain, Northern France, Belgium, and the North German plain, eastwards to Bohemia and Silesia. This vast northern basin was the theatre of a remarkable succession of geological revolutions. While its eastern portions, during the earlier part of the Cretaceous period, were submerged under the sea, its western tracts were the site of the delta of a great river, probably descending from the land that still lay massed towards the north. During the later ages of the period, the whole of this area formed a broad and long gulf or inlet, the southern margin of which seems to have been defined by the ridge of old rocks that runs from the headlands of Brittany through Central France, the Black Forest, and the high grounds of Bohemia. South of that ridge lay the open ocean which extended all over Southern Europe and the north of Africa, and spread eastwards into Asia.

Bearing in mind this peculiar disposition of sea and land, we can understand why the development of the Cretaceous system, alike in regard to its deposits and its fossils, should be so different in the area of the northern basin from that of the southern regions. In the one case we meet with the local and changing accumulations of a comparatively shallow and somewhat isolated portion of the sea-bed, wherein are mingled abundant traces of the proximity of land. In the other we are presented with evidence of a wide open sea, where the same kinds of deposits and the same forms of marine life extend with little change over vast distances. Obviously, it is not the local type of the northern basin, but the more general and widespread type of Southern Europe that should be taken for the distinctive characteristics of the Cretaceous system. But the northern basin was the first to be systematically explored, and is still the best known, and hence its features have not unnaturally usurped the place of importance which ought properly to be assigned to the other wider area.

Regarding the period as a whole, let us first consider the general character of its distinguishing flora and fauna, and then pass on to trace the history of the period as revealed by the succession of strata. The plants of the Cretaceous system show that the vegetable kingdom had now made a most important advance in organisation. In the lower half of the system the fossil plants yet found are on the whole like those of the Jurassic rocks—that is, they include some of the same genera of ferns, cycads, and conifers which these rocks contained. But already the ancestors of our common trees and flowering plants must have made their appearance, for in the upper half of the system their remains occur in abundance. This earliest dicotyledonous flora numbered among its members species of maple, alder, aralia, poplar, myrica, oak, fig, walnut, beech, plane, sassafras, laurel, cinnamon, ivy, dogwood, magnolia, gum-tree, ilex, buckthorn, cassia, credneria, and others.   The modern aspect of this assemblage of plants is in striking contrast to the more antique look of all the older floras.   There were likewise species of pine (*Pinus*), Californian pine (*Sequoia*), juniper, and other conifers, various

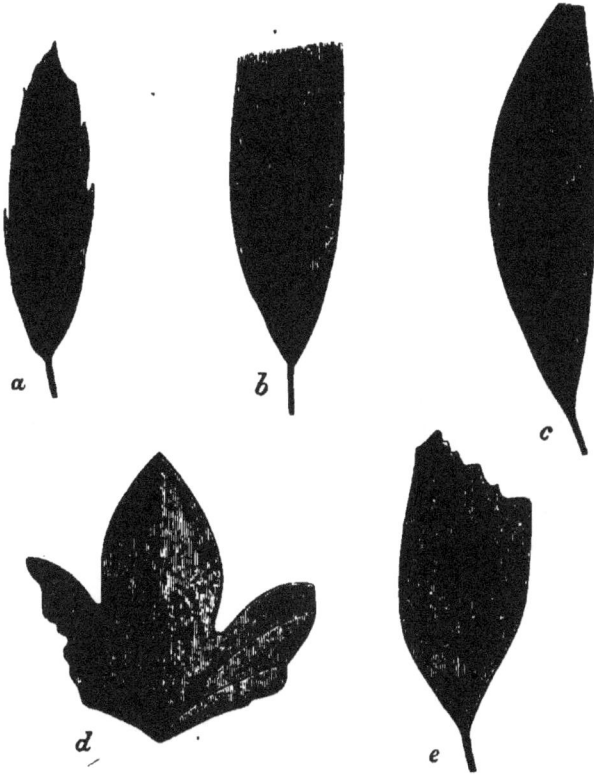

FIG. 184.—Cretaceous Plants.   (a) *Quercus rinkiana* (⅔); (b) *Cinnamomum sezannense* (⅔); (c) *Ficus atavina* (⅔); (d) *Sassafras recurvata* (⅔); (e) *Juglans arctica* (⅓).

cycads, forms of screw-pine (*Pandanus*), palms (*Sabal*), and numerous ferns (*Gleichenia, Asplenium,* etc.)   This flora spread over the land, surrounding the northern Cretaceous basin, and extended northwards even as far as North Greenland, from which nearly 200 species of Cretaceous plants have been obtained.   The inference may be deduced that the climate of the globe must then have been much warmer than at present.   The luxuriant vegetation disinterred from the Cretaceous rocks of North Greenland includes more than forty kinds of ferns, besides laurels, figs, magnolias, and other plants, which show that, though the winters were no doubt dark, they must have been extremely mild. There

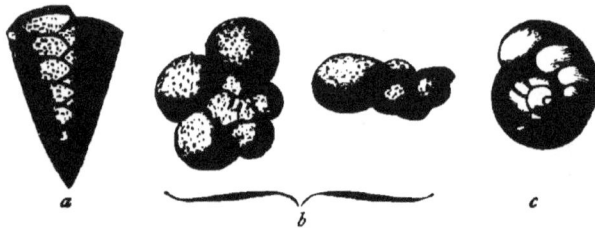

FIG. 185.—Cretaceous Foraminifera. (a) *Textularia baudouiniana* ($\frac{20}{1}$); (b) *Globigerina cretacea* ($\frac{17}{1}$); (c) *Rotalina voltziana* ($\frac{20}{1}$).

could have been no perpetual frost and snow in these Arctic latitudes in Cretaceous times.

Foraminifera abound in some of the Cretaceous limestones, indeed, in some places they form almost the only constituent of these rocks. They are plentiful in the white chalk of England, France, and Belgium, one of the more frequent genera being *Globigerina* (Fig. 185) which still lives in enormous numbers in the Atlantic, and forms at the bottom of that ocean a grey ooze not unlike chalk (Fig. 33). Sponges lived in great numbers in the Cretaceous sea. Their minute siliceous spicules are abundant in the Chalk, and even entire sponges enveloped in flint are not uncommon (*Ventriculites*, Fig. 186). Sea-urchins are among the most familiar fossils of the Chalk, and must have lived in great numbers on the Cretaceous sea-bottom. Some of their genera are still living, and have been dredged up in recent years from great depths in the ocean. Among the more characteristic Cretaceous types are *Ananchytes*, *Holaster*, *Micraster*, and *Echinoconus* (Fig. 187). The brachiopods were still represented chiefly by the ancient genera *Terebratula* and *Rhynchonella*. Lamellibranchs abounded, especially the genera *Ostrea*, *Exogyra*, *Inoceramus* (Fig. 188), *Lima*,

FIG. 186.—Cretaceous Sponge (*Ventriculites decurrens*, $\frac{1}{2}$).

Pecten, and the various forms of Hippuritids. These last (*Hippurites*, *Radiolites*, *Caprina*, etc., Fig. 189) are specially characteristic, being, so far as we know, confined to the Cretaceous system ; hence their occurrence serves to indicate the Cretaceous

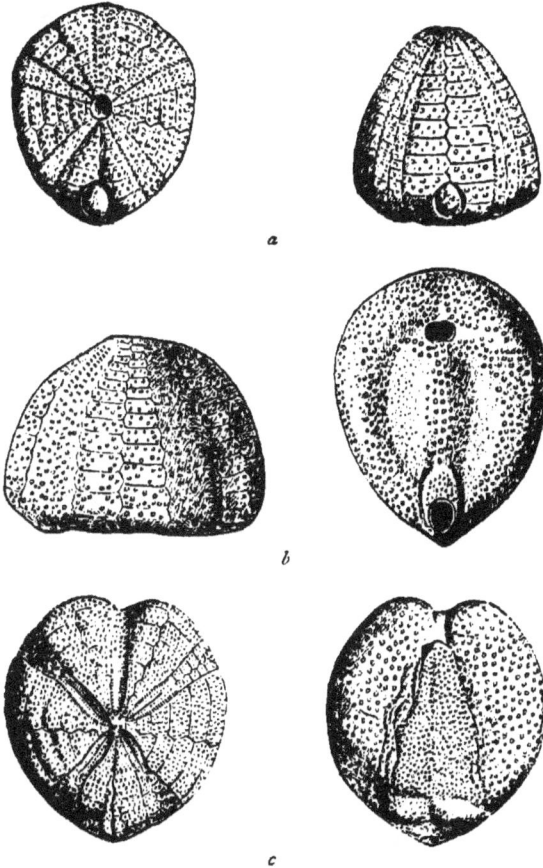

FIG. 187.—Cretaceous Sea-urchins. (*a*) *Echinoconus conicus*, ⅔ (= *Galerites albo-galerus*), under surface and side view ; (*b*) *Ananchytes ovatus* (½), side view and under surface ; (*c*) *Micraster cor-arguinum* (½), upper and under surface.

age of the rock containing them. They have been imbedded in such numbers in the limestones of the south of Europe as to give the name of hippurite-limestone to these rocks. They are comparatively infrequent in the strata of the northern Cretaceous basin.

Probably the most distinctive feature in the molluscan life of the Cretaceous seas was the extraordinary variety in the develop-

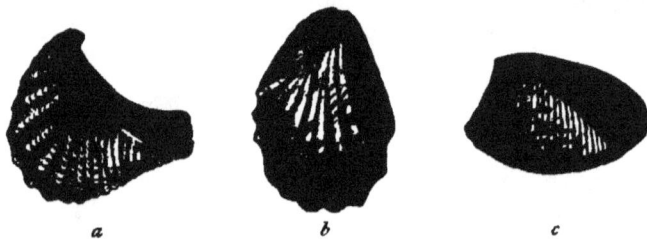

FIG. 188.—Cretaceous Lamellibranchs. (*a*) *Trigonia aliformis* (½); (*b*) *Inoceramus sulcatus* (½); (*c*) *Nucula bivirgata* (natural size).

ment of the cephalopods.   This is all the more remarkable from the fact that before the next geological period the great majority

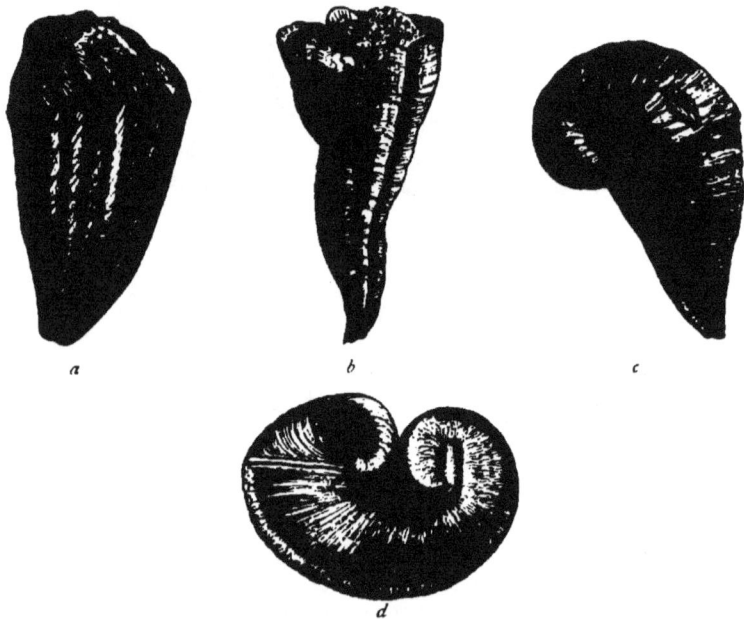

FIG. 189.—Cretaceous Lamellibranchs (Hippurites). (*a*) *Radiolites acuticostata* (½); (*b*) *Hippurites toucasiana* (½); (*c*) *Caprina Aguilloni* (½); (*d*) *Caprotina toucasianus* (½).

of these types appear to have become extinct.   The ammonites and belemnites, which played so important a part in the fauna of Mesozoic time, died out about the close of that long succession of periods. At least in Europe, while their remains continue to

Fɪɢ. 190.—Cretaceous Cephalopods. (a) *Baculites anceps* (⅓); (b) *Ptychoceras emerici-anus* (½); (c) *Toxoceras bituberculatus* (⅔); (d) *Hamites rotundus* (½); (e) *Ancylo-ceras renauxianus* (¹⁄₁₂); (f) *Scaphites æqualis* (⅔); (g) *Crioceras villiersianus* (⅓); (h) *Helicoceras annulatus*; (i) *Ammonites inflatus* (½); (k) *Turrilites catenatus* (½).

present themselves up to the top of the Cretaceous system, they disappear entirely from the overlying strata. It is curious to observe that while these important tribes were about to vanish, other cephalopods of new and varied types flourished contemporaneously with them. Never before or since, indeed, have the cephalopodan types been so manifold (Fig. 190). For instance, *Baculites* is a straight chambered shell reminding us of the ancient *Orthoceras*. In *Toxoceras* the shell is bent into the form of a bow. In *Hamites* it is long, tapering, and curved upon itself like a hook. In *Ancyloceras* it is coiled at the posterior end, the other being bent back upon itself; while in *Scaphites* the coils are adherent. In *Ptychoceras* the shell is long, tapering, and bent once back on itself, the two portions being in contact. In *Crioceras* it is coiled, and the coils are not adherent, as they are in the ammonites. In *Helicoceras* the shell is coiled spirally, the coils remaining free, while in *Turrilites* they are adherent.

The fishes of the Cretaceous period are chiefly known by teeth belonging to various genera of sharks (*Otodus, Lamna, Oxyrhina*).

FIG. 191.—Cretaceous Fish (*Beryx lewesiensis*, ¼).

But they also include the earliest known representatives of the modern osseous or teleostean fishes, such as the herring, salmon, and cod (*Osmeroides, Enchodus, Beryx*, Fig. 191, *Syllæmus*, etc.)

Already reptilian life seems to have been on the decline, at least there is much less variety and abundance of it in the Cretaceous system than in that which immediately preceded it. Turtles and tortoises continued to haunt the low shores of the time. Ichthyosaurs, plesiosaurs, pterosaurs, and deinosaurs still lived, but in diminishing numbers, and they are not known to have survived the Cretaceous period. One of the most remarkable of the deinosaurs, and interesting from being one of the last of its

race, is that known as *Iguanodon* (Fig. 192). Only scattered teeth
and bones of this animal were known, until a few years ago the
fortunate discovery of a number of entire skeletons in Belgium enabled its structure to be almost completely made known, and threw
much fresh light on the osteology of the deinosaurs.   It was a herb-
ivorous and probably amphibious creature, able, no doubt, to walk
along the shores, with an unwieldy gait, on its long hind legs, and
balancing itself by its strong massive tail, which was doubtless a
powerful instrument of propulsion through the water.   Its extraordinary fore legs, with the strong spurs on the digits, must have

FIG. 192.—Cretaceous deinosaur (*Iguanodon*, about $\frac{1}{60}$).

been formidable weapons of defence against its carnivorous contemporaries.   Another gigantic reptile, the *Mosasaurus*, believed
to have been 75 feet long, was furnished with fin-like paddles for
swimming.   Several kinds of crocodiles have also been disinterred
from Cretaceous rocks in Europe.

Still more remarkable is the assemblage of remains of animal
life exhumed from corresponding rocks in the Western Territories
of North America.   Among these the *Discosaurus* was a snake-
like animal some 40 feet long, with a swan-like neck supporting
a slim head which it could raise 20 feet out of the water, or dart
to the bottom and catch its prey.   The pythonomorphs or sea-
serpents were especially numerous.

The remains of true birds have been obtained from the
Cretaceous rocks both of Europe and North America.   They are

Y

related to the living ostrich, but some of them were furnished with teeth set in a continuous groove (*Hesperornis*), others had large teeth in distinct sockets (*Ichthyornis*).

The following are the principal subdivisions of the Cretaceous system in Europe in descending order. The stages are based upon more or less well-marked fossil evidence, but they are also for the most part to be distinguished by lithological characters :—

Upper	Danian	Pisolitic limestone of Paris basin ; Chalk of Hainault, Ciply, Maestricht, Faxoe in Denmark, and the south of Sweden ; absent in England (*Belemnitella, Baculites*).
	Senonian	Chalk-with-flints of Norwich, Brighton, Flamborough Head, and Dover, north of France (*Belemnitella, Marsupites, Micraster*); sandstones of Westphalia and Saxony.
	Turonian	Chalk-without-flints of Dover and north of France (*Holaster planus, Terebratulina gracilis, Inoceramus labiatus*); sandstones, limestones, and marls of Saxony and Bohemia ; Hippurite limestone of Southern France and Mediterranean basin.
	Cenomanian	Grey Chalk of Folkestone (*Holaster subglobosus, Belemnitella plena*), Chalk-Marl, red chalk of Hunstanton, Glauconitic Marl and Upper Greensand (*Ammonites rostratus, Pecten asper*); Chalk of Rouen ; earthy limestones and marls in Hanover replaced southwards by plant-bearing sandstones, clays, and thin coal-seams ; Hippurite limestones of Southern Europe.
	Albian	Gault (*Ammonites cristatus, A. denarius, A. auritus*).
Lower	Neocomian	In Southern England a fluviatile (partly marine) succession of sands and clays (Wealden), surmounted by sands, clays, and limestones (Lower Greensand) ; in Northern England a series of clays and limestones, with marine fossils (upper part of Speeton clay) ; limestones and marls of Neuchâtel ; compact crystalline limestones in Provence (*Ammonites Dehayesi, A. nisus* in upper division ; abundant *Ancyloceras* with *Pecten cinctus* in middle ; *Ammonites noricus, A. astieramus, Ostrea Couloni* in lower).

It will be remembered that towards the close of the Jurassic period the floor of the sea in the western part of the European area was gently raised, some of the younger Jurassic marine lime-stones being ridged up into islets or low land, with lakes or estuaries in which the Purbeck beds were deposited. This terrestrial condition of the geography was maintained and extended in the same region during the early part of the Cretaceous period. The geological history of Europe as revealed by the various subdivisions in the foregoing Table may be briefly given.

Neocomian (from Neocomum, the old Latin name of Neuchâtel in Switzerland). This stage in the south of England and thence eastwards across Hanover consists of a mass of sand and clay sometimes 1800 feet thick, representing the delta of a river. Only a portion of this delta remains, but as it extends in an east and west direction for at least 200, and from north to south for perhaps 100 miles, its total area may have been 20,000 square miles, indicating a large river comparable with the Quorra of the present day. This stream not improbably descended from the north or north-west. It carried down the drifted vegetation of the land, with occasional carcases of the iguanodons and other terrestrial or amphibious creatures of the time. From their great development in the Weald of Sussex these delta-deposits have been called Wealden. They attain there a thickness of more than 1800 feet and consist of the following subdivisions in descending order.

Weald Clay	.	1000 feet.
Hastings Sand group, comprising—		
3. Tunbridge Wells Sand .	.	140 to 380 ,,
2. Wadhurst Clay . . .	.	120 to 180 ,,
1. Ashdown Sand . . .	.	400 to 500 ,,

Beyond the area overspread by the sand and mud of the delta, the ordinary marine sediments accumulated, with their characteristic organic remains. We find these sediments in Yorkshire (upper part of Speeton clay), which must then have lain beyond the estuary of the river. They stretch thence eastwards through North-Western Germany, and are found at the base of Cretaceous system through France into Switzerland. The Lower Greensand which overlies the Wealden group in the south of England contains marine fossils, and points to the submergence of the delta.

Albian (from the department of the Aube in France). In England this stage nearly corresponds to the band of dark, stiff, blue clay known as the Gault. Extending over the Wealden sands and clays, the Gault (100 to 200 feet or more in thickness), with its abundant marine fossils, shows how thoroughly the Wealden delta was now submerged beneath the sea.

Cenomanian (from Coenomanum, the old Latin name of the town of Mans in the department of Sarthe, France). This stage comprises a group of impure chalky, glauconitic, and sandy deposits lying at the base of the Chalk in England and the north

of France. The subdivisions of this stage in England are shown
in the following table in descending order :—

> Grey chalk forming the base of the Chalk.
> Chalk Marl (Red Chalk of Hunstanton).
> Glauconitic Marl.
> Upper Greensand.

Certain sandy portions of this group have been called the Upper
Greensand. The Glauconitic (or Chloritic) Marl is an impure, dull white, or yellowish chalk, with grains of glauconite and
phosphatic nodules. The Chalk-Marl is an impure band of
chalk sometimes overlain by a zone of Grey Chalk which forms
the base of the true Chalk-without-flints. ' All these deposits
indicate the accumulations of a shallow sea, probably not far from
land. Traced eastwards into Germany, they undergo great
changes in their lithological characters, passing at last in Saxony
and Bohemia into sandstones and clays full of remains of terrestrial vegetation, and even including some thin seams of coal. It
is in these beds that the oldest dicotyledonous plants in Europe
have been found. It is evident that land existed in the heart of
Germany during this stage of the Cretaceous period. In Southern
France, on the other hand, the corresponding strata are massive
hippurite-limestones which sweep through the great Mediterranean
basin, and show how large an area of Southern Europe then lay
under the sea.

Turonian (from Touraine). This stage includes the lower
part of the Chalk, above the Grey Chalk. The thick mass of
white crumbly limestone known as the Chalk has been referred to
as the most conspicuous member of the Cretaceous system in the
west of Europe. It has long been grouped into two parts, a lower
band of "Chalk-without-flints," and an upper band of "Chalk-
with-flints." The former corresponds, on the whole, with the
Turonian stage. The Chalk is a remarkably pure limestone, composed chiefly of crumbled foraminifera, urchins, molluscs, and
other marine organisms. It must have been laid down in a sea
singularly free from fine sediment ; but there is no evidence that
this sea was one of great depth. On the contrary, though the
Chalk itself resembles the Globigerina ooze of the deeper parts of
the Atlantic Ocean, the characters of its foraminifera and other
organic remains indicate comparatively shallow-water conditions.
The basin in which it was laid down shallowed eastwards, where,
from the evidence of sandstones, coal-seams, and plants, there
was land at the time ; while, probably, towards the west there

was connection with the open sea. The total thickness of the Chalk, including the Cenomanian, Turonian, and Senonian stages, exceeds 1200 feet.

Senonian (from Sens, in the department of Yonne). This stage corresponds generally with the original English subdivision of Upper Chalk, or Chalk-with-flints. Its most conspicuous feature is the presence of the layers of nodules or irregular lumps of black flint which mark the stratification of the Chalk. The origin of these concretions has been the subject of much discussion among geologists, and it cannot be said to have been even yet satisfactorily solved. Some marine plants (diatoms) and animals (radiolarians, sponges, etc.) secrete silica from sea-water, and build it up into their framework. But the flints are not mere siliceous organisms, though organic remains may often be observed enclosed within them. They are amorphous lumps of dark silica, containing a little organic matter. By some process, not yet well understood, these aggregations of silica have gathered usually round organic nuclei, such as sponges, urchins, shells, etc. The decomposition of organic matter on the sea-floor may have been the principal cause in determining the abstraction and deposition of silica. Not infrequently an organism, such as a brachiopod or echinus, originally composed of carbonate of lime, has been completely transformed into flint.

The Chalk is well exposed along the sea-cliffs of the east and south of England. It forms the promontories of Flamborough Head, Dover, Beachy Head, and the Needles in the Isle of Wight. The white cliffs of Kent are repeated on the opposite coast of France, where the Chalk with all its lithological and palæontological characters reappears, and whence it extends through Northern France into Belgium.

Danian (from Denmark). This stage has not been recognised in England. Its component chalky strata occur in scattered patches over Northern France, Belgium, and Denmark, to the south of Sweden.

The Cretaceous hippurite-limestones of Southern Europe and the basin of the Mediterranean are prolonged through Asia Minor into Persia, where they cover a vast area. They have been found likewise on the flanks of the Himalaya Mountains, so that the open Cretaceous sea must have stretched right across the heart of the Old World. In the Indian Deccan a great extent of country, estimated at 200,000 square miles, lies buried under horizontal or nearly horizontal sheets of lava, which have a united thickness of from 4000 to 5000 feet, and were erupted during the Cretaceous period. These eruptions, from the presence of interstratified layers containing remains of fresh-water shells, land-plants, and insects, are believed to have taken place on land and not under the sea.

Cretaceous rocks cover an enormous area in North America. They attain no great thickness in the Eastern States, but they thicken southwards, until in Texas they present massive lime-stones indicative of deeper and clearer water than elsewhere in that region. They attain gigantic proportions in Colorado, Utah, and Wyoming, whence they are prolonged northwards into the British territories, with a maximum thickness of 11,000 to 13,000 feet. They have yielded a remarkably abundant and varied series of organic remains. In their upper parts (Laramie group) they contain a large assemblage of land-plants, half of which are allied to still living American trees, and in some places these plants are aggregated into valuable seams of coal. The numerous reptilian and bird remains found in these strata have been already noticed.

Rocks assigned to the Cretaceous system cover a wide region of Queensland, and also attain a considerable thickness in New Zealand.

CHAPTER XXV

THE Cretaceous system closes the long succession of Secondary or ·Mesozoic formations. The rocks which come next in order are classed as Tertiary or Cainozoic. When these names were originally chosen, geologists in general believed not only that the divisions into which they grouped the stratified rocks of the earth's crust corresponded on the whole with well-defined periods of time, but that the abrupt transitions, so often traceable between systems of rocks, served to mark geological revolutions, in which old forms of life as well as old geographical conditions disappeared and gave place to new. One of the most notable of such breaks in the record was supposed to separate the Cretaceous system from all the younger rocks. This opinion arose from the study of the geology of Western Europe, and more especially of South-Eastern England and North-Western France. The top of the Chalk, partly worn down by denudation, was found to be abruptly succeeded by the pebble-beds, sands, and clays of the lower Tertiary groups. No species of fossils found in the Chalk were known to occur also in the younger strata. It was quite natural, therefore, that the hiatus at the top of the Cretaceous system should have been regarded as marking the occurrence of some great geological catastrophe and new creation, and, consequently, as one of the great divisional lines of the Geological Record.

More detailed investigation, however, has gradually overthrown this belief. In Northern France, Belgium, and Denmark various scattered deposits (Danian, p. 322) serve to bridge over the gap that was supposed to separate Mesozoic and Cainozoic formations. In the Alps no satisfactory line could be found to separate un- doubtedly Cretaceous strata from others as obviously Tertiary. And in various parts of the world, especially in Western North

America, other testimony was gradually obtained to show that no general convulsion marked the end of the Secondary and beginning of the Tertiary periods, but that the changes on the earth's surface proceeded in the same orderly connection and sequence as during previous and subsequent geological ages. The break in the continuity of the deposits in Western Europe only means that in that part of the world the record of the intervening ages has not been preserved. Either strata containing the record were never deposited in the region in question, or, having been deposited, they have subsequently been removed.

Bearing in mind, then, that such geological terms are only used for convenience of classification and description, and that what is termed Mesozoic time glided insensibly into what is called Cainozoic, we have now to enter upon the consideration of that section of the earth's history comprised within the Tertiary or Cainozoic periods. The importance of this part of the geological chronicle may be inferred from the following facts. During Tertiary time the sea-bed was ridged up into land to such an extent as to give the continents nearly their existing area and contour. The crust of the earth was upturned into great mountain ranges, and notably into that long band of lofty ground stretching from the Pyrenees right through the heart of Europe and Asia to Japan. Some portions of the Tertiary sea-bed now form mountain peaks 16,000 feet or more above the sea. The generally warm climate of the globe, indicated by the world-wide diffusion of the same species of shells in Palæozoic and less conspicuously in Mesozoic time, now slowly passed into the modern phase of graduated temperatures, from great heat at the equator to extreme cold around the poles. At the beginning of the Tertiary or Cainozoic periods the climate was mild even far within the Arctic Circle, but at their close it became so cold that snow and ice spread far southward over Europe and North America.

The plants and animals of Tertiary time are strikingly modern in their general aspect. The vegetation consists, for the most part, of genera that are still familiar in the meadows, woodlands, and forests of the present day. The assemblage of animals, too, becomes increasingly like that of our own time as we follow the upward succession of strata in which the remains are preserved. In one strongly marked feature, however, does the Tertiary fauna stand contrasted alike with everything that preceded and followed it. If the Secondary periods could appropriately be grouped together under the name of the "Age of Reptiles," Tertiary time may not less fitly be called the "Age of Mammals." As the manifold reptilian types died out, the mammals, in ever-increasing complexity of organisation, took their place in the animal world. By the end of the Tertiary periods they had reached a variety of type and a magnitude of size altogether astonishing, and far surpassing what they now present. The great variety of pachyderms is an especially marked feature among them.

The rocks embraced under the terms Cainozoic or Tertiary have been classified according to a principle different from any followed with regard to the older formations. When they began to be sedulously studied, it was found that the percentage of recent species of shells became more numerous as the strata were followed from older to newer platforms. The French naturalist Deshayes determined the proportions of these species in the different Tertiary groups of strata, and the English geologist Lyell proposed a scheme of classification based on these ratios. His names, with modifications as to their application, have been generally adopted. They are compounds of the Greek καινος, recent, with affixes denoting the proportion of living species.

To the oldest Tertiary deposits, containing only about 3 per cent of living species of shells, the name Eocene (dawn of the recent) was given. The next series, containing a larger number of living species, has received the name of Oligocene (few recent). The third division in order is named Miocene, to indicate that the living species, though in still larger proportions, are yet a minority of the whole shells. The overlying series forming the uppermost of the Tertiary divisions is termed Pliocene (more recent), because the majority are now living species. The same system of nomenclature has been retained for the next overlying group, which forms the lowest member of the Post-tertiary or Quaternary series. This group is called Pleistocene (most recent), and all the species of shells in it are still living at the present time. It must not be supposed that the mere percentage of living or of extinct species of shells in a deposit always affords satisfactory evidence of geological age. Obviously, there may have been circumstances favourable or unfavourable to the existence of some shells on the sea-bottom which that deposit represents, or to the subsequent preservation of their remains. The system of classification by means of shell-percentages must be used with some latitude, and with due regard to other evidence of geological age.

EOCENE.

In Europe great geographical changes took place at the close of the Cretaceous period. The wide depression in which the Chalk had been deposited was gradually and irregularly elevated, and over its site a series of somewhat local deposits of clay, sand, marl, and limestone was laid down, partly in small basins of the sea-floor, and partly in estuaries, rivers, or lakes. In Southern Europe, however, the more open sea maintained its place, and over its floor were accumulated widespread and thick sheets of limestone which, from the crowded nummulites which they contain, are known as Nummulitic Limestone. These characteristic rocks extend all over the basin of the Mediterranean, stretching far into Africa and sweeping eastwards through the Alps, Carpathians, and Caucasus, across Asia to China and Japan. In North America the rocks classed as Eocene present two contrasted types. Down the eastern and western borders of the Continent, from the coast of New Jersey into the Gulf of Mexico on the one side, and along the coast ranges of California and Oregon on the other, they are marine deposits, though occasionally presenting layers of lignite with terrestrial plants. Over the vast plateaux which support the Rocky Mountains, however, they are of lacustrine origin, and show that in what is now the heart of the Continent the bed of the Cretaceous sea was upraised into a succession of vast lakes, round which grew a luxuriant vegetation. In these lakes a total mass of Eocene strata, estimated at not less than 12,000 feet, was deposited, entombing and preserving an extraordinarily abundant and varied record of the plant and animal life of the time.

The flora of Eocene time points to a somewhat tropical climate. Among its plants are many which have living representatives now in the hotter parts of India, Australia, Africa, and America. Above the ferns (*Lygodium*, *Asplenium*, etc.) which clustered below, rose clumps of palms, cactuses, and aroids; numerous conifers and other evergreens gave the foliage an umbrageous aspect, while many deciduous trees—ancestors of some of the familiar forms of our woodlands—raised their branches to the sun. Among the conifers were many cypress-like trees (*Callitris*, *Cupressinites*), pines (*Pinus*, *Sequoia*), and yews (*Salisburia* or Ginko). Species of aloe (*Agave*), sarsaparilla (*Smilax*), and amomum were mingled with fan-palms (*Sabal*,

Chamærops) and screw-pines (*Pandanus, Nipa*), together with early forms of fig (*Ficus*), elm (*Ulmus*), poplar (*Populus*), willow (*Salix*), hazel (*Corylus*), hornbeam (*Carpinus*), chestnut (*Castanea*), beech (*Fagus*), plane (*Platanus*), walnut (*Juglans*), liquidambar, magnolia, proteaceous plants (Fig. 193) resembling those of Australia and the Cape, water-bean (*Nelumbium*), water-lily (*Victoria*), maple (*Acer*), gum-tree (*Eucalyptus*), cotoneaster, plum (*Prunus*), almond (*Amygdalus*), laurel (*Laurus*), cinnamon tree (*Cinnamomum*), and many more.

Fig. 193.—Eocene Plant (*Petrophiloides Richardsonii*), natural size.

The fauna likewise points to the extension of a warm climate over regions that are now entirely temperate. This is particularly noticeable with regard to the mollusca. The species are, with perhaps a few exceptions, all extinct, but many of the genera are still living in the warmer seas of the globe. Some of the most

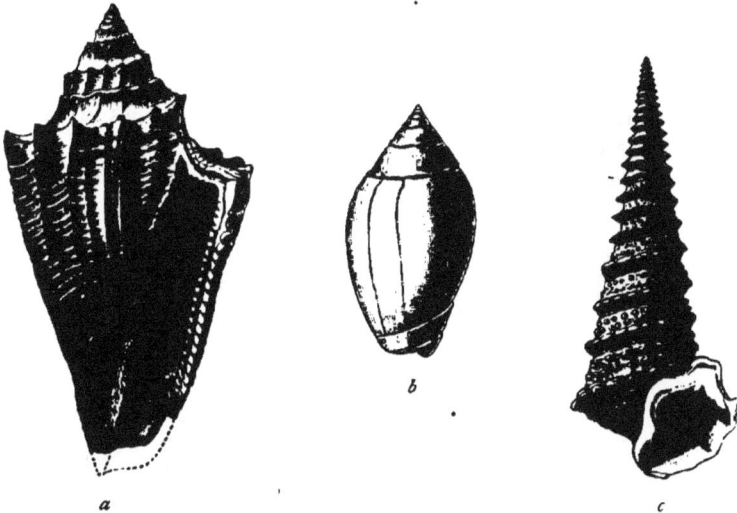

Fig. 194.—Eocene Molluscs. (*a*) *Voluta luctatrix* (⅔); (*b*) *Oliva Branderi* (natural size); (*c*) *Cerithium tricarinatum* (⅔).

characteristic forms are species of *Nautilus, Oliva, Voluta, Conus, Mitra, Cyrena, Cytherea, Chama.* The genus of Foraminifera, called *Nummulites* from the resemblance of the organism to a

piece of money, is enormously abundant in the limestones above
referred to as nummulitic limestones.   It must have flourished in
vast profusion over the floor of the sea, which in older Tertiary
time spread across the heart of the Old World from the Atlantic
to the Pacific Oceans.   Some of the most common fish-remains
found in the Eocene strata belong to the genera *Lamna, Otodus,
Myliobates, Pristis.*   Reptilian life, which enjoyed such a pre-
ponderance during the Mesozoic ages, is conspicuously diminished
in the Eocene deposits alike in number of individuals and variety
of structure.   The genera are chiefly turtles, tortoises, crocodiles,
and sea-snakes, presenting in their general assemblage a decidedly
modern aspect compared with the reptilian fauna of the Secondary
rocks.   Remains of birds are comparatively rare as fossils.   We
have seen that the earliest known type has been obtained in the
Jurassic system, and that others have been found in the Cretaceous
rocks.   Still more modern forms occur in Eocene strata ; they
include one (*Argillornis*) which may have been a forerunner of
the living gannet ; another, of large size (*Dasornis*), akin to the
gigantic extinct ostrich-like moa (*Dinornis*) of New Zealand ; a
third (*Agnopterus*) shows an affinity with the flamingo ; while the
buzzard, woodcock, quail, pelican, ibis, and African hornbill are
represented by ancestral forms.   That the early type which linked
birds with reptiles was still living is shown by the remains of one
curious genus (*Odontopteryx*) which had serrated jaws in which
the teeth were projections of the bony substance.

   But, as stated above, it was chiefly in higher forms of life that the

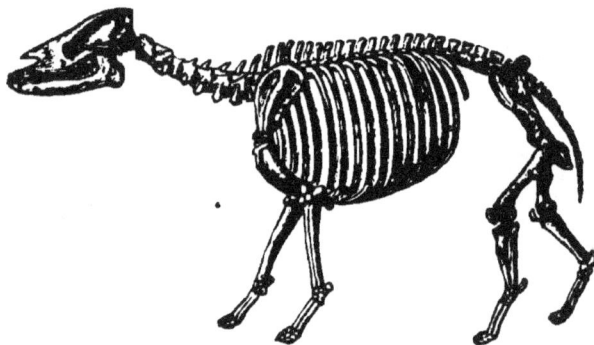

FIG. 195.—Eocene Mammal (*Palæotherium magnum*, ⅕.)

fauna of early Tertiary time stood out in strong contrast with that
of previous ages of geological history.   The mammalia now took

the leading place in the animal world, which they have retained
ever since.   Among the Eocene mammals reference may here be
made to the numerous tapir-like creatures which then flourished
(*Coryphodon*, *Palæotherium*, Fig. 195, *Anchilophus*, etc.)   Some
of the forms were intermediate in character between tapirs and
horses, and included the supposed ancestors of the modern horse
—small pony-like animals, with three, four, and even traces of
five toes on each foot.   Many of the mammals of Eocene time

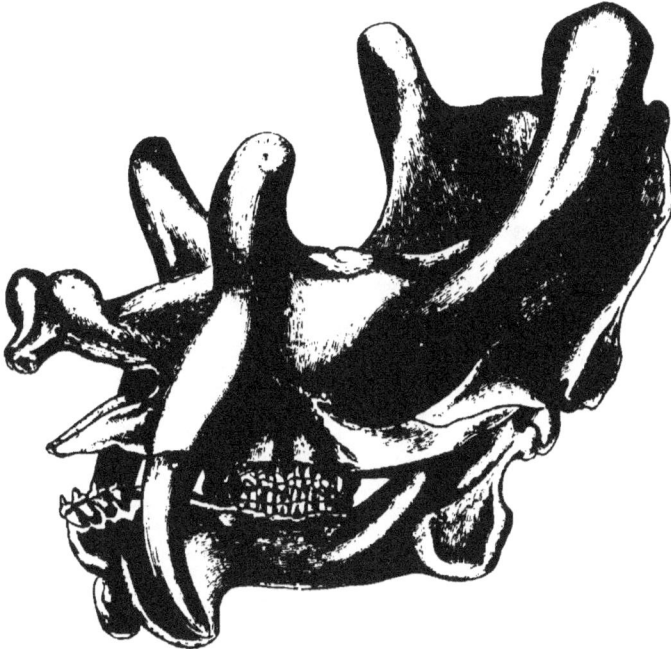

FIG. 196.—Skull of *Tinoceras ingens* (about ₁⁄₇).

presented more or less close resemblances to wolves, foxes,
wolverines, and other modern forms.   There were likewise true
opossums.   Numerous herds of hog-like animals (*Hyopotamus*)
and of hornless deer and antelopes wandered over the land,
while in the woodlands lived early ancestors of our present
squirrels, hedgehogs, bats, and lemurs.   Among these various
tribes which recall existing genera, others of strange and long-
extinct types roamed along the borders of the great lakes in
Western North America.   The Tillodonts were a remarkable
order, in which the characters of the ungulates, rodents, and

carnivores were curiously combined. These animals, perhaps rather less in size than the living tapir, had skeletons resembling those of carnivores, but with large prominent incisor teeth like those of rodents, and with molar teeth possessing grinding crowns like those of ungulates. Still more extraordinary were the forms to which the name of Deinocerata has been given (*Deinoceras*, *Tinoceras*, Fig. 196). These were somewhat like elephants in size, and like rhinoceroses in general build, but the skull bore a pair of horn-like projections on the snout, another pair on the forehead, and one on each cheek.

The Eocene rocks of England are confined to the south-eastern part of the country, from the coast of Hampshire into Norfolk. They vary in character from district to district, sands and gravels being replaced by clays according to the conditions in which the sediment was accumulated. They are prolonged into the north of France and Belgium. Arranged in tabular form, they may be grouped as follows :—

	ENGLAND.	FRANCE AND BELGIUM.
Upper.	Barton Clay and Upper Bagshot Sands.	Marine gypsum and marls of Paris ; sands and calcareous sandstones of Belgium (Wemmelian). Sands (marine), with estuarine and fresh-water limestones, etc. (Sables Moyens).
Middle.	Bracklesham Beds (leaf-beds of Alum Bay and Bournemouth), Middle Bagshot beds.	Calcaire - grossier divided into (3) upper limestones, with marine and fresh - water fossils ; (2) middle limestones, with marine shells and terrestrial vegetation ; (1) lower glauconitic marine limestones and sands. Sandstones and sands (Bruxellian) of Belgium.
Lower.	Lower part of Bagshot sands. London Clay. Oldhaven Beds. Woolwich and Reading Beds. Thanet Beds.	Paniselian sands of Belgium. Ypresian clays and overlying sands of Belgium. Absent in Paris basin. Landenian gravels and sands of Belgium. Sands of Bracheux (Paris basin), Heersian beds of Belgium, marls of Meudon ; fresh-water limestones of Rilly and Sézanne. Limestone of Mons in Belgium.

In striking contrast with these comparatively thin and locally developed deposits are those of the Alps, Southern Europe, and the basin of the Mediterranean. Masses of nummulitic limestone and sandstone, several thousand feet thick, have been upraised, folded, and fractured, and now form important parts of the great mountain chains which run through Europe and the north of Africa. Similar rocks have been uplifted along the flanks of the great chain of heights that sweeps through the heart of Asia, reaching in the Himalaya range a height of 16,500 feet above the sea-level. We thus learn not only that a large part of the existing continents lay under the sea during Eocene time, but that the principal mountain-chains of the Old World have been upheaved to their present altitudes since the beginning of the Tertiary periods. The great Eocene lake-basins of North America—so remarkable a feature in the geography—survived till a much later part of Tertiary time.

OLIGOCENE.

Under this name geologists have placed a group of strata usually of comparatively insignificant thickness, chiefly of fresh-water and estuarine, but partly also of marine origin, which, in Western and Central Europe, show how the bays and shallow seas of that region in the Eocene period were gradually obliterated and replaced by land and by sheets of fresh water. They attain in Switzerland a thickness of several thousand feet of sandstones, conglomerates, and marls, almost entirely of lacustrine origin, and forming a group of massive mountains (Rigi, Rossberg). A large lake occupied their site and continued to be an important feature in the geography of Central Europe during this and the following geological period. Other sheets of fresh water were scattered over the west of Europe. One of the largest of these lay in Central France, over the old district known as the Limagne d'Auvergne. In Germany, lacustrine and terrestrial deposits, including numerous seams of lignite or brown coal, are separated by a group of strata full of marine shells, foraminifera, etc., showing how the lakes and woodlands were submerged beneath the sea. In the Paris basin, and in the Isle of Wight, the strata are chiefly of fresh-water origin, but contain occasional marine intercalations. Evidently the Oligocene period, throughout the European area, was one of considerable oscillation in the earth's crust. During this time, too, the volcanic eruptions took place whereby the great sheets of basalt forming the terraced hills of the north of Ireland and Western Islands of Scotland were thrown out. •

An epoch of frequent change in the relative positions of sea and land is one in which there may be exceptional facilities for the preservation of a record of the plants and animals of the time. Oligocene strata have accordingly a peculiar interest from the abundant remains they contain of the contemporaneous terrestrial plants and animals. The land flora of that period is probably better known than that of any other section of the Geological Record, chiefly from the extraordinary abundance of its remains which have been preserved in the sediments of the ancient Swiss lake. Judging of it from these remains, we learn that it was in great measure made up of evergreens, and in various ways resembled the existing vegetation of tropical India and Australia and that of sub-tropical America. Its fan-palms, feather-palms, conifers, evergreen oaks, laurels, and other evergreen trees, gave a peculiarly verdant umbrageous character to the landscape in all seasons of the year, while numerous proteaceous shrubs glowed with their bright blooms on the lower grounds.

Of the terrestrial fauna numerous remains have been found in the lacustrine deposits of the time. We know that the borders of

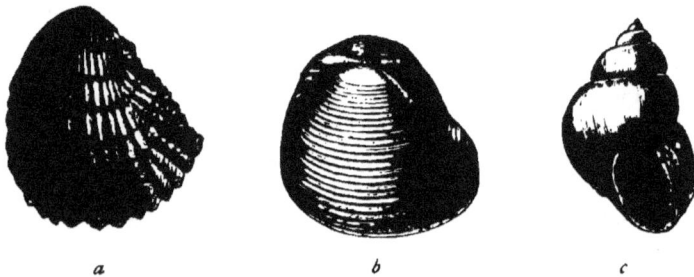

FIG. 197.—Oligocene Molluscs. (a) *Ostrea ventilabrum* (½); (b) *Corbula subpisum* (⅔); (c) *Paludina lenta* (natural size).

the lakes in Central France were frequented by many different kinds of birds—paroquets, trogons, flamingoes, ibises, pelicans, maraboots, cranes, secretary birds, eagles, grouse, and other forms. This association of birds recalls that around the lakes of Southern Africa at the present time. The mammals appeared in still more numerous and abundant types. Pachyderms abounded, including the *Anoplotherium*—a slender, long-tailed animal, about the size

of an ass, with two toes on each foot ; certain transitional types of
ungulates, with affinities to the pigs, peccaries, and chevrotains
(*Anthracotherium, Charopotamus, Hyopotamus*, etc.) ; various
forms of the tapir family, and of dogs, civets, martens, marmots,
bats, moles, and shrews.   The carnivora still presented marsupial characters, and in not a few of the animal types features
of structure were combined which are now only found in distinct
genera.   The Eocene palæotheres and the Oligocene anoplotheres
appear to have died out before the end of the Oligocene period.
The fresh water teemed with molluscs, belonging chiefly to genera
that still live in our rivers and lakes, such as *Unio, Cyrena,
Paludina, Planorbis, Limnæa, Helix*, and others (Fig. 197).
    In the Isle of Wight the highest Eocene strata were followed
by a group of fresh-water, estuarine, and marine deposits, formerly
classed as Upper Eocene, but placed here in the Oligocene division.   They are arranged in the following manner in descending
order :—

> Hamstead series—clays, marls, and shelly layers, with fresh-water and
> estuarine shells and land plants.   About 260 feet.
> Bembridge series—marls and limestone, with fresh-water shells below, and
> estuarine shells above.   About 110 feet.
> Osborne series—clays, marls, sands, and limestones, with abundant fresh-
> water shells.   About 100 feet.
> Headon series—consisting of an upper and lower division, containing fresh
> and brackish water fossils and a middle group in which marine shells
> and corals occur.   About 150 feet.

    These Isle of Wight strata, having a total depth of more than
600 feet, were for many years the only known examples in Britain
referable to this portion of the Geological Record, and they are
still the only beds in this country which in their abundant
molluscs allow a comparison to be made between them and corresponding rocks on the Continent.   But at Bovey Tracey in
Devonshire a small lake-basin has been discovered, the deposits
of which have yielded a large number of terrestrial plants comparable with those found in the Oligocene strata of Switzerland
and Germany.   Between the great sheets of basalt, also, that
form the plateaux of Antrim and the Inner Hebrides, numerous
remains of a similar vegetation have been discovered.   There
can be no doubt that these volcanic rocks were poured out over
the surface of the land, and that the plants, whose remains have
been disinterred from the intercalated layers of tuff and hardened
clay, grew upon that land.   The basalts and other lavas, even after

the great denudation which they have undergone, are still in some places more than 3000 feet thick. They were poured out in wide-spreading sheets that completely buried the previous topography and extended as vast lava-plains, like those of younger date, which form so impressive a feature in the scenery of Montana, Idaho, and Oregon, in Western North America.

In the Paris basin, the Oligocene strata follow immediately upon the Eocene group described on p. 334. They consist of (1) a lower division of gypsum (65 feet) and marls, with terrestrial shells, and remains of palæotheres and anoplotheres; (2) a middle band of marl, limestone, and sand, with lacustrine and estuarine shells; and (3) an upper division, in which the most conspicuous members are the sands and hard siliceous sandstone of Fontainebleau.

In Northern Germany the subjoined succession of strata in descending order has been noted.

Upper	Marine marls, clays, and sands. Brown coal of the Lower Rhine, with abundant terrestrial vegetation and some marine bands.
Middle	Sands and Septaria-clay, with abundant marine fauna; occasionally a brown-coal group occurs.
Lower	Marine beds of Egeln, with marine shells and corals. Amber beds of Königsberg, containing 4 or 5 feet of glauconitic sand, with abundant pieces of amber, which is the fossil resin of different species of coniferous trees. A large number of species of insects has been enclosed and preserved in the amber. Lower brown coal—sands, sandstones, clays, and conglomerates, with interstratified seams of brown coal and an abundant terrestrial flora, in which coniferæ are prominent.

CHAPTER XXVI

MIOCENE—PLIOCENE

THE geological period at which we are now arrived, one of the most important in the history of the configuration of the existing continents, embraced that portion of geological time during which the great mountain-chains of the globe were uplifted into their present commanding positions. There is good reason to believe that these lines of elevation are of great geological antiquity, and that they have again and again been pushed upward during great terrestrial disturbances. But the intervals between these successive upthrusts were probably often of immense duration, so that the mountains, being exposed to continuous and prolonged denudation, were worn down, sometimes perhaps almost to the very roots. In all probability the nucleus of the line of the Alps, for example, dates back to a remote geological period. But only in Tertiary time did it attain its present dimensions. We have seen that, during the Eocene period, the sea of the nummulitic limestone extended over at least a considerable part of the Alpine region, and that, as the limestone now forms crumpled and dislocated mountainous masses, the great upheaval of the chain must have taken place after Eocene time. Not improbably the process was a prolonged one, advancing in successive uplifts with intervals of rest. The final upheaval that gave the Alps their colossal bulk did not take place until the Miocene period or later, for the Miocene strata have been involved in the earth-movements, and have been thrust up, bent, and broken. Nor were the terrestrial convulsions confined to Central Europe, all over the globe there seem to have been extensive disturbances. The Eocene sea-bed with its thick accumulations of nummulite-limestone was ridged up into land, and portions of it were carried upward on the flanks of the mountains, in the Himalayas to a height of 16,500 feet above the sea.

While these revolutions were taking place in its topography, Europe continued to enjoy a climate which, to judge from the remains of plants and animals preserved in the Miocene rocks, must still have been of a somewhat tropical character. The flora that clothed the slopes of the Alps was not unlike that of the forests of India and Australia at the present time. Palms of various kinds still flourished all over Central and Western Europe, mingled with conifers, laurels, evergreen oaks, magnolias, myrtles, mimosas, acacias, sumachs, figs, oaks, and various still living

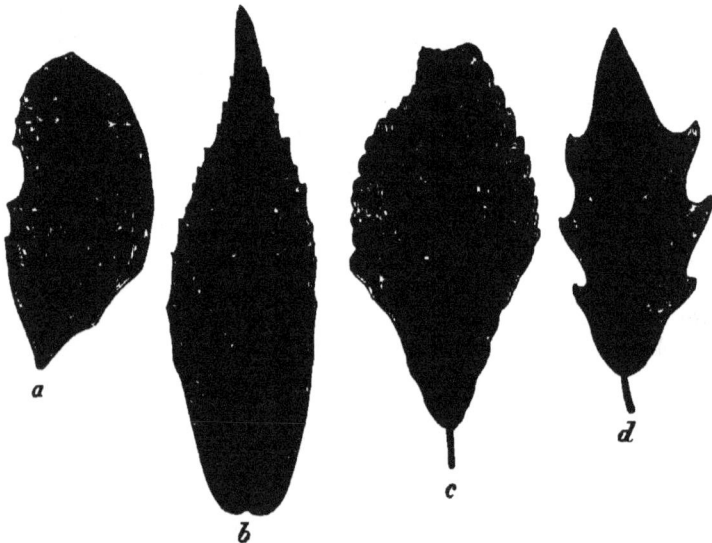

FIG. 198.—Miocene Plants; (a) *Magnolia Inglefieldi* (½); (b) *Rhus Meriani* (natural size); (c) *Ficus decandolleana* (½); (d) *Quercus ilicoides* (⅔).

genera of proteaceous shrubs (Fig. 198). But there is evidence of the incoming of a more temperate climate, for, in the higher parts of the Miocene series of strata, the vegetation was characterised by the abundance of its beeches, poplars, hornbeams, elms, laurels, pondweeds, etc.

Remains of the terrestrial fauna have been well preserved in the deposits that gathered over the floors of the lakes. We know, for instance, that in the woodlands surrounding the large Miocene lake of Switzerland insect life was remarkably abundant. From the proportions of the different kinds that have been exhumed, it has been inferred that the total insect population was then more varied in some respects than it is now in any part of Europe, wood-beetles being especially numerous and large.   In the thick underwood, frogs, toads, lizards, and snakes found their food. Through the forests there roamed antelopes, deer, and three-toed horses, while opossums, apes, and monkeys (*Pliopithecus, Dryo-pithecus, Oreopithecus*) gamboled among the branches.   Wild cats, bears (*Hyænarctos*), and sabre-toothed lions (*Machairodus*) were among the prominent carnivores of the time.   But the most striking denizens of these scenes were undoubtedly the huge proboscidian creatures among which the Mastodon and Deino-

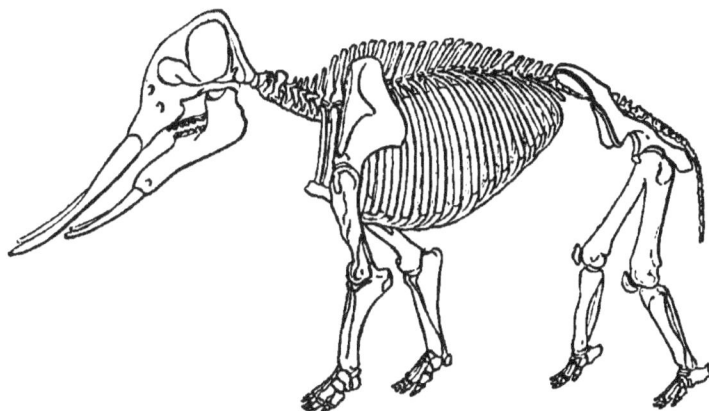

FIG. 199.—*Mastodon augustidens* ($\frac{1}{12}$).

therium took the lead.   The mastodon (Fig. 199) was a large and long-extinct form of elephant, which, besides tusks in the upper jaw, had often also a pair in the lower jaw.   The deino-therium (Fig. 200) possessed two large tusks in the lower jaw which were curved downwards.   This huge animal probably frequented the rivers of the time, using its powerful curved tusks to dig up roots, and perhaps to moor itself to the banks. Contemporaneous with these colossal pachyderms were species of rhinoceros, hippopotamus, and tapir.   The rivers were haunted by crocodiles, turtles, beavers, and otters ; while the seas were tenanted by ancestors of our living morse, sea-calf, dolphin, and lamantin.   It is strange to reflect that such an assemblage of animals should once have found a home all over Europe.

The deposits referable to the Miocene period in Europe indi-cate a great change in the geography of the region since Eocene

and Oligocene times.   While most of the Continent remained
land, with large lakes scattered over its surface, certain tracts had
subsided beneath shallow seas which penetrated here and there
by long arms into the very heart of the region.   Britain continued
to be a land surface, and as such was continuously exposed to
denudation.   Instead of the formation of new deposits, there was
an uninterruped waste of those already existing.   So vast indeed
has been the destruction of the Tertiary strata of Britain that it
has evidently been in progress for an enormous period of time.
Much of it, no doubt, took place during the long interval required
elsewhere for the accumulation of the Miocene series of rocks.

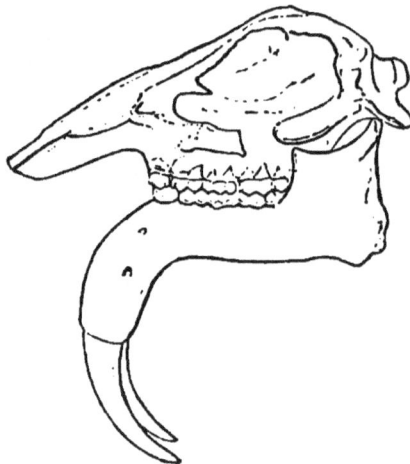

Fig. 200.—Skull of *Deinotherium giganteum* (reduced).

Not only were the soft sands and clays of the older Tertiary
groups of south-eastern England worn away from hundreds of
square miles which they originally covered, but even the hard
basalt-sheets of Antrim and the Inner Hebrides were so cut down
by the various agents of denudation that wide and deep valleys
were carved out of them, and hundreds of feet of solid rock were
gradually removed from their surface.

While Britain remained land, arms of the sea spread over what
is now Belgium, and the basins of the Loire, Indre, and Cher,
stretching across Southern France to the Mediterranean, passing
along the northern base of the Alps, running into the valley of
the Rhine as far north as Mainz, sweeping eastwards round the
eastern end of the Alps, and expanding into the broad gulf of
Vienna among the submerged heights of Austria and Hungary.

The strata that tell this story of submergence contain an
abundant assemblage of marine shells, many of which belong to
genera that now live in warmer seas than those which at present
bathe the coasts of Europe. Among them are *Cancellaria, Cyprœa,
Mitra, Murex, Strombus, Arca, Cardita, Cytherea, Pectunculus,
Spondylus,* together with genera, such as *Ostrea, Pecten, Cardium,
Tapes, Tellina,* which are familiar in the Northern seas.

The district of France, formerly called Touraine, is largely
overspread with shelly sands and marls, rarely more than 50 feet
thick, and locally known as " Faluns." These deposits represent
the floor of the shallow Miocene strait which extended across
France. They have yielded upwards of 300 species of shells, the
general character of which marks a warmer climate than now
exists in Southern Europe. The tableland of Spain, with its
northern mountainous border, rose along the southern margin of
this strait which connected the Atlantic and the Mediterranean.
Through this broad passage the large cetaceans of the time passed
freely from sea to sea, for their bones are found in the upraised
sea-bottom. The carcases of the mammals that then lived
among the Pyrenees—mastodons, rhinoceroses, lions, giraffes, deer,
apes, and monkeys—were likewise swept down into the sea. The
deposits of the shallow Miocene straits and bays thus supply us
with evidence of the position of the land and the character of its
inhabitants. Eastwards the sea appears to have deepened over
the region now occupied by the gulf of Genoa and the encircling
mountain ranges, for the Miocene deposits of that part of the basin
of the Mediterranean, consisting almost wholly of blue marls, are
said to reach the great thickness of more than 10,000 feet.
Beyond that depression the sea once more shallowed across the
site of South-Eastern Europe. In the Vienna basin its deposits
are well developed and consist of two divisions : (1) a lower
group (Mediterranean or marine stage) of limestones, marls, clays,
and sands, containing an abundant assemblage of shells, some of .
which belong to species still living in the present Mediterranean
Sea, or off the west coast of Africa, and also numerous remains of
land-plants which again recall the living floras of India and
Australia ; and (2) an upper group (Sarmatian or Cerithium
stage) of sands, gravels, and clays in which the shells and terrestrial plants point to a much more temperate climate than that
indicated by the lower beds.

On the northern side of the Swiss Alps, the lake which was
formed by the uplifting of the Eocene sea-floor, and in which so

thick a succession of Oligocene strata was laid down, eventually disappeared among the terrestrial movements that submerged so much of Europe beneath the Miocene sea. Marine bands containing undoubted Miocene shells extend across Switzerland : but among them there are such abundant remains of terrestrial vegetation as to show that the land was not far off. No doubt the Alps, not yet uplifted to their ultimate height, rose along the southern borders of the strait that ran across Central Europe, and bore on their slopes luxuriant forest-growths. In Switzerland, however, we learn that before the close of the Miocene period the sea was once more excluded from the district, and another lake made its appearance. The marls, limestones, and sandstones accumulated in this lake (Œningen Beds) are among the most interesting geological deposits in Europe, from the great number and perfect preservation of the plants, insects, fishes, and mammals which have been obtained from them. A large part of our knowledge regarding the terrestrial vegetation and animal life of the Miocene period has been derived from these strata.

Passing beyond the European area, we find that some of the characteristic vegetation of Miocene time spread northwards far within the Arctic Circle. In Spitzbergen and in North Greenland, an abundant series of plant-remains has been discovered, including a good many which occur also as fossils in the Miocene deposits of Central Europe. More than half of them are trees, among which are thirty species of conifers, also beeches, oaks, planes, poplars, maples, walnuts, limes, and magnolias. This flora has been traced as far as 81° 45′ north latitude, where the last expedition sent out from England found a seam of coal 25 to 30 feet thick, covered with black shales full of plant-remains.

Miocene deposits occupy a considerable area in North America. In the Eastern States, they are of marine origin and follow generally the tract of the underlying Eocene beds. In the Western States and Territories they are lacustrine, and show that the lakes which covered so wide an expanse in early Tertiary time still existed, but in greatly diminished proportions. They have preserved many interesting relics of the terrestrial life of the period—three-toed horses, tapiroid animals, hogs as large as rhinoceroses, true rhinoceroses, huge elephant-like creatures allied to deinoceras and tapir, stags, camels, beavers, wolves, bears, and lions. In India, also, thick masses of sedimentary rock occur containing remains of mastodon, deinotherium, and other Miocene animals.

PLIOCENE

The last division of the Tertiary series of formations lays before
us the history of the geological changes that brought about the
present general distribution of land and sea, and completed the
existing framework of the continents.   Contrasted with the previous
Tertiary groups, it is, on the whole, insignificant in thickness and
extent, and it probably records the passing of a much less period
of time, during which the amount of terrestrial revolution was
comparatively trifling.   Only in the basin of the Mediterranean
are there any European Pliocene strata worthy of note on account
of their thickness.   The floor of that sea slowly subsided until
sands, clays, and accumulated shell-beds had been piled up to a
depth of several thousand feet.   An important volcanic episode
then took place.   Etna, Vesuvius, and the other volcanoes of
Central Italy began their eruptions.   Thick masses of Pliocene
sediments were ridged up on both sides of the Apennines, and in
Sicily were upheaved to a height of nearly 4000 feet above the
present sea-level.   This elevation of the Pliocene sea-bed in the
Mediterranean area was not improbably connected with other
movements within the European region.   The shallow firths and
bays which still indented the Continent were finally raised into dry
land, and the Alps may then have received their final uplift.
While the European Pliocene deposits have their maximum thickness in the Mediterranean basin, they elsewhere represent the
sediments of shallow seas and of lakes and rivers.

The flora of the Pliocene period affords evidence of the continued advance of a more temperate climate.   The tropical types
of vegetation one by one retreated southwards in the European
region, leaving behind them a vegetation that partook of the
characters of those of the present Canary Islands, of North
America, and of Eastern Asia and Japan, but which, as time wore
on, approached more and more to the present European flora (Fig.
201). It included species of bamboo, sarsaparilla (*Smilax*), glyptostrobus, taxodium, sequoia, magnolia, tulip tree (*Liriodendron*),
maple (*Acer*), buckthorn (*Rhamnus*), sumach (*Rhus*), plum
(*Prunus*), laurel (*Laurus*), cinnamon-tree (*Cinnamomum*), sassafras, fig (*Ficus*), elm (*Ulmus*), willow (*Salix*), poplar (*Populus*),
alder (*Alnus*), birch (*Betula*), liquidambar, oak (*Quercus*),
evergreen oak (*Quercus ilex*), plane (*Platanus*), walnut (*Juglans*),
hickory (*Carya*), and other now familiar trees.

The fauna presented likewise evidence that the climate, during at least the earlier part of the Pliocene period, still continued warm enough to permit tribes of animals to roam over Europe,

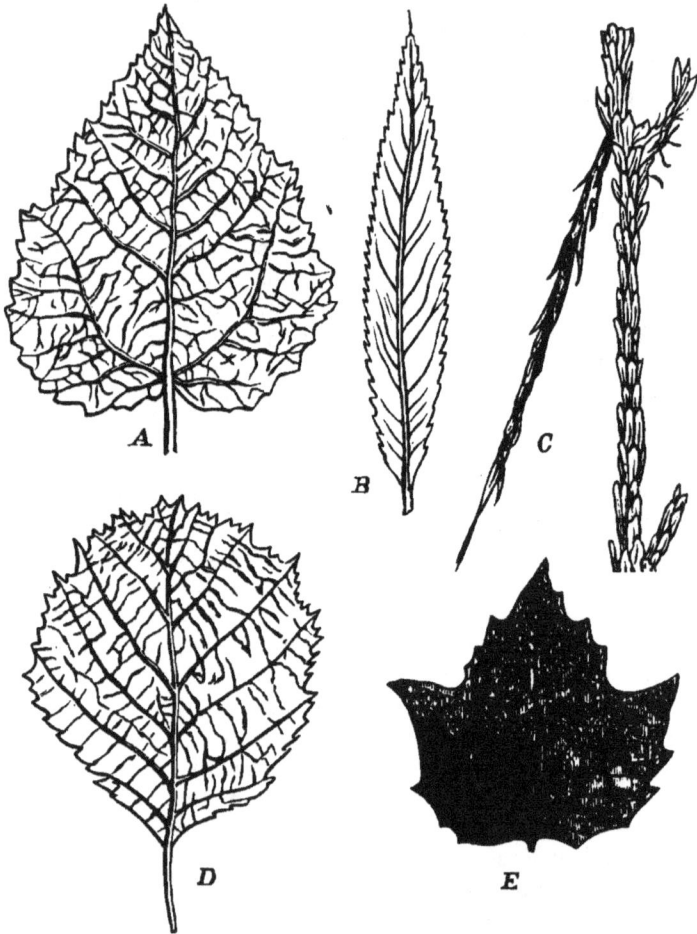

FIG. 201.—Pliocene Plants. (A) *Populus canescens*; (B) *Salix alba*; (C) *Glyptostrobus europæus*; (D) *Alnus glutinosa*; (E) *Platanus aceroides* (all natural size except E, which is ½).

the descendants of which are now confined to regions south of the Mediterranean basin. Some of the huge mammalian types that had survived from an earlier time now died out; such was the

case with the deinotherium and mastodon.   Herds of pachyder-
matous animals formed a distinguishing feature of the fauna—
rhinoceroses, hippopotamuses, and elephants, with troops of herb-
ivorous quadrupeds—gazelles, antelopes, deer, giraffes, horses,
oxen, and strange long-extinct types linking together genera that
are now quite distinct.   There were, likewise, carnivores (wild-cats,
bears, hyænas, etc.), and many monkeys.   The remains of
monkeys have been found fossil in Europe 14° farther north than
their descendants now live.

The shells of the Pliocene deposits afford important evidence
regarding the gradual change of climate.   The great majority of

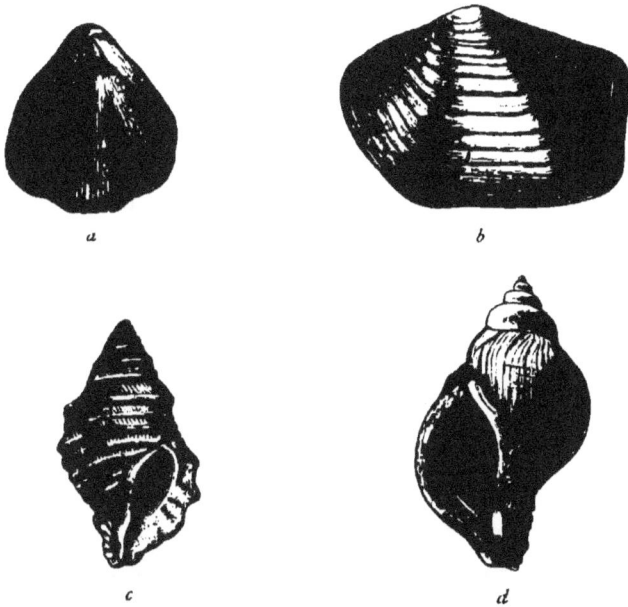

FIG. 202.—Pliocene Marine Shells.   (a) *Rhynchonella psittacea* (natural size); (b)
Panopæa norvegica (½); (c) *Purpura lapillus* (½); (d) *Trophon antiquum* (½).

them belong to still living species (Fig. 202).   They consequently
supply an excellent basis for comparison with the existing distribution of the same species.   When the deposits containing them
are examined with reference to the present habitats of the species,
it is found that the percentage of what are now northern shells
increases from the lower to the higher parts of the series.   In
Pliocene time, each species no doubt flourished only in that part
of the sea where it found its congenial temperature and food.

We infer that its requirements are still the same at the present day, in other words, that the temperature of the regions within which the species is now confined afford, on the whole, an indication of the temperature of the areas within which it lived in the Pliocene seas. On this basis of comparison, the inference has been drawn that the climate in the northern hemisphere, after becoming temperate, passed on to a more rigorous stage. In the end thoroughly Arctic conditions spread over most of Europe and a large part of North America, during the period that succeeded the Pliocene (p. 352).

In Britain Pliocene deposits are almost entirely confined to the counties of Norfolk and Suffolk. They consist of various shelly sands, gravels, and marls, which have long been known as " Crag." Arranged in descending order, the following are the recognised subdivisions :—

Forest-Bed Group — Upper fresh-water, estuarine, and Lower fresh-water sands and silts, with layers of peat, having a total depth of 10 to 70 feet. Among the terrestrial plants are cones of Scotch fir (*Pinus sylvestris*) and spruce (*Abies*), leaves of water-lily (*Nymphæa alba*), yellow pond-lily (*Nuphar luteum*), hornwort (*Ceratophyllum*), blackthorn (*Prunus spinosa*), bog-bean (*Menyanthes trifoliata*), oak, and hazel, with land and fresh-water shells, and many mammals, including species of wolf, fox, machairodus, hyæna, glutton, bear, seal, horse, rhinoceros, hippopotamus, pig, ox, musk-sheep, deer, beaver, trogontherium (a huge extinct kind of beaver), mole, elephant (*E. antiquus, E. meridionalis, E. primigenius*), etc. This group of strata is found at the base of the sea-cliff of boulder-clay in Norfolk, and extends under the present sea.

Chillesford Group — Sands and clays occurring as a thin local deposit in Suffolk, 6 to 16 feet thick, with marine shells, about two-thirds of which still live in Arctic waters (*Mya truncata, Cyprina islandica, Astarte borealis, Tellina obliqua*).

Norwich (fluvio-marine or mammaliferous) Crag — Shelly sand and gravel, 5 to 10 feet thick, containing 93 per cent of still living species of shells and bones and teeth of mastodon, elephant (*E. meridionalis, E. antiquus*), hippopotamus, rhinoceros, etc. The proportion of northern shells is 14.6 per cent, and the following species are included—*Rhynchonella psittacea, Scalaria grœnlandica, Panopœa norvegica, Astarte borealis*. About twenty species of land or fresh-water shells also occur.

Red Crag	A local and inconstant accumulation, 25 feet thick, of red and dark brown ferruginous shelly sand, with numerous species of shells of which 10.7 per cent are northern forms. Some of the characteristic shells of the deposit are— *Trophon antiquum, Voluta Lamberti, Purpura lapillus, Pectunculus glycimeris, Cardium edule.*
Lenham Beds (Diestian)	Sands and ironstones filling hollows of the Chalk of the North Downs, more than 600 feet above the sea, and containing nearly 200 species of fossils, all of which, save 22, have been found in the Coralline Crag.
St. Erth Beds	A local deposit of clays and gravels found at St. Erth in Cornwall, with abundant and well-preserved shells, probably of older Pliocene age, about 40 per cent being of extinct species.
White (Suffolk or Coralline) Crag	Shelly sands and clays containing 84 per cent of still living shells, whereof 5 per cent are northern species. One of the characteristics of the deposit is the large number (140 species) of coral-like polyzoa (corallines or bryozoa), whence one of the names given to this subdivision.

On the Continent the youngest Tertiary deposits cover comparatively small areas and mark some of the last tracts occupied by the sea. Thus, in the Vienna basin there is evidence that the sea, shut off from the main ocean, and partly converted into an inland sea, like the Caspian, was gradually filled up with sediment and raised into land. Along the northern borders of the Mediterranean Sea, thick masses of marine Pliocene strata show the prolonged depression of that region during Pliocene time, and its subsequent elevation. In the south of France these strata, lying unconformably on everything older than themselves, reach a height of 1150 feet above the sea. Along both sides of the Apennine chain, Pliocene blue marls, clays, and sands, known as the sub-Apennine beds, have been uplifted into a range of low hills. These deposits swell out southwards, reaching their greatest thickness (2000 feet or more) in Sicily, which was probably the region of maximum subsidence during Pliocene time. Here and there, in the Italian strata of this period, remains of terrestrial vegetation and land-animals are abundantly preserved. One of the most noted localities for these fossils is the upper part of the valley of the Arno.

Perhaps the most curious and interesting assemblage of the land-fauna of Europe during Pliocene time has been found in some hard red clays, alternating with gravels, at Pikermi in Attica.

Thirty-one genera of mammals have there been obtained, of which twenty-two are extinct. The ruminants, specially well represented among these remains, include species of giraffe, helladotherium (Fig. 203), antelopes, gazelles, and other forms allied to, but distinct from, any living genera. There are likewise the bones of gigantic wild boars, several species of rhinoceros, mastodon,

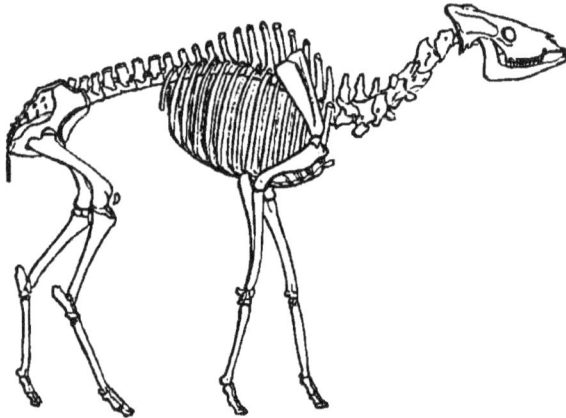

FIG. 203.—*Helladotherium Duvernoyi* ($\frac{1}{50}$)—a gigantic animal belonging to the same family as the living giraffe, Pikermi, Attica.

deinotherium, porcupine, hyæna, various extinct carnivores, and a monkey.

In India a somewhat similar fauna has been obtained from a massive series of fresh-water sandstones, known as the Siwalik group. A large proportion of the remains belong to existing genera of animals, such as macaque, bear, elephant, horse, hippopotamus, giraffe, ox, porcupine, goat, sheep, and camel. Various extinct types were contemporary with these animals, two of the most extraordinary of them being the *Sivatherium* and *Bramatherium*—colossal, four-horned creatures allied to our living antelopes and prong-bucks.

CHAPTER XXVII

WE have now arrived at the last main division of the Geological
Record, that which is named POST-TERTIARY or QUATERNARY, and
which includes all the formations accumulated from the close of
the Tertiary periods down to the present day. But no sharp line
can be drawn at the top of the Tertiary groups of strata. On the
contrary, it is often difficult, or indeed impossible, satisfactorily to
decide whether a particular deposit should be classed among the
younger Tertiary or among the Post-tertiary groups. In the latter,
all the molluscs are believed to belong to still living species, and
the mammals, although also mostly still of existing species, include
some which have become extinct. These extinct forms are
numerous in proportion to the antiquity of the deposits in which
they have been preserved. Accordingly, a classification of the
Quaternary strata has been adopted, in which the older portions,
containing a good many extinct mammals, have been formed into
what is termed the Pleistocene, Post-pliocene, or Glacial group,
while the younger deposits, containing few or no extinct mammals,
are termed Recent.

The gradual refrigeration of climate which is revealed to us by
the shells of the crag was prolonged and intensified in Post-tertiary
time. Ultimately the northern part of the northern hemisphere
was covered with snow and ice, which extended into the heart of
Europe and descended far southward in North America. The
previous denizens of land and sea were in large measure driven
out or even in many cases wholly extirpated by the cold, while
northern forms advanced southward to take their places. The
reindeer, for instance, roamed in great numbers across Southern
France, and Arctic vegetation spread all over Northern and

Central Europe, even as far as the Pyrenees. After the cold had reached its climax, the ice-fields began to retreat, and the northern flora and fauna to retire before the advance of the plants and animals which had been banished by the increasingly severe temperature. And at last the present conditions of climate were reached. The story of this Ice Age is told by the Pleistocene or Post-pliocene formations, while that of the changes which immediately led to the establishment of the present order of things is made known in the Recent deposits.

PLEISTOCENE, POST-PLIOCENE, OR GLACIAL.

The evidence from which geologists have unravelled the history of the Ice Age or cold episode which came after the Tertiary periods in the northern hemisphere may here be briefly given. All over Northern Europe and the northern part of North America the solid rocks, where of hardness sufficient to retain it, are found to present a characteristic smoothed, polished, and striated surface. Even on crags and rocky bosses that have remained for long periods exposed to the action of the weather, this peculiar worn surface may be traced ; but where they have been protected by a covering of clay, these markings are often as fresh as when they were first made. The groovings and fine striæ do not occur at random, but in every district run in one or more determinate directions. The faces of rock that look one way are rounded off, smoothed, and polished ; those that face to the opposite quarter are more or less rough and angular. The quarter to which the worn faces are directed corresponds with that to which the striæ and grooves on the rock-surfaces point. There can be no doubt that all this smoothing, polishing, grooving, and striation has been done by land-ice ; that the trend of the striæ marks the direction in which the ice moved, those faces of rock which looked towards the ice being ground away, while those that looked away from it escaped. By following out the directions of the rock-striæ we can still trace the march of the ice across the land (see Chapter VI).

As the ice travelled, it carried with it more more or less detritus, as a glacier does at the present day. Some of this material may have lain on the surface, but probably most of it was pushed along at the bottom of the ice. Accordingly, above the ice-worn surfaces of rock, there lies a great deposit of clay and boulders, evidently the debris that accumulated under the ice-sheet and was left on the surface of the ground when the ice retired. This deposit, called Boulder-Clay or Till, bears distinct corroborative testimony to the movement of the ice. It is always more or less local in origin, but contains a variable proportion of stones which have travelled for a greater or less distance, sometimes for several hundred miles. When these stones are traced to their places of origin, which are often not hard to seek, they are found to have come from the same quarter as that indicated by the striation of the rocks. If, for example, the ice-worn bosses of rock show the ice to have crept from north to south, the boulders will be found to have a northern origin. The height to which striated rock-surfaces and scattered erratic blocks can be traced affords some measure of the depth of the ice-sheet.

From this kind of evidence it has been ascertained that the whole of Northern Europe, amounting in all to probably not less than 770,000 square miles, was buried under one vast expanse of snow and ice. The ice-sheet was thickest in the north and west, whence it thinned away southward and eastward. Upon Scandinavia it was not improbably between 6000 and 7000 feet thick. It has left its mark at heights of more than 3000 feet in the Scottish Highlands, and over North-Western Scotland it was probably not less than 5000 feet thick. Where it abutted upon the range of the Harz Mountains, it appears to have been still not far short of 1500 feet in thickness.

This vast mantle of ice was in continual motion, creeping outward and downward from the high grounds to the sea. The direction taken by its principal currents can still be followed. In Scandinavia, as shown by the rock-striæ and the transport of boulders, it swept westward into the Atlantic, eastward into the Gulf of Bothnia, which it completely filled up, and southward across Denmark and the low grounds of Northern Germany. The basin of the Baltic was completely choked up with ice ; so also was that of the North Sea as far south as the neighbourhood of London. From the' same evidence we know that the ice which streamed off the British Islands moved eastward from the slopes of Scotland into the hollow of the North Sea, part of it turning to the left to join the south-western margin of the Scandinavian sheet, and move with it northwards and westwards across the Orkney and Shetland Islands into the Atlantic, and another branch bending southwards and moving with the southerly expansion of the Scandinavian ice along the floor of the North Sea and the low grounds of the east of England ; and that on the west side of Scotland the ice filled up and crept down all the fjords, burying

the Western Islands under its mantle and marching out into the
Atlantic.   The western margin of the ice-fields, from the south-
west of Ireland to the North Cape of Norway, must have presented a vast wall of ice some 2000 miles long, and probably
several hundred feet high, breaking off into icebergs which floated
away with the prevailing currents and winds.   The Irish Sea was
likewise filled with ice, moving in a general southerly direction.

Northern Europe must thus have presented the aspect of
North Greenland at the present time.   The evidence of rock-striæ
and ice-borne blocks enabled us to determine approximately the
southern limit to which the great ice-cap reached.   As even the
southern coast of Ireland is intensely ice-worn, the edge of the
ice must have extended some distance beyond Cape Clear, rising
out of the sea with a precipitous front that faced to the south.
Thence the ice-cliff swung eastwards, passing probably along the
line of the Bristol Channel and keeping to the north of the valley
of the Thames.

That the northern ice moved down the bed of the North Sea
is shown by the boulder-clays and transported stones of the eastern
counties of England, among which fragments of well-known Norwegian rocks are recognisable.   Its southern margin ran across
what is now Holland, and skirted the high grounds of Westphalia,
Hanover, and the Harz, which probably there arrested its southward extension.   There is evidence that the ice swept round into
the Lowlands of Saxony up to the chain of the Erz, Riesen, and
Sudeten Mountains, whence its southern limit turned eastward
across Silesia, Poland, and Gallicia, and then swung round to the
north, passing across Russia by way of Kieff and Nijni Novgorod
to the Arctic Ocean.

In Europe no distinct topographical feature appears to mark
the southern limit reached by the ice-sheet; this limit can only
be approximately fixed by the most southerly localities where
striated rocks and transported blocks have been observed.   In
North America, however, the margin of the great ice-cap is
prominently defined by a mound or series of mounds of detritus
which seem to have been pushed in front of the ice.   These
mounds, beginning on the coast of Massachusetts, run across the
Continent with a wonderful persistence for more than 3000 miles.
They form what American geologists call the "terminal moraine."

The detritus left by the ice-sheet consists of earthy, sandy, or
clayey material (Boulder-Clay, Till) more or less charged with
stones of all sizes up to blocks weighing many tons.   For the

most part it is unstratified, and bears witness to the irregular way in which it was tumbled down by the ice. In some districts, it has been more or less arranged in water, and then assumes a stratified character. The stones in the detritus, more especially where they are hard and are imbedded in a clayey matrix, present smooth striated surfaces, the striæ usually running along the length of the stone, but not infrequently crossing each other, the older being partially effaced by a newer set (Fig. 24). This characteristic striation points unmistakably to the slow creeping motion of land-ice.

But the boulder-clays, earths, and gravels left by the great ice-sheet are not simply one continuous deposit. On the contrary, they contain intercalations of stratified sand, clay, and even peat. In these included strata organic remains occur, for the most part those of terrestrial plants and animals, showing that the ice again and again retreated, leaving the country to be covered with vegetation, and to be tenanted by land animals ; but that after longer or shorter periods of diminution it once more advanced southward over its former area. These intervals of retreat are known as "interglacial periods." Probably they were of prolonged duration, the climate becoming comparatively mild and equable while they lasted. The occurrence of boulder-clays above the interglacial deposits shows a subsequent lowering of the temperature, with a consequent renewal of glacial conditions.

The Pleistocene deposits thus reveal to us a prolonged period of cold broken up by shorter intervals of milder climate. The fossils which they contain throw curious and interesting light on these oscillations of temperature. Among the plants, leaves of Arctic species of birch and willow are found far to the south of their present limits ; on the other hand, remains of plants now confined to temperate latitudes are found fossil in Siberia, and others, now living in more genial climates than those of Central Europe, are associated in interglacial deposits, with the remains of the still indigenous vegetation.

To the same effect, but still more striking, is the testimony of the Pleistocene fauna, with its strange mingling of northern and southern forms. The marine shells imbedded in the glacial clays of Scotland, though chiefly belonging to species that still live in the adjoining seas, include a few that are now restricted to more northern latitudes (*Pecten islandicus*, *Leda lanceolata*, *Tellina lata*, etc., Fig. 204). Turning to the terrestrial mammals, we find among the Pleistocene deposits the remains of the last of the huge pachyderms which, through Tertiary time, had been so striking a feature of the animal population of Europe. The hairy mammoth (*Elephas primigenius*, Fig. 205) and the woolly rhinoceros (*R. tichorhinus*) now roamed all over the Continent and across Britain, which had not yet become an island. During the retreat of the snow and ice, they found their way into the forests

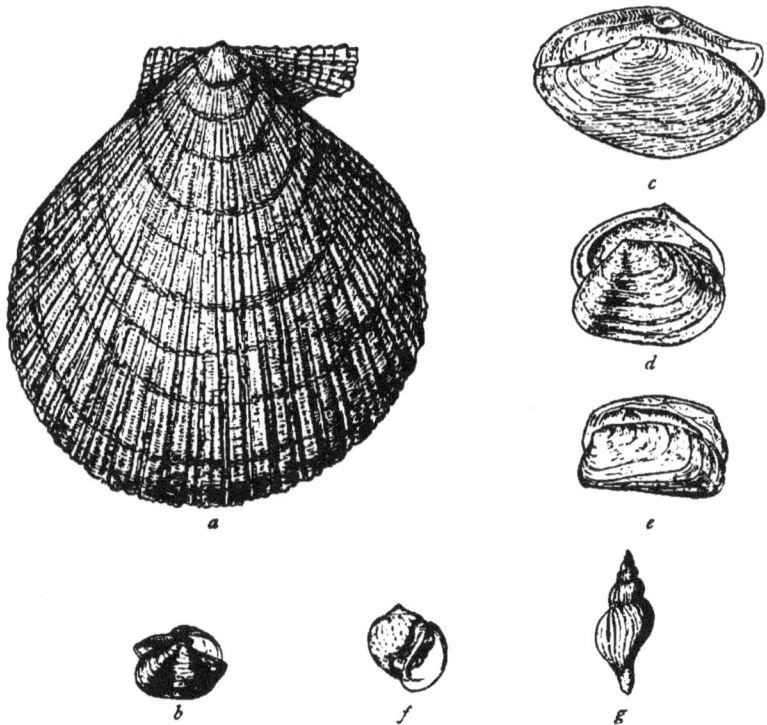

FIG. 204.—Pleistocene or Glacial Shells. (*a*) *Pecten islandicus* (½); (*b*) *Leda truncata* (½); (*c*) *Leda lanceolata* (½); (*d*) *Tellina lata* (½); (*e*) *Saxicava rugosa* (½); (*f*) *Natica clausa* (½); (*g*) *Trophon scalariforme* (½).

and pastures of Northern Siberia. Driven southwards when the cold increased, they were accompanied by numerous Arctic animals which have not yet become extinct. Herds of reindeer (*Cervus tarandus*) sought the pastures of Central France and Switzerland ; the glutton (*Gulo luscus*) came to the South of England and to Auvergne ; the musk-sheep (*Ovibos moschatus*, Fig. 206) and Arctic fox (*Canis lagopus*) wandered southward to the Pyrenees. But as each oscillation of climate slowly brought in a milder temperature, and pushed the snow and ice north-ward, animals of southern types made their way into Southern and Central Europe. Among these immigrants were the porcupine

FIG. 205.—Mammoth (*Elephas primigenius*) from the skeleton in the Musée Royal, Brussels.

(*Hystrix*), leopard (*Felis pardus*), African lynx (*Felis pardina*), lion (*Felis leo*), hyæna, elephant, and hippopotamus, the bones of which have been found in the Pleistocene deposits.

After the height of the cold period or Ice Age had been reached and the general tempera-ture of the northern hemisphere began to rise again, the ice retreated from the low grounds, but still con-tinued among the mountains. The existing snow-fields and glaciers of the Alps, the Pyrenees, and Scan-dinavia are the lineal descendants of those vaster ice-sheets which formerly overspread so much of Europe. The glaciers of the Alps, large though they are, can be shown to be merely the relics of

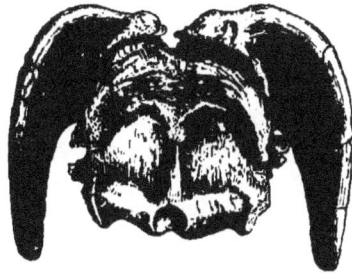

FIG. 206.—Back view of skull of Musk-sheep (*Ovibos moschatus*, ¼), Brick-earth, Crayford, Kent.

their former size. The glacier of the Rhone, for example, as is proved by rock-striæ and transported blocks, once extended 170 miles in direct distance from its modern termination, and rose hundreds of feet above its present surface, burying the valleys and overflowing considerable ridges of hills. The glacier of the Aar stretched once as far as Berne—a distance of about 70 miles from its present termination; and, judging from the marks it has left on the mountains, it must have been not less than 4000 feet thick at the Lake of Brienz.

Though elsewhere in Europe the glaciers have long ago vanished from most of the high grounds, they have left unmistakable traces of their former presence. Thus in hundreds of valleys among the Highlands of Scotland, in the Lake District, and North Wales, admirably ice-worn bosses of rock and beautifully perfect moraines may be seen. We can even trace, in the succession of moraines that become smaller as they approach the head of a valley, the stages of retreat of the original glacier as it shrank before the increasing warmth, till at last it disappeared together with the snow-basin that fed it.

Other relics of the retirement of the ice-sheet are supplied by the long mounds and heaps of gravel and sand, so abundantly strewn over many Lowlands of Northern Europe. These sometimes form ridges, rising 20 or 30 feet above the ground on either side of them, and running for a number of miles. Elsewhere they are heaped together irregularly, often enclosing pools of water. They are known as *Ösar* in Sweden, *Kames* in Scotland, and *Eskers* in Ireland.

During the later stages of the Ice Age the level of the land in Western Europe was lower than it is now. When elevation began, the upward movement continued with long intervals of rest until the land reached its present position. These pauses during the prolonged upheaval are marked by lines of raised beach (p. 112), well seen along both sides of Scotland, and also along the sea margin of Norway.

So slowly and gradually did the great cold disappear that the Ice Age insensibly passed into the Recent or existing period. There can be no doubt that man appeared in Europe before the climate had become as mild as it now is, for his flint-flakes and bone implements are found associated with the bones of Arctic animals in Central France, and traces of his presence in rudely chipped stone instruments occur in deposits which point to frozen rivers. Indeed, in a certain sense, it may be said that the Ice Age still exists among the snow-fields and glaciers of Europe.

Arranged in chronological order, the evidence from which the history of the Pleistocene period is determined may be given as follows :—

Last traces of local glaciers ; terminal and lateral moraines.

Marine terraces or raised beaches, sometimes with moraines resting upon them ; rock-shelves cut probably by waves and floating ice, and marking former levels of the sea.   These beaches and shelves indicate pauses during the last upheaval of the land.   Marine clays with Arctic shells.

Erratic blocks chiefly transported by the great ice-sheet, but partly also by floating ice during the rise of the land, and by valley-glaciers.

Sands and gravels (kames) arranged in heaps, mounds, and ridges, and due in some way to the melting of the edges of the ice-sheet, often associated with lacustrine deposits formed in their hollows, and containing lake-shells and terrestrial plants and animals.

Boulder-clay, till, or bottom-moraine of the great ice-sheet ; the upper part sometimes rudely stratified, and in some regions separated from the lower part by a series of "middle sands and gravels"; the lower part quite unstratified and full of transported stones and boulders.   Finely laminated clays, sands, layers of peat, and traces of terrestrial surfaces occur at different levels in the boulder-clay, and mark "interglacial periods" of milder climate.

Polished and striated surfaces of rock, ground down by the movement of the ice-sheet.

RECENT.

The insensible gradation of what is termed the Pleistocene into the Recent series of deposits affords a good illustration of the true relations of the successive geological formations to each other.   We can trace this gradual passage because it is so recent that there has not yet been time for those geological revolutions, which in the past have so often removed or concealed the evidence that would otherwise have been available to show that one period or group of formations merged insensibly into that which followed it.

The recent formations are those which have been accumulated since the present general arrangement of land and sea, the present distribution of climate, and the present floras and faunas of the globe were established.   They are particularly distinguished by traces of the existence of man.   Hence the geological age to which they belong is spoken of as the Human Period.   But, as has already been pointed out, there is good evidence that man had already appeared in Europe during Pleistocene time, so that the discovery of human relics does not afford certain evidence that the deposit containing them belongs to the Recent series. Nevertheless, it is in this series that vestiges of man become abundant, and that the proofs of his advancing civilisation are contained.

Man differs in one notable respect from the other mammals whose remains occur in a fossil state. Comparatively seldom are any of his bones discovered as fossils ; but he has left behind him other more enduring monuments of his presence in the form of implements of stone, metal, bone, or horn. These relics are in a sense more valuable than his bones would have been, for while they afford us certain testimony to his existence, they give at the same time some indication of his degree of civilisation and his employments. His handiwork thus comes to possess much geological value ; his stone-hatchets, flint-flakes, bone-needles, and other pieces of workmanship are to be regarded as true fossils, from which much regarding his early history has to be determined.

In the river-valleys of the north-west of France and south-east of England human implements have been found in the higher alluvial terraces. After careful exploration, it has been ascertained that these objects have not been buried there subsequently, but must have been covered up at the time the gravel was being formed. The higher terraces are of course the older deposits of the rivers, which have since deepened their valleys until they now flow at a much lower level (p. 38). The excavation of valleys must have been a slow process. Within a human lifetime it is impossible to detect any appreciable lowering of the ground from this cause. Even during the many centuries of which we have authentic human records, we can hardly anywhere detect proof of such a change. How vast then must have been the interval between the time when the rivers flowed at the level of the upper terraces and the present day ! Other evidence of the great age of these higher alluvia is to be found in the number of extinct animals whose remains are buried in them. The human implements likewise bear their testimony in support of the antiquity of the terraces, for they are extremely rude in design and construction, indicative of a race not yet advanced beyond the early stages of barbarism. In the lower and therefore younger terraces, and in other deposits which may also be regarded as belonging to a later date, the articles of human fabrication exhibit evidence of much higher skill and more tasteful design, whence they have been inferred to be the workmanship of a subsequent period when men had made considerable progress in the arts of life. Accordingly, a classification has been adopted, based upon the amount of finish in the stone weapons and implements, the ruder workmanship being assumed to mark the higher antiquity. The older deposits, with coarsely chipped and roughly finished human stone implements, are termed Palaeolithic, and the younger deposits with more artistically finished works in stone, bone, or metal are known as Neolithic. It will be understood that this arrangement is one rather for convenience of description than for a determination of true chronological sequence. It is quite probable, for example, that some of the palæolithic gravels date back to the Pleistocene Ice Age, while other deposits containing similar weapons and a similar assemblage of extinct mammals may belong to a much later time, when the ice had long retreated to the north. It is obvious, too, that we know nothing of the relative progress made in the arts of life by the early races of man. One race may have continued fashioning the palæolithic type of implement long after another race had already learnt to make use of the neolithic type. Even at the present day we see some barbarous races employing rude weapons of stone not unlike those of the palæolithic gravels, while others fabricate stone arrow-heads and implements of bone exactly resembling those of the neolithic deposits. It would hardly be incorrect to say that in some respects certain tribes of mankind are still in the palæolithic or neolithic condition of human progress.

1. Palæolithic.

The formations included under this term are distinguished by containing the rudest shapes of human stone implements, associated with the remains of mammals, some of which are entirely extinct, while others have disappeared from the districts where their remains have been found. These deposits may be conveniently classed under the heads of alluvium, brick-earth, cavern-beds, calcareous tufas, and loess.

Alluvium.—Reference has just been made to the upper river-terraces, which, rising sometimes 80 or 100 feet above the present level of the rivers, belong to a very ancient period in the history of the excavation of the valleys, and yet contain rude human implements. The mammalian bones, found in the sands, loams, and gravels of these terraces, include extinct species of elephant, rhinoceros, hippopotamus, and other animals. The human tools are roughly chipped pieces of flint or other hard stone, and their abundance in some river-gravels has suggested the belief that they were employed when the rivers were frozen over, for breaking the ice and other operations connected with fishing.

The high river-gravels of the Somme and of the valleys in the south-east of England have been specially prolific in these traces of early man.

Brick-earths.—On gentle slopes and on plains, the slow drifting action of wind and rain transports the finer particles of soil and accumulates them as a superficial layer of loam or brick-earth. In the south-east of England, considerable tracts of

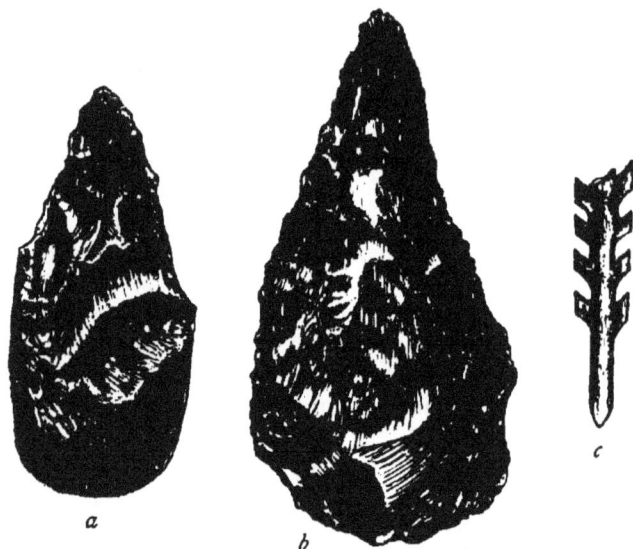

FIG. 207.—Palæolithic Implements. (a) Flint implement, Reculver (⅓), chipped out of a rounded pebble; (b) Flint implement (⅓) from old river-gravel at Biddenham, Bedford, where remains of cave-bear, reindeer, mammoth, bison, hippopotamus, rhinoceros, and other mammalia have been found; (c) Bone harpoon-head (⅓) from the red cave-earth underlying the stalagmite floor of Kent's Cavern (a and b reduced from Mr. Evans's "Ancient Stone Implements").

country have been covered with a deposit of this nature. It is still in course of accumulation, but, as already stated (p. 20), its lower parts must date back to a high antiquity, for they contain the bones of extinct mammals, together with human implements of palæolithic type.

Cave-earth and stalagmite.—The origin of caverns in limestone districts was described in Chapter V, and reference was made to the formation of stalagmite on their floors, and to the remarkably perfect preservation of animal remains in and beneath that deposit. Many of these caves were dens tenanted by hyænas or other beasts of prey (p. 56). Some of them were inhabited by man. In certain cases, they have communicated with the ground above by openings in their roofs, through which the bodies of animals have fallen or been washed by floods. The stalagmite, by covering over the bones left on the floor of the caverns, or in the earth deposited there by water, has preserved them as a singularly interesting record of the life of the time.

Calcareous Tufa.—Here and there, the incrustation of tufa formed round the outflow of calcareous springs has preserved the remains of the vegetation and of the land-animals of the palæolithic time (compare Fig. 21).

Loess.—This is the name given to a remarkable accumulation of pale yellowish calcareous sandy earth which occurs in some of the larger river valleys of Central Europe, especially in those of the Rhine and the Danube ; it likewise covers vast regions of China, and is found well developed in the basin of the Mississippi. It is unstratified and tolerably compact, so that it presents steep slopes or vertical walls along some parts of the valleys, and can be excavated into chambers and passages. In China subterranean villages have been dug out of it, along the sides of the valleys which it has filled up. It contains remains of terrestrial plants and snail-shells, also occasional bones of land-animals. It bears little or no relation to the levels of the ground, for it crosses over from one valley to another, and even mounts up to heights several thousand feet above the sea and far above the surrounding valleys. Its origin has been the subject of much discussion among geologists and travellers. But the result of much careful investigation bestowed upon it goes to show that the loess is probably a sub-aerial deposit formed by the long-continued drifting of fine dust by the wind. It was probably accumulated during a comparatively dry period when the climate of Central Europe, after the disappearance of the ice-sheet, resembled that of the steppes of the south-east of Russia. The assemblage of animals whose bones have been found in it closely resembles that of these steppes at the present time ; for it includes species of jerboa, porcupine, wild horses, antelopes, etc. Among its fossils, however, there occur also the bones of the mammoth, woolly rhinoceros, musk-sheep, hare, wolf, stoat, etc., together with palæolithic stone implements.

Thus the association of animals in the palæolithic formations shows a commingling of the denizens of warmer and colder climates, like that already noticed as characteristic of the Ice Age, and hence the inference above alluded to that the palæolithic gravels may themselves be interglacial. Among the animals distinctively of more southern type mention may be made of the lion, hyæna, hippopotamus, lynx, leopard, Caffer cat ; while among the northern forms are the glutton, Arctic fox, reindeer, Alpine hare (*Lepus variabilis*), Norwegian lemming (*Myodes torquatus*), and musk-sheep.   The animals which then roamed over Europe, but are now wholly extinct, included the mammoth, woolly rhinoceros, and other species of the genus, Irish elk (*Megaceros hibernicus*),

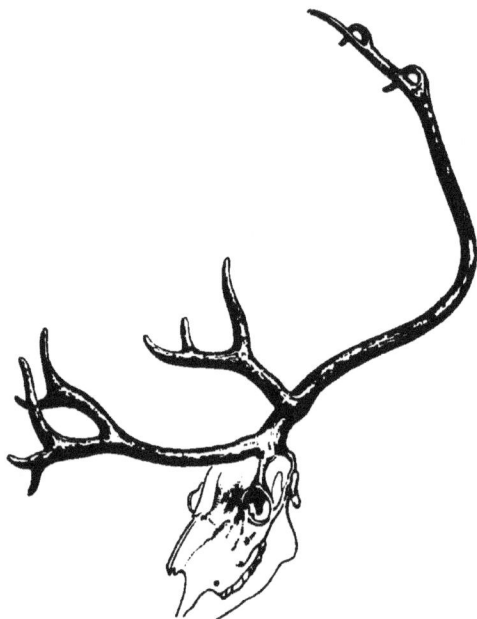

FIG. 208.—Antler of Reindeer (⅕) found at Bilney Moor, East Dereham, Norfolk.

and cave-bear (*Ursus spelæus*).   The traces of man consist almost entirely of pieces of his handiwork ; only rarely are any of his bones to be seen.   Besides the rude chipped flints, he has left behind him, on tusks of the mammoth and horns and bones of the reindeer and other animals, preserved in the stalagmite of cavern-floors, vigorous incised outline-sketches and carvings representing the species of animals with which he was familiar, and some of which have long died out.   He was evidently a hunter and fisher, living in caves and rock-shelters, and pursuing with flint-tipped arrow and javelin the bison, reindeer, horse, mammoth, rhinoceros, cave-bear, and other wild beasts of his time.

2 Neolithic

In this division, the human implements indicate a considerable advance in the arts of life, and the remains of the mammoth, rhinoceros, and other prevalent extinct forms of the palæolithic series are absent. The deposits here included consist of river-gravels, cave-floors, peat-bogs, lake-bottoms, raised beaches, sand-hills, pile-dwellings, shell-mounds, and other superficial accumulations in which the traces of human occupation have been preserved.

After the extinction of the huge pachyderms, the European fauna assumed the general character which it now presents, but with the presence of at least one animal, the Irish elk, that has since become extinct, and of others, such as the reindeer, elk, wild ox or urus, grizzly bear, brown bear, wolf, wild boar, and beaver, which, though still living, have long been extirpated from many districts wherein they were once plentiful. This local extinction has no doubt, in many if not in most cases, been the result, directly or indirectly, of human interference. But man not only drove out or annihilated the old native animals. As tribe after tribe of human population migrated into Europe from some region in Asia, they carried with them the animals they had domesticated—the hog, horse, sheep, goat, shorthorn, and dog. The remains of these creatures never occur among the palæolithic deposits; they make their appearance for the first time in the neolithic accumulations, whence the inference has been drawn that they never formed part of the aboriginal fauna of Europe, but were introduced by the human races of the neolithic period.

The stone articles of human workmanship found in neolithic deposits consist of polished celts and other weapons, hammers, knives, and many other implements of domestic use. Knives, needles, pins, and other objects were made out of bone or horn. There is evidence also that the arts of spinning, weaving, and pottery-making were not unknown. The discovery of several kinds of grain shows that the neolithic folk were also farmers. Vast numbers of these various relics have been found at the pile-dwellings of Switzerland and other countries. For purposes of security these people were in the habit of constructing their wooden dwellings in lakes on foundations of beams, wattled-work, stones, and earth. Sometimes these erections were apt to be

destroyed by fire, as well as to decay by age. And their places
were taken by new constructions of a similar kind built on their
site. Hence, as generation after generation lived there, all kinds
of articles dropped into the lakes were covered up in the silt that
slowly gathered on the bottom. And now, when the lakes' are
drained, or when their level is lowered by prolonged drought,
these accumulated droppings are laid open for the researches of

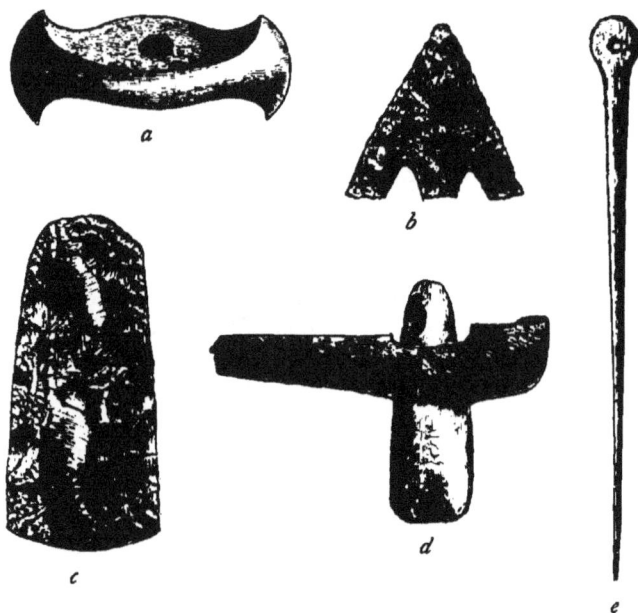

FIG. 209.—Neolithic Implements. (*a*) Stone axe-head (⅓); (*b*) Barbed flint arrow-head
(natural size); (*c*) Roughly-chipped flint celt (⅓); (*d*) Polished celt (⅓), with part of
its original wooden hand still attached, found in a peat-bog, Cumberland; (*e*) Bone-
needle (natural size), Swiss Lake Dwellings; *a, b, c, d*, reduced from Mr. Evans's
"Ancient Stone Implements."

antiquaries and geologists. Many important relics of neolithic
man have likewise been obtained from the floors of caverns
and rock-shelters—places that from their convenience would continue to be used as in palæolithic time. Interesting evidence,
also, of the successive stages of civilisation reached by early man
in Europe, is supplied by the older Danish peat-bogs, in the lower
parts of which remains of the Scotch fir (*Pinus sylvestris*), a tree
that had become extinct in that country before the historic period,
are associated with neolithic implements. In a higher layer of

the peat, trunks of common oak are found, together with bronze implements, while in the uppermost portion, the beech-tree and iron weapons take their place.

Between the neolithic and the present period no line can be drawn. They shade insensibly into each other, and the materials from which the history of their geographical and climatal vicissitudes, their changes of fauna and flora, and their human migrations and development, form a common ground for the labours of the archæologist, the historian, and the geologist.

During the Recent period the same agencies have been and are at work as those which have been in progress during the vast succession of previous periods. In the foregoing pages we have followed in brief outline each of these great periods, and after this survey we are led back again to the world of to-day with which the first chapters of this book began. In this circle of observation no trace can anywhere be detected of a break in the continuity of the evolution through which our globe has passed. Everywhere in the rocks beneath our feet, as on the surface of the earth, we see proofs of the operations of the same laws and the working of the same processes.

Such, however, have been the disturbances of the terrestrial crust that, although undoubtedly there has been no general interruption of the Geological Record, local interruptions have almost everywhere taken place. The sea-floor of one period has been raised into the dry land of another, and again, the dry land, with its chronicles of river and lake, has been submerged beneath the sea. Each hill and ridge thus comes to possess its own special history, which it will readily reveal if questioned in the right way.

We are surrounded with monuments of the geological past. But these monuments are being slowly destroyed by the very same processes to which they owed their origin. Air, rain, frost, springs, rivers, glaciers, waves, and all the other connected agents of demolition, are ceaselessly at work wherever land rises above the sea. It is in the course of this demolition that the characteristic features of the scenery of the land are carved out. The higher and harder parts are left as mountains and hills, the softer parts are hollowed out into valleys, and the materials worn away from them are strewn over plains. And as it is now, so doubtless has it been through the long ages of geological history. Decay and renovation in never-ending cycles have followed each other since the beginning of time.

But amid these cycles there has been a marvellous upward progress of organic being. It is undoubtedly the greatest triumph of geological science to have demonstrated that the present plants and animals of the globe were not the first inhabitants of the earth, but that they have appeared only as the descendants of a vast ancestry,—as the latest comers in a majestic procession which has been marching through an unknown series of ages. At the head of this procession we ourselves stand, heirs of all the progress of the past and moving forward into the future wherein progress towards something higher and nobler must still be for us, as it has been for all creation, the guiding law.

APPENDIX

THE VEGETABLE KINGDOM

I. CRYPTOGAMS OR FLOWERLESS PLANTS.[1]

THESE bear spores that differ from true seeds in consisting only of one or
more cells without an embryo. They include the following classes :—

Algæ—Fungi.—These embrace the smallest and simplest forms of
vegetation—fresh-water confervæ, desmidiæ, mushrooms, lichens, sea-
weeds, etc. Some of them secrete carbonate of lime and form a stony
crust, as in the case of the marine nullipores (p. 84), others secrete
silica, as in the frustules of diatoms (p. 83, Fig. 31). These hard
parts are most likely to occur as fossils ; but impressions of some of
the larger kinds of sea-weeds may be left in soft mud or sand (pp. 242,
249). Fungi are not well adapted for preservation, but traces of them
have been noticed even in rocks of the Carboniferous period.

Characeæ are fresh-water plants, some of which abstract carbonate of
lime from the water and deposit it as an incrustation on their surface.
Hence their calcified nucules or spiral seed-like bodies [gyrogonites]
and stems may accumulate at the bottom of lakes.

Muscineæ, mosses, and liverworts afford little facility for fossilisation.
But some of the mosses (sphagnum, etc.) form beds of peat (p. 82).

Filices, ferns, bearing fronds on which are placed the sporangia or
spore-cases. Many of them possess a tough tissue which can for
some time resist decomposition. Traces of ferns are consequently
abundant among the fossiliferous rocks (Figs. 131, 141, 159).

Ophioglossaceæ, adder's tongues and moonworts.

Rhizocarpeæ, pepperworts.

Equisetaceæ, horse-tails, with hollow striated siliceous jointed stems or
shoots (Figs. 143, 164). These stems possess considerable durability,
and where buried in mud or marl may retain their forms for an indefinite period. Allied plants [*Calamites*, Fig. 143] have been
abundantly preserved among some of the older geological formations
(Old Red Sandstone, Carboniferous, Permian).

Lycopodiaceæ, club-mosses, plants with leafy branches like mosses,
growing in favourable conditions into tree-like shrubs that might be

---

[1] Names placed within square brackets ([ ]) are fossil forms.

mistaken for conifers. Their dichotomous stems and their fertile
branches, which resemble cones and bear spore-cases, offer themselves
for ready preservation as fossils. The spores are highly inflammable,
and it is worthy of notice that similar spores have been detected in
enormous abundance in the Carboniferous system. *Lycopodium* and
Selaginella are familiar living genera. (For extinct forms see Fig.
142, p. 273).

II. PHANEROGAMS OR FLOWERING PLANTS.

i. GYMNOSPERMS or plants with naked seeds ; that is, seeds not enclosed
in an ovary.

Cycadeæ, small plants resembling both palms and tree-ferns. The
pinnate leaves are hard and leathery, and have been frequently preserved as fossils. *Cycas* and *Zamia* are two typical genera (Figs.
164, 171).

Coniferæ, the Pine family.—The stiff hard leaves and the hard seed-
cones may be looked for in the fossil state (Figs. 159, 164). The
resinous wood also sometimes long resists decomposition, and may be
gradually petrified. Trunks of pine are often met with in peat-mosses.
The Coniferæ have been subdivided into the following families :—

1. Cupressineæ, cypresses, including *Juniperus* (Juniper), *Libo-
cedrus, Thuja, Thujopsis, Cupressus, Taxodium, Glyptostrobus.*
2. Abietineæ, pines and firs, including *Pinus, Abies, Cedrus,
Araucaria* (p. 299), *Dammara, Cunninghamia, Sequoia.*
3. Podocarpeæ, trees growing in New Zealand, Java, China,
Japan, etc., bearing a succulent fruit or a thick fleshy stalk.
4. Taxineæ, yews, plants with fleshy fruit, including the genera
Taxus, Salisburia, Phyllocladus.

Gnetaceæ, joint-firs, small trees or shrubs with jointed stems (*Gnetum,
Ephedra, Welwitschia*).

ii. ANGIOSPERMS, or plants bearing their seed within an ovary. They
are subdivided into two great classes—the Monocotyledons or Endogens, and
the Dicotyledons or Exogens.

Monocotyledons, so called from their having only one cotyledon or
seed-lobe. They are also known as "Endogens," from the fact that
they chiefly increase in diameter by growth in the interior, whereby
the exterior layers are pushed outwards. Their seeds are usually
enclosed in strong sheaths or shells, of which the cocoanut is a striking
example. The following are some of the families :—

Lemnaceæ (duck-weeds) ; Potamogetoneæ (pond-weeds) ; Panda-
naceæ (screw-pines) ; Palmaceæ (palms) ; Typhaceæ (typhads,
marshy plants) ; Cyperaceæ (sedges) ; Gramineæ (grasses) ; Junca-
ceæ (rushes) ; Liliaceæ (lilies) ; Irideæ (irises) ; Dioscoreaceæ
(yams) ; Taccaceæ (tacca) ; Musaceæ (plantains and bananas) ;
Zingiberaceæ (gingerworts) ; Orchideæ (orchids).

Dicotyledons or plants that have two cotyledons or seed-lobes ; also
called "Exogens," because their stems increase by successive layers
added to the exterior. This division includes the most highly organised
members of the vegetable kingdom. Our common flowers and hard-
wood trees belong to it. The sections, orders, and families into
which it has been partitioned are so numerous that only some of the

more interesting or important to the geologist can be inserted here. It is chiefly the leaves and seeds that occur in the fossil condition and furnish means of recognising the plants (Figs. 184, 198, 201). Urticaceæ (nettles); Platanaceæ (planes); Cannabineæ (hemps); Ulmaceæ (elms); Betulaceæ (birches); Nelumbiaceæ (lotus plants); Nymphæaceæ (water-lilies); Ranunculaceæ (crowfoots); Anonaceæ (custard-apples); Berberideæ (barberries); Laurineæ (laurels); Myristicaceæ (nutmegs); Papaveraceæ (poppies); Fumariaceæ (fumitories); Cruciferæ (plants with cross-shaped flowers, such as wallflower, *Brassica*, which is the original genus from which our cultivated cabbage, cauliflower, broccoli, and turnip are derived, *Sinapis* or mustard, cress, radish, etc.); Convolvulaceæ (bindweeds); Solanaceæ (nightshades, potato); Bignoniaceæ (trumpet flowers); Plantagineæ (ribworts or plantains); Labiatæ (plants with labiate flowers, such as mint, sage, lavender); Oleaceæ (olives); Jasminiaceæ (jasmines); Gentianaceæ (gentians); Valerianaceæ (valerians); Cucurbitaceæ (cucumbers and gourds); Campanulaceæ (bell-flowers); Compositæ (plants with compound flowers); Primulaceæ (primroses); Ericaceæ (heaths); Rhamnaceæ (buckthorns); Sapindaceæ (soap-trees); Balsamineæ (balsam tribe); Geraniaceæ (cranesbills, geraniums); Euphorbiaceæ (spurge tribe); Araliaceæ (ivy tribe); Cornaceæ (dogwood tribe); Saxifragaceæ (saxifrage tribe); Proteaceæ (found principally in Australia and Cape of Good Hope, see Fig. 193); Papilionaceæ (plants bearing flowers like those of the pea, bean, clover, etc.); Pomeæ (apple tribe); Rosaceæ (rose tribe); Amygdaleæ (almond tribe); Myrtaceæ (myrtle tribe); Cactaceæ (Indian figs, cactus tribe); Myricaceæ (galewort tribe); Juglandeæ (walnut tribe).

THE ANIMAL KINGDOM.

I. INVERTEBRATES.

I. PROTOZOA.—Animals simple in structure and usually minute in size, with bodies composed of a structureless jelly-like substance (sarcode) which, in some cases, secretes siliceous or calcareous needles or shells which serve to protect them. It is only these hard parts which have any chance of being preserved as fossils.

CLASS i. RHIZOPODS, having generally a calcareous shell or siliceous skeleton; divided into the following three orders :—

Foraminifera—having usually a calcareous shell pierced with fine pores through which slender thread-like processes protrude from the jelly-like body. These minute creatures live in enormous abundance in various parts of the ocean, and their shells gather as a deposit of ooze at the bottom (*Globigerina, Lagena, Nummulina*). Their remains are also found in the geological formations, sometimes constituting masses of limestone (Figs. 146, 185).

Heliozoa—fresh-water forms sometimes with a radial siliceous skeleton (*Acanthocystis, Clathrulina*).

Radiolaria—marine creatures with radial siliceous skeleton, which usually consists of small siliceous needles or spicules united together. They occur in vast numbers on some parts of the sea-floor, where their remains form a siliceous ooze (*Thalassicolla, Polycistina*, p. 84).

CLASS ii. INFUSORIA—protozoa living chiefly in fresh water, and having a definite form enclosed within an external membrane, and usually with a mouth and anus. From their perishable nature these animals are not met with in a fossil state.

II. SPONGIDA (sponges), chiefly marine forms possessing an internal skeleton of horny fibres or of calcareous or siliceous spicules. The horny sponges are illustrated by the common sponge of domestic use which is the skeleton of a Mediterranean genus, composed of a close network of horny fibres. Such forms are too perishable to be looked for as fossils. The siliceous sponges secrete minute siliceous spicules which are dispersed in a network of sponge-fibres, sometimes in a glassy framework of six-rayed spicules (*Hexactinellidæ*). The calcareous sponges, as their name implies, secrete carbonate of lime as the substance of which their spicules consist (see Fig. 186).

III. CŒLENTERATA (zoophytes), radially symmetrical animals with a body composed of cells arranged in an outer and an inner layer enclosing a body-cavity.

Hydrozoa, including the fresh-water *Hydra*, and the marine jelly-fishes, millepores, *Campanularia, Sertularia*, etc. Most of these animals offer little facility for preservation as fossils; but some of them possess horny or calcareous structures which have been preserved in sediment-ary deposits. Among these an extinct type of Hydrozoa, known as Graptolites, occurs abundantly in some of the older parts of the Geological Record (Figs. 117, 121).

Ctenophora—spherical or cylindrical Medusæ, including the Venus Girdle of the Mediterranean, and the Beroe of Northern waters.

Actinozoa (corals), Polypes having a cavity in the body divided by vertical partitions into a number of compartments. The common Actinia or sea-anemone is an example; but it is an exception to the general rule that the internal parts are strengthened by a secretion of carbonate of lime. It is this calcareous skeleton which forms the familiar part of corals.

Rugosa (Tetracoralla), the older forms of coral in which the calcareous partitions are arranged in multiples of four with trans-verse partitions [*Zaphrentis, Cyathophyllum, Amplexus*, etc., Figs. 122, 137, 147].

Alcyonaria (Octocoralla), including *Alcyonia, Pennatula*, and *Gorgonia*, animals with eight-plumed tentacles and calcareous bodies (sclerodermites) which form the foundation of a calcareous or horny skeleton (Fig. 122).

Zoantharia (Hexacoralla), including the more modern forms of corals, wherein the tentacles are either six or some multiple of six. Among the families comprised in this Order are the soft-bodied *Actinidæ*; also *Turbinolidæ, Oculinidæ, Astræidæ* or star-corals, *Fungidæ* or mushroom corals, *Madreporidæ* (Fig. 172).

IV. ECHINODERMATA—animals possessing usually a symmetrical fivefold grouping of parts, and enclosed in a skin which is strengthened by hard calcareous granules, spicules, or close-fitting plates.

Crinoidea—globular or cup-shaped, with jointed arms, and usually fixed

APPENDIX 373

by a jointed calcareous stalk. Most of the Crinoids are now extinct. Among the living forms are *Pentacrinus, Rhizocrinus, Bathycrinus,* and *Comatula* (see Figs. 149, 165, 173). Allied to the Crinoids are the extinct Cystideans (Fig. 123) and the Blastoids (Fig. 150), found in Palæozoic formations.

Asteroidea (star - fishes).—The parts of these animals most readily preserved as fossils are the calcareous plates which run along the five rays of the star. These have been found in the marine deposits of many geological periods (Fig. 123). In the Brittle Stars (Ophiuroidea) the arms are flexible, cylindrical, and quite sharply marked off from the central disc.

Echinoidea (sea-urchins), spherical, heart-shaped, or disc-shaped, with a usually immovable skeleton of calcareous plates which encloses the body like a shell and bears calcareous movable spines. They comprise the regular echinoids (*Cidaris, Echinus,* etc., Figs. 148, 174) and the irregular echinoids either compressed into the form of a shield (*Clypeaster*) or of a heart-shape (*Spatangus,* Fig. 187).

Holothuroidea—worm-like elongated animals, with a leathery body in which the calcareous secretion is confined to isolated particles, scales, or spicules. These calcareous bodies have been found abundantly in the Carboniferous system, and are the only evidence of the existence of this division of the echinoderms at so ancient a period.

V. VERMES, comprising the various forms of worms. Comparatively few of these animals occur in the fossil state. Many of them are worms or flukes living in the intestines or other parts of the body. The only important class to the student of geological history are the annelides or segmented worms.

Errantia, free-swimming predaceous sea-worms. The only hard parts of these creatures capable of surviving as fossils are the horny jaws which have been met with in some numbers even in ancient geological formations. But as many of the species live in and crawl over mud they leave behind them in their burrows and trails evidence of their presence. Such markings remain abundantly in many ancient rocks (Fig. 124).

Tubicolæ—sedentary worms, living within tubes within which they can withdraw for protection. This tube may remain as the only permanent relic of their existence. Sometimes it is a leathery substance ; in other cases it consists of grains of sand or other particles cemented by a glutinous secretion, or of solid carbonate of lime. The most familiar example of this Order is the *Serpula* which may so frequently be seen encrusting dead shells thrown up upon the beach.

Oligochæta—earthworms and aquatic worms—are not found as fossils. But the common earthworm is an important agent in mixing the soil and bringing up its fine particles within reach of rain and wind (p. 18).

VI. ARTHROPODA (Articulata). These differ from the worms in having jointed appendages attached to the body, which serve as organs of locomotion. They possess a tough chitinous skin which usually becomes hardened by the deposit of calcareous matter. The articulate animals are divided into four great classes as follows :—

i. **Crustacea**—chiefly aquatic forms with two pairs of antennæ and numerous paired legs. They include the Phyllopods, remarkable for their compressed bivalve shell which is frequently found in the fossil state (Fig. 126); the Ostracods, small forms enclosed in a bivalve shell, and with seven pairs of appendages, the minute shells being abundant in the fossil state (*Cypris*); the Cirripedes or barnacles, so commonly seen encrusting shore rocks; the Amphipods; the Isopods; the Decapods, which are either macrurous (long-tailed), as in prawns and lobsters (Fig. 178), or brachyurous (short-tailed), as in sea-crabs and land-crabs. A remarkable group of extinct Crustaceans is comprised in the Order Eurypterida (Fig. 135). The Xiphosura are still found living in the form of Limulus or King-crab; but they date back to the Carboniferous period. The earliest forms of Crustaceans belong to another extinct Order, the Trilobites (Figs. 118, 125, 136, 151).

  ii. **Arachnida**—air-breathing arthropods, with two pairs of jaws and four pairs of ambulatory legs, including mites, spiders, and scorpions. Some of these animals (scorpions) have chitinous integuments which resist decomposition and have been abundantly preserved in the rocks (pp. 215, 256, 273).

  iii. **Myriapoda**, including the chilopods or centipedes, feeding entirely on animals which they bite and kill with their poisonous secretion; and the chilognaths or millipedes and galley-worms which live in damp places and feed on vegetable and dead animal matters.

  iv. **Insecta.**—Among the orders of insects of most interest in geological history are—

> *Orthoptera*, with two usually unequal pairs of wings (earwigs, cockroaches, praying-insect, grasshoppers, locusts, crickets, book-lice, termites or white ants, ephemeridæ or mayflies, dragonflies).
> *Neuroptera*, insects with wings in which the nervures form a network (*Corydalis*, camel-neck flies, ant-lions, phryganidæ or spring-flies).
> *Hemiptera*, including lice, cochineal insect, plant-lice, cicadas, bugs, water-bugs, water-scorpions.
> *Diptera*, with large glassy front wings, including the various kinds of flies, such as the house-fly, dung-fly, gad-fly, gnat, gall-fly, and flea.
> *Lepidoptera*, butterflies and moths.
> *Coleoptera*, beetles, the most durable parts of which are the horny wing-covers (elytra), so often to be found in woods and peat-mosses. They include lady-birds, stag-beetles, tiger-beetles, etc.
> *Hymenoptera*, with four membranous wings, having few nervures, comprising ants, wasps, bees.

VII. MOLLUSCOIDA—under this division may be grouped the Tunicaries, Polyzoa, and Brachiopoda.

  i. **Tunicata**, sea-squirts—simple or compound, fixed or free organisms which have been named from the leathery integument within which they are enclosed. Though some of them abstract carbonate of lime from sea-water, they present no hard parts for fossilisation, and the class is not known in the fossil state.

  ii. **Polyzoa** (Bryozoa), sea-mats and sea-mosses—composite animals, each enclosed in a horny or calcareous case, and united into colonies which are generally attached to some foreign body and often resemble plants in outer form. The calcareous colonies form durable objects which have been abundantly preserved as fossils. Polyzoa are met with among the oldest fossiliferous formations and still abound in the present sea. The common lace-like *Flustra*, so frequently to be seen encrusting the fronds of sea-weeds or dead shells, is a familiar example of them. Among the fossil forms (many of which have long been extinct) some of the most important genera are *Fenestella* ("lace-coral," Fig. 152), *Polypora, Retepora, Glauconome, Hippothoa, Heteropora, Fascicularia*.

iii. **Brachiopoda,** lamp-shells—molluscous animals, having bivalve, calcareous, or horny shells, one valve placed on the back, the other on the front of each individual, and taking their name from two long ciliated arms which proceed from the sides of the mouth and create the currents that bring their food. They are grouped in two Orders : (1) The Inarticulata, in which the two valves are not united along the hinge line (*Lingula*, Fig. 119, *Discina, Crania*); and (2) the Articulata, in which the two valves are hinged together with teeth (*Terebratula, Rhynchonella*, Figs. 127, 138, 153, 160, 202). The brachiopods attained their chief development during the earlier periods of geological time, and are now represented by comparatively few living forms. The shells are equal sided, but the ventral is usually larger than the dorsal valve, and is prolonged into a prominent beak, by which it fixes itself, or through which the pedicle passes whereby it is attached to the sea-floor. The following are characteristic genera :—

> *Terebratula* (still living), *Stringocephalus* (Devonian), *Thecidium* (Trias to present time), *Spirifera* (chiefly Palæozoic), *Atrypa* (Palæozoic), *Rhynchonella* (Lower Silurian to present time), *Pentamerus* (Silurian), *Orthis, Strophomena, Productus* (Palæozoic), *Leptæna* (Palæozoic to Lias), *Crania, Discina, Lingula* (from early Palæozoic to present time).

VIII. MOLLUSCA—animals with soft bodies, inclosed in a muscular envelope which is usually covered with a strong calcareous shell. These hard shells are durable objects, and when covered up in sediment remain for an indefinite period as evidence of the existence of the animals to which they belonged. The great abundance of the mollusca also in the sea and in terrestrial waters gives a peculiar value to their remains. They (with the Brachiopoda) furnish by far the most valuable data to the geologist for the identification and comparison of marine sedimentary deposits of all ages. They are divided into the following classes :—

i. **Lamellibranchiata**—ordinary bivalves like the cockle, mussel, and oyster, in which the valves are placed on the right and left sides of the body. The following are the more important families :—

> Ostreidæ, oysters—including among other genera *Ostrea* (Fig. 197), *Anomia, Pecten* (Figs. 166, 204), *Lima, Plicatula* (Fig. 175), [*Gryphæa*, Fig. 175, *Exogyra, Aviculopecten*, Fig. 154].
> Aviculidæ, wing-shells—*Avicula* (Fig. 166), [*Posidonomya, Bakevellia*, Fig. 161, *Gervillia*], *Perna*, [*Inoceramus*, Fig. 188], *Pinna*.
> Mytilidæ, mussels—*Mytilus* (mussel), *Modiola* (horse-mussel), *Lithodomus, Dreissena*, [*Orthonota*, Fig. 128].
> Arcadæ, including among other genera *Arca, Cucullæa* (Fig. 139), *Pectunculus, Nucula* (Fig. 188), *Leda* (Fig. 204).
> Trigoniadæ—*Trigonia* (Figs. 175, 188), [*Myophoria*, Fig. 166, *Schizodus*, Fig. 161, *Axinus*].
> Unionidæ—*Unio* (river-mussel), *Anodon*, [*Anthracosia*].
> Chamidæ—*Chama*, [*Diceras, Requienia*]
> [Hippuritidæ, Rudistes—*Hippurites, Radiolites, Caprina, Caprotina*, all confined to the Cretaceous system, Fig. 189].

Tridacnidæ—*Tridacna*.
Cardiadæ—*Cardium* (cockle, Figs. 32, 166).
Lucinidæ—*Lucina, Corbis*.
Cycladidæ—*Cyclas, Cyrena*, fluviatile and estuarine shells.
Cyprinidæ — *Cyprina, Astarte, Isocardia, Cypricardia, [Megalodon, Cardinia], Cardita*.
Veneridæ—*Venus, Cytherea, Artemis, Tapes*.
Mactridæ—*Mactra, Lutraria*.
Tellinidæ—*Tellina* (Fig. 204), *Psammobia, Sanguinolaria, Donax*.
Solenidæ—*Solen* (razor-shell).
Myacidæ—*Mya* (gaper), *Corbula* (Fig. 197), *Panopœa* (Fig. 202), *Glycimeris*.
Anatinidæ—*Anatina, Thracia, Pholadomya* (almost now extinct), [*Myacites, Edmondia*, Fig. 154].
Gastrochænidæ—*Saxicava* (Fig. 204).
Pholadidæ—*Pholas, Xylophaga, Teredo*.

ii. **Gasteropoda** or snails are named from the way in which they creep on the broad foot-like expansion of the lower part of the body.   They are almost all protected by a univalve shell which is usually spiral.

The carnivorous gasteropods possess a respiratory siphon and are all marine.   Among them are the genera *Strombus, Rostellaria, Murex, Pyrula, Fusus, Buccinum* (whelk), *Nassa* (dog-whelk), *Purpura* (Fig. 202), *Cassidaria, Oliva* (Fig. 194), *Conus, Pleurotoma, Voluta* (Fig. 194), *Mitra, Cyprœa* (cowry).

There is another group the members of which possess no respiratory siphon, and mostly live on plants.   Among their more prominent genera are *Natica* (Fig. 204), *Chemnitzia*, [*Loxonema*], *Cerithium* (Fig. 194), *Potamides, Nerinœa, Aporrhais, Melania, Turritella, Scalaria, Littorina* (periwinkle), *Rissoa, Paludina* (Fig. 197), *Nerita, Neritina, Turbo, Trochus*, [*Euomphalus*, Fig. 155], *Haliotis*, [*Pleurotomaria, Murchisonia*], *Fissurella, Calyptrœa, Pileopsis,* *Patella* (rock-limpet), *Dentalium, Chiton*.

The pulmoniferous or air-breathing gasteropods include the land-snails, and have a broad foot and usually a spiral shell.   The following are among the more important genera : *Helix, Vitrina, Succinea, Bulimus, Pupa, Clausilia, Limax* (slug), *Limnæa* (pond-snail), *Ancylus* (river-limpet), *Planorbis, Cyclostoma, Cyclophorus*.

The sea-slugs possess either no shell or one so small and thin as to be wholly or partially concealed by the animal, and therefore unlikely to be preserved in a fossil state.   Some, however, occur as fossils, particularly the genera *Tornatella, Bulla*, and *Cylichna*.

The heteropod gasteropods are animals inhabiting the open sea, in which they are fitted to swim by a peculiar expansion of the foot into a fin-like tail or a fan-shaped ventral fin.   Their more important living genera are *Firola, Carinaria, Atlanta*, while of extinct genera *Bellerophon* (Figs. 129, 155), *Maclurea*, and *Ophileta* may be mentioned.

iii. **Pteropoda**—a small group of molluscs swarming in the open sea, in which they swim by means of two wing-like fins proceeding from the sides of the mouth.   They are all small, but extinct forms of much larger size are found fossil in rocks of all ages, even in some of the most ancient.   The shell when present in the living forms is glassy and translucent; but in some of the fossil genera it is thicker. The living genera *Hyalea* and *Cleodora* occur also fossil among the Tertiary rocks. The more important fossil genera are *Theca, Pterotheca, Hyolithes, Tentaculites,* and *Conularia* (Fig. 156), all of which occur in the Palæozoic formations.

iv. **Cephalopoda**—the cuttle-fishes, squids, and pearly nautilus—are the highest division of the mollusca, being distinguished by the long muscular arms placed round their mouth, and the plume-like gills by which they breathe. The living forms are nearly all destitute of an outer shell, which, however, is possessed by the pearly nautilus and argonaut or paper nautilus. Some of them have a horny or calcareous internal bone (cuttle-bone). They also possess powerful horny or partly calcareous jaws like a parrot's beak. It is only these hard parts that can be expected to occur as fossils.

In former periods the cephalopods were enormously more abundant than they are now, and as most of them possessed an outer shell their remains have been abundantly preserved among the rocks. During the earlier ages of geological history, they appear to have been the magnates of the sea.

According to the number of their breathing gills, cephalopods are grouped into Dibranchiate or two-gilled, and Tetrabranchiate or four-gilled.

The Dibranchiate forms now living include the *Argonauta* or paper-sailor, the *Octopus*, the calamaries or squids (*Loligo, Sepiola, Sepioteuthis,* etc.), *Sepia,* and *Spirula.* Some of these occur also in the fossil state, but the family of the *Belemnites* (Fig. 177), so abundant in Jurassic and Cretaceous time, died out at the close of Secondary time.

The Tetrabranchiate genera are protected by an external chambered shell with siphuncle. They attained their chief development in Palæozoic and Mesozoic time, and are now almost extinct. The shell in the fossil forms is sometimes quite straight (Orthoceras), and from this simplest form successive stages of curvature may be observed till it becomes a flat coil (Ammonites). The more important fossil genera are *Nautilus* (Fig. 167), the only living genus, found also fossil as far back as early Palæozoic deposits, *Lituites* (Fig. 130), *Clymenia* (Fig. 139), *Orthoceras* (Figs. 130, 157), *Phragmoceras, Cyrtoceras, Ammonites* (Fig. 176), *Goniatites* (Fig. 157), *Ceratites* (Fig. 167), *Crioceras, Toxoceras, Ancyloceras, Scaphites, Turrilites, Hamites, Baculites.* (For some of the leading varieties of type, see Fig 190.)

VERTEBRATA.

PISCES, fishes.—The parts most likely to be preserved in a fossil state are the bones of the skeleton, especially the teeth; also bony scales and external plates and spines. Those types of fishes which possess these hard parts have accordingly been abundantly preserved among the stratified rocks of the earth's crust. The following are the four sub-classes into which the great class of fishes has been divided :—

i. **Leptocardii**—animals possessing neither brain nor heart, ribs nor jaws, and so lowly an organisation that their claim to be ranked among the vertebrata is disputed.

ii. **Cyclostomata**—animals with a cartilaginous skeleton, the skull not separate from the body, and no real jaws or ribs. Some of them have horny denticles on the mouth, and these are the only hard parts that offer any facilities for fossilisation. The living Lamprey and the Hag-fish are examples. Certain tooth-like bodies called "Conodonts," which occur in Silurian rocks, have been supposed to be teeth of Cyclostomes.

iii. **Teleostei**—embrace the vast majority of the living fishes of the present day. They possess a bony skeleton, and hence are often spoken of as the osseous fishes. The vertebræ are usually biconcave, each face showing a deep conical hollow. Most of them possess teeth which are usually isolated and pointed. They are for the most part covered with overlapping horny scales, but sometimes they have dermal plates of true bone, or are encased in a calcareous cuirass (see Fig. 191). As examples may be cited—the perch, mullet, bream, sword-fish, John Dory, sun-fish, mackerel, tunny, lump-sucker, goby, blenny, stickleback, wrass, cod, whiting, haddock, hake, ling, flat-fishes, carp, pike, salmon, trout, herring, pilchard, eel.

iv. **Palæichthyes**—including Elasmobranchs, Chimæroids, and Ganoids.

1. Elasmobranchs, with a cartilaginous skeleton and a skin which may be naked, and is never covered with scales as in the osseous fishes, but may bear small prominences which harden by the secretion of carbonate of lime and become tooth-like, or where small and close-set form shagreen. These calcareous portions sometimes form dermal plates or tubercules, and also spines which commonly rise in front of the dorsal fins. It is these dermal defences which are so common in the fossil state under the name of Ichthyodorulites (Fig. 158). The Elasmo-branchs include the Sharks and Rays.

2. Chimaeroids, represented in the living Chimæra and by the fossil *Rhynchodus* (Devonian), *Ischiodus* (Mesozoic), *Edaphodon* (Cretaceous and Eocene).

3. Ganoids—these fishes have a cartilaginous or ossified skeleton. They are usually covered with bony plates or scales. At present this order is almost extinct, only a few forms having survived. But in former times it embraced by far the largest part of the vertebrate life of the globe, and from the durability of the external bony plates and scales the remains of the extinct genera have been plentifully preserved. Eight sub-orders have been recognised, viz. (1) Placoderms, entirely extinct, but well represented by *Scaphaspis, Cephalaspis, Coccosteus*, and other Palæozoic genera (Fig. 134) ; (2) Acanthodians, also extinct, *Acanthodes* (Fig. 133), *Cheiracanthus* ; (3) Dipnoi, represented by the living *Lepidosiren* of the Amazon and *Ceratodus* of Queensland rivers ; and by *Dipterus*, and *Phaneropleuron* of the Old Red Sandstone ; (4) Chondrosteans, of which the living sturgeon is a type, and which is represented in the fossil state by *Palæoniscus* (Permian) and *Chondrosteus* (Lias) ; (5) Polypteroideans, represented by the modern *Polypterus* of the Nile, and by many extinct genera, as the Palæozoic *Diplopterus, Megalichthys, Osteolepis* (Fig. 133).

Cœlacanthus, Holoptychius, Strepsodus, etc. ; (6) Pycnodontoideans, entirely extinct, represented among the Mesozoic formations by *Pleurolepis, Gyrodus, Pycnodus,* and other genera ; (7) Lepidosteans, of which the living lepidosteus or "gar-pike" of North America is the type. Large numbers of extinct genera have been met with in Palæozoic and Mesozoic rocks. Among these are *Pholidophorus* (Fig. 179), *Lepidotus, Cheirolepis, Amblypterus, Eurynotus, Platysomus* (Fig. 162) ; (8) Amiodians, represented by the living *Amia,* a mud-fish of the fresh waters of the United States, and by several Mesozoic and Tertiary extinct genera, as *Caturus, Leptolepis.*

AMPHIBIA—newts, frogs, salamanders, etc.—are divisible into four orders as follows :—

i. **Urodela** or tailed amphibians—animals with elongated bodies and relatively short limbs, devoid of scales or pectoral plates. They comprise the living newts (*Triton*), salamanders, and mud-eels (*Siren*). Traces of forms supposed to be allied to some of these animals have been met with in Permian rocks, but it is only in Tertiary strata that undoubted remains of Urodela have been found.

ii. **Anura,** or tailless amphibians—animals with relatively short and broad bodies and two pairs of limbs, of which the hinder are longer and stronger. Though there are no scales nor pectoral plates, portions of the skin of the back are in some cases ossified. They are typified by the frogs and toads. They are only found fossil in younger Tertiary deposits.

iii. **Peromela,** or snake-like amphibians—animals with serpentiform bodies without limbs. They are not known as fossils.

iv. **Labyrinthodonta**—animals now entirely extinct which possessed bodies somewhat like those of the salamanders, with relatively weak limbs and long tail. The head was defended by hard plates of bone, the breast by three sculptured bony plates, and the lower side of the body by an armour of oval plates or scales. The feet appear to have been five-toed. The footprints of these creatures were first found ; but more or less perfect skeletons of them have since been obtained. Their name is taken from the labyrinthine structure of their large teeth. The earliest known Labyrinthodonts are from the Carboniferous system ; they formed the magnates of the world until they were supplanted in early Mesozoic time by the great development of Reptilians (Fig. 163).

REPTILIA or true reptiles, with horny scales or bony scutes, are represented now by turtles, tortoises, snakes, lizards, and crocodiles, but flourished formerly in many remarkable forms which have long been extinct. Embracing all the known living and extinct types in one view, we may group them into the following orders :—

i. **Chelonia**—the tortoises and turtles, distinguished for the most part by the bony case or box in which the body is enveloped. As many of these animals are of aquatic habits their hard parts must often be covered up and preserved in sedimentary deposits. They are not uncommon in the fossil state, as far back as the Jurassic series.

ii. **Ophidia**—snakes and serpents, covered with horny scales, and remarkable for the number of their vetebræ (which, in some pythons, amount to more than 400), and for the want of limbs. They are not known fossil except in the Tertiary formations.

iii. **Lacertilia**—lizards, chamæleons. The oldest forms occur in the Permian system (*Protorosaurus*); in the Triassic period lived the *Rhyncosaurus*, *Hyperodapedon*, and *Telerpeton* (Fig. 168); in the Cretaceous, the long-necked *Dolichosaurus* and the gigantic *Mosasaurus*.

iv. **Crocodilia**—the crocodiles, alligators, and gavials form the highest type of living reptiles. The earliest trace of them in a fossil state is in the *Stagonolepis* of the Trias (Fig. 169). They abounded in the Jurassic seas, the genera *Teleosaurus* and *Steneosaurus* being conspicuous, while in Cretaceous time the *Goniopholis* abounded. None of the modern crocodiles, however, are truly marine.

The following orders of reptiles are now wholly extinct :—

v. **Ichthyosauria** —animals somewhat resembling whales in shape, the head being joined to the body with no distinct neck, and the body tapering away behind. They appear to have been covered merely with skin, and moved through the water by means of two pairs of paddles. In the huge head the most conspicuous feature was the large eye-orbits filled with a circle of bony plates that remain often well preserved. The typical genus, *Ichthyosaurus*, is abundant in the Lias (Fig. 180).

vi. **Plesiosauria**—distinguished for the most part by the disproportionate length of the neck and the smallness of the head. Like the ichthyosaurs, the plesiosaurs appear to have had no bony covering upon their skin ; they had two pairs of paddles, those behind being largest, and a comparatively short tail. The earliest plesiosaurs are found in Triassic rocks (*Nothosaurus*, *Simosaurus*, *Pistosaurus*), but they are most characteristic of the Jurassic formations (*Plesiosaurus*, *Pliosaurus*).

vii. **Dicynodontia**—lizard-like animals with crocodilian vertebræ and tortoise-like jaws, which were probably cased in a horny beak. They have been found in certain supposed Triassic strata in Scotland, South Africa, and India (*Dicynodon*, *Oudenodon*).

viii. **Pterosauria,** flying reptiles or Pterodactyls—distinguished by the length of their heads and necks, and the proportionately great size of their fore-limbs, on which the outer finger was enormously elongated to support a wing-like membrane. These animals no doubt flew from tree to tree, and hopped or shuffled along the ground. They appear to have been confined to the Mesozoic periods. The important genera are *Pterodactylus* (Fig. 181), *Rhamphorhynchus*, *Dimorphodon*, and *Pteranodon*.

ix. **Deinosauria**—a remarkable group of animals, mostly of enormous size, which presented structures linking them with birds. Some of them had a covering only of naked skin, others possessed an armour of bony plates like those of the crocodile. The hind-limbs are usually enormously developed in comparison with the fore-limbs, showing that the animals probably walked on their hind feet. The deinosaurs abounded during the Mesozoic ages and in many diverse types. Some of the more important genera are *Iguanodon* (Fig. 192), *Hylæosaurus*, *Cetiosaurus*, *Megalosaurus*, and *Compsognathus*. The largest animal yet known, the *Atlantosaurus*, has been found in the Jurassic rocks of North America (p. 307).

x. **Thecodontia**—a group of carnivorous reptiles, remarkable for possessing teeth which have been classed by Professor Owen as incisor, canine, and molar. Their remains have been found in supposed Triassic rocks in South Africa.

AVES, birds.—This important section of the animal kingdom has been but sparingly found in the fossil state. The facility with which birds can escape by flight from the destruction that befalls other land-animals will no doubt suffice to explain why their fossil remains should be so infrequent. The oldest known birds had curious reptilian affinities, being furnished with jaws and teeth. Taking all the known forms of birds, recent and fossil, they may be grouped in the following subdivisions :—

I. **Saururæ**—in this singular extinct group the vertebral column was prolonged into a long lizard-like tail, each vertebra of which, however, bore a couple of quill-feathers. The only known example is the *Archæopteryx* of the Jurassic system—the oldest bird yet discovered (Fig. 182).

II. **Odontornithes** or toothed birds. Some of these (Odontolcæ) were diving birds with rudimentary wings, ratite sternum, powerful legs, a strong tail for steering, and jaws with numerous conical teeth sunk in a deep continuous groove (*Hesperornis*). Others (Odontotormæ) were provided with strong wings and carinate sternum, and had their teeth sunk in separate sockets, as in the crocodiles. The toothed birds have long been extinct. They have been found most abundantly in the Cretaceous rocks of Kansas.

III. **Ratitæ**—the cursores or running birds, such as the ostrich, cassowary, rhea, emeu, and apteryx. These have not with certainty been found fossil in strata older than the Tertiary series. The gigantic extinct *Dinornis* of New Zealand belongs to this class.

IV. **Carinatæ**—generally possessing powers of flight. These include most of the birds of the present day. The arrangement of this great sub-class into definite orders has not been satisfactorily accomplished. The student, however, may find some advantage in making himself acquainted with the following names which, though in process of being superseded, are still in common use. N a t a t o r e s—swimmers or palmipeds, with short legs placed behind and provided with webbed feet. These include gulls, penguins, geese, ducks, swans, cormorants, etc. Remains of this order are found in Cretaceous and Tertiary strata. G r a l l a t o r e s—waders, chiefly found by the shores of rivers, lakes, or the sea, distinguished by the length of their legs, which are not completely webbed. They include plovers, cranes, flamingoes, storks, herons, snipes, etc. They have been found fossil in Cretaceous and Tertiary rocks. R a s o r e s— scratchers or gallinaceous birds, including the various tribes of fowls and pigeons. They are found fossil in Tertiary strata. S c a n s o r e s—climbers, including the parrots, cuckoos, toucans, and trogons. They are only found fossil in Tertiary and Post-tertiary rocks. I n s e s s o r e s—perchers, passerine birds include by far the largest number of living birds, and all the ordinary song-birds. They have not been met with in a fossil state in rocks older than the Tertiary series. R a p t o r e s or birds of prey, comprising birds with strong, curved, sharp-edged, and pointed bills, and strong talons, as the eagles, hawks, falcons, vultures, and owls. This order also has not been obtained fossil except in Tertiary and Post-tertiary rocks.

MAMMALIA.—The highest class of the vertebrata is represented chiefly on the land, the marine representatives being few in number, though often of large size (whales, dolphins, porpoises, manatee, seals, morse). In marine deposits, therefore, we need not expect to find mammalian remains abundant at the present time. Doubtless from the time of their first appearance mammals have always been, on the whole, terrestrial animals ; their fossil remains consequently occur but sparingly among ancient geological formations. The earliest known examples belong to the Marsupial type, and have been found in the Triassic and Jurassic rocks of Europe and North America.

I. PROTOTHERIA or ORNITHODELPHIA—including the two types of Ornithorhynchus and Echidna.

II. METATHERIA or DIDELPHIA—Marsupial animals. Comprising the Opossums (Didelphidæ), Dasyures, Myrmecobius, Perameles, Kangaroos, and Wombats. As just mentioned, it is representatives of this section of the vertebrates that occur fossil among the Mesozoic rocks [*Microlestes*, Fig. 170, *Dromatherium, Amphitherium, Phascolotherium,* Fig. 183].

III. EUTHERIA or MONODELPHIA—including the vast majority of living and extinct mammalia. They may be grouped as follows :—

    Edentata—sloths, ant-eaters, armadilloes, pangolins, and African ant-eaters. Some enormous extinct types of Edentates have been found in America [*Megatherium, Mylodon, Moropus, Glyptodon*].

    Sirenia—aquatic fish-like animals including the manatee, dugong, sea-cow. The last named is now extinct, the last having been killed so recently as 1768. Numerous fossil remains of Sirenians occur in Miocene and Pliocene deposits of Europe [*Halitherium*].

    Cetacea—whales including the Balænidæ or whalebone whales, Delphinoidea or toothed whales (Sperm whale, Ziphius, Narwhal, Porpoise, Ca'ing whale, Grampus, Dolphin). Cetacean ear-bones and other bones are not infrequent in Tertiary and Post-tertiary strata.

    Insectivora—small terrestrial mammals like the shrews, moles, myogale. No fossil insectivores older than Eocene times are known, except perhaps *Stereognathus* of the Stonesfield slate.

    Cheiroptera—animals with the fore-limbs adapted for flight, including the tribe of bats. Fossil representatives are found as far back as Eocene rocks.

    Rodentia—small terrestrial plant-eating mammals, distinguished by their large chisel-shaped incisor teeth, specially adapted for gnawing, and by the absence of canines. Among them are squirrels, marmots, beavers, dormice, rats, mice, voles, lemmings, jumping mice, jerboas, porcupines, chinchillas, cavies, rabbits, and hares. Fossil rodents belonging to most of the existing families have been met with in Tertiary and Recent strata, together with some extinct types.

    Ungulata or hoofed animals include the Hyrax, the Proboscideans (elephants, Fig. 205, and the extinct types of *Mastodon*, Fig. 199, *Deinotherium*, Fig. 200, etc.) and the extinct type of the *Deinocerata* (Fig. 196); the perissodactyl or odd-toed group (tapirs, rhinoceroses, horses, [*Palæotherium*], Fig. 195), and the artiodactyl or even-toed group (hippopotamus, peccary, swine, llama, camel, chevrotains, the true ruminants, such as deer, antelopes, giraffes, [*Helladotherium*, Fig. 203], and all bovine animals). The earliest known forms are of Eocene age.

    Carnivora—so named from the majority of them subsisting on animal food and being eminently beasts of prey. They are divided into (1) Fissipedes or true carnivores, generally adapted for life on land, comprising (a) the Æluroids or cat-like forms (lions, tigers, cats, puma, jaguar, cheetah, civet-cat, ichneumon, hyæna, and various fossil forms found in Tertiary and Post-tertiary deposits); (b) the Cynoids or dog-like forms (dogs, wolves, foxes); and (c) the Arctoids or bears and their allies (otters, badgers, weasels, raccoons, panda); (2) Pinnipedes or aquatic carnivores, divisible into three well-marked families: (a) Otariids or sea-bears; (b) Trichechids or walruses; (c) Phocids or true seals.

Primates, the highest division of vertebrate life, comprising (1) the Lemuroid animals; (2) the Hapalids or marmosets; (3) the Cebids or American monkeys; (4) the Cercopithecids, the monkeys of the Old World, exclusive of the apes; (5) the Simiids or man-like apes (*Troglodytes, Gorilla, Simia,* and *Hylobates*); (6) Man.

INDEX

2 C

Basalt rocks, 164, 166*
Basaltic structure, 165, 166* ·
Bases, 117
Basic rocks, 164
Bath Oolite, 309
Bathonian, 232, 309, 310
Bat, fossil, 333, 337
Beaches, raised, 112, 358, 359
Bear, fossil, 341, 344, 347, 348, 350
Beaver, fossil, 341, 344, 348
Bed (in stratigraphy), 142, 172, 173, 222
Beech, fossil, 314, 331, 340, 344
Beetles, fossil, 304, 341
Belemnite, 302, 303, 318
Belemnitella, 322
Bellerophon, 246, 255, 256*, 280, 281*
Bembridge Series, 337
Bermuda, recent limestone formations of, 85, 151
Beryl in granite, 164
Beryx, 320*
Betula, 345
Biotite, 132
Birch, climate of, 219 ; fossil, 345, 355
Birds, fossil, 307*, 321, 332 ; with teeth, 221, 322, 332
Bison hunted by early man, 364
Black Jura, 309
Blackthorn, fossil, 348
Blastoids, 221, 277, 278*
Blattidæ (cockroaches), fossil, 256, 274
Blende, 209
Blocks, erratic, 64, 359 ; volcanic, 102*, 154, 205*, 206
Blow-holes, 70, 71*
Boars, fossil, 350, 365
Bog-bean, fossil, 348
Bog-iron ore, 47, 129, 157
Bog-manganese, 130
Bogs ; see Peat
Bombs, volcanic, 153
Bone-beds, 160, 296, 298
Bone-breccia, 161
Bone-caves, 56, 90, 160, 362
Bosses, eruptive, 201, 202*
Bottom-moraine, 359
Boulder-clay, 353, 354, 359
Bracheux, sands of, 334
" Brachiopods, Age of," 246

Brachiopods, fossil, 246*, 253, 255*, 264*, 265, 275, 279, 280*, 287*, 301, 316, 347*, 348
Brachymetopus, 278
Bracklesham Beds, 334
Bradford Clay, 309
Bramatherium, 350
Branchiosaurus, 289
Breakers ; see Waves
Breaks in succession, 222
Breccia, 149*, 154
Brick-clay, 152
Brick-earth, 19, 20*, 362
Britain, ancient glaciers and ice-sheets of, 65, 353, 358 ; lava plains of, 110
Brittle-stars, fossil, 252
Bronteus, 263*, 264
Brontosaurus, 306
Brown coal, 159, 338
Brown colour of rocks, usual cause of, 123, 129, 151
Bruxellian, 334
Buckthorn, fossil, 315, 345
Bunter or Lower Trias, 232, 297
Burrows of animals, 81, 212, 244
Butterfly, earliest known, 304
Buzzard, ancestral forms of, 332

CACTUS, fossil, 330
Caffer cat, fossil, 364
Cainozoic time, 231, 232, 327
Cairngorm stones, 128
Calamites, 259, 272, 274*, 287
Calc-sinter, 57, 157
Calcaire-grossier, 334
Calcareous, 151
Calcareous sand, 158
Calcareous tufa, 57 ; fossils from, 363
Calceola, 263*, 265
Calceola group, 266
Calcification, 217
Calcite, 124*, 126, 134*, 135*, 209
Calcium, 116, 121
Calcium-carbonate (see Lime, carbonate of), 25, 26, 52, 55, 58, 84, 86, 89, 119, 121, 124, 134, 151, 154, 158, 215, 217
Calcium-phosphate, 120, 137
Calcium-silicates, decomposition of, 53

Earthworms, influence of, in removal of soil, 18, 81
Echidna, 296
Echinoconus, 316, 317*
Echinodermata, fossil, 243, 252*, 277, 278*, 294, 301*, 316, 317*
Edmondia, 279, 280*
Egeln, marine beds of, 338
Elements, most important simple, 116
Elephant, climate indicated by bones of, 219, 220
Elephants, fossil, 348, 350, 356, 357*, 361
Elevation, proofs of, 112, 339
Elk, fossil, 365 ; Irish, 83, 364, 365
Ellipsocephalus, 244*
Elm, fossil, 331, 340, 345
Enaliosaurs, 305
Enchodus, 320
Encrinurus, 253
Encrinus, 294
Endo-skeleton, 215
Energy, solar, 93 ; terrestrial, 93
Eocene, 231, 329, 330
Eophyton, 242*
Eoscorpius, 273
Eozoon, 235
Ephemera, fossil, 260, 274
Equisetaceæ, fossil, 272, 273*, 293*, 299
Equus, 221
Erosion, base-level of, 74 ; by running water, 28
Erratic blocks, 64, 359
Eruptive, definition of, 143
Eruptive rocks, chief varieties of, 161 ; their place in the architecture of the earth's crust, 200
Eskers, 358
Estuaries, deposit of mud in, 76
Eucalyptus, 331
Euchirosaurus, 290
Euomphalus, 279, 281*
Europe, glaciation of, 351, 353
Eurypterids, 262*, 265, 279, 303
Evergreen oaks, fossil, 336, 340, 345
Exogyra, 302, 308, 316
Exo-skeleton, 215

Fagus, 331
Fairy stones, 140*, 141, 178
False-bedding, 37, 174, 175*

Faluns, 343
Famennian, 266
Fan-palms, 330, 336
Fans of alluvium, 35
Fault-rock, 195*
Faults, 195*, 207*
Favosites, 251, 277
Feather-palms, fossil, 336
Felsite, 163
Felsitic structure, 144, 163
Felspars, 131, 161, 164*
Fenestella, 279*
Ferns, fossil, 259*, 272*, 286*, 293, 314, 315, 330
Ferric oxide, 59, 123, 185
Ferrous carbonate, 135, 141, 157, 160
Ferrous oxide, 122
Ferruginous, 151
Fibrous minerals, 127
Ficus (fig), fossil, 314, 315*, 331, 340*, 345
Fire-clay, 152 ; is an ancient soil, 102, 173, 206, 269
Fir, Scotch, fossil, 348
Fishes, fossil, 239, 255, 260*, 261*, 281, 282*, 288*, 289, 294, 304*, 320*, 332
Fishes, sudden destruction of, 262, 290
Fissures, 209 ; volcanic, 103, 108 ; formed by earthquakes, 111
Fjords, 45
Flamingo, fossil, 336
Flint, 160, 178, 325
Flint implements, 362*, 366*
Flood-plain of rivers, 39
Flow-structure, 145*, 146, 162
Fluid-cavities in crystals, 143*
Fluorides, 124, 137
Fluorine, 116, 120
Fluor-spar, 120, 125, 137, 138*, 209
Fluvio-marine Crag, 348
Fluxion-structure, 145*, 146, 162.
Foliated, definition of, 147
Fontainebleau Sandstone, 338
Footprints in rocks, 176, 212, 294, 295, 307
Foraminifera, fossil, 249, 276*, 316*, 331
Forest-bed, 348
Forest-marble, 309
Forests, protective influence of, 82 ; disappearance of, 89

Marl Slate, 232, 285, 289
Marlstone, 309
Marine animals more likely than ter-
restrial to occur as fossils, 213
Marmots, fossil, 337
Marsupials, fossil, 296, 308*, 311
Marsupites, 322
Martens, fossil, 337
Massive, definition of, 143
Mastodon, 221, 341*, 343, 347, 348,
350
Mastodonsaurus, 294
Mayfly, fossil, 260, 274, 304
May Hill sandstone, 257
Mechanical deformation of rocks, 146,
147, 168, 169, 170
Mediterranean stage (Miocene), 343
Megaceros, 364
Megalichthys, 275
Megalosaurus, 306
Megaptilus, 275
Melanerpeton, 290
Menevian group, 233, 247
Menyanthes, 348
Mesozoic time, 231, 232, 291
Metalloids, most important in Earth's
crust, 116, 117
Metals, most important in Earth's
crust, 116, 121
Metamorphic rocks, 168
Metamorphism, 146, 147, 168, 193.
194*, 198*, 202, 203*, 248
Meteoric iron, 122 ; in deep sea de-
posits, 78
Meteorites, 228
Mica, 132, 162, 164*
Mica-schist, 169
Mica-slate, 169, 194, 198, 203
Micaceous, 151
Mice, influence of in baring soil, 18
Micraster, 316, 317*, 322
Microcline, 132
Microfelsitic, 144
Microlestes, 296*
Microscope, use, of in Geology, 140,
143, 144, 162
Millipedes, fossil, 260, 274
Millstone Grit, 232, 283
Mimosa, fossil, 340 •
Minerals, definition of, 123 ; most
important, 123 ; different modes of
origin of, 126

Mineral veins, 209
Miocene, 231, 329, 339
Mississippi, sediment transported by,
28
Mitra, 331, 343
Moa, 332
Modiola, 279
Modiolopsis, 246, 254
Molasse, 218
Moles, influence of burrowing habits
of, 18, 81
Moles, fossil, 337, 348
Mollusca, importance of, as fossils, 244
Monkeys, fossil, 341, 343, 347
Monoclinic, 126*
Monograptus, 250*
Monometric crystals, 125*
Monotis, 294
Mons, limestone of, 334
Montlivaltia, 300
Moon, former distance of, from the
Earth, 229
Moraines, 63, 149 ; terminal, of ice-
sheet, 354
Morse, fossil, 341
Mosasaurus, 321
Moselle, gorge of, 31 ; volcanic vents
on the, 104
Moulds and casts of organisms, 216
Mountains, upheaval of, at various
times, 112 ; determination of rela-
tive dates of, 183*
Mud, 152 ; depth of water indicated
by, 219 ; inimical to some kinds of
marine life, 172
Mudstone, 153
Murchison on Silurian rocks, 240,
248
Murchisonia, 255
Murex, 343
Muschelkalk, 232, 297
Muscovite, 132
Musk-sheep, climate of, 219 ; fossil,
348, 363, 364 ; former southward
range of, 356
Mya, 348
Myliobates, 332
Mylonitic, defined, 147
Myophoria, 294, 295*
Myriapods, fossil, 273
Myrica, fossil, 314
Myrtle, fossil, 340

2 D

THE END

Printed by R. & R. CLARK, LIMITED, *Edinburgh.*

www.ingramcontent.com/pod-product-compliance
Lightning Source LLC
Chambersburg PA
CBHW021347210326
41599CB00011B/782